2nd International Conference on Applied Sciences (ICAS 2021)

Held online

E3S Web of Conferences Volume 348 (2022)

Bogor, Indonesia
8 – 9 September 2021

Editors:

O. Farobie
T. Soma
Y. Arkeman
A. Daryanto,
M. Komatsuzaki
G. Duteurtre
I. Hermadi
B. Purwanto
A. Azhar
G. Saefurahman
S. Irawan
H. Wijaya

ISBN: 978-1-7138-5227-8

Printed from e-media with permission by:

Curran Associates, Inc.
57 Morehouse Lane
Red Hook, NY 12571

Some format issues inherent in the e-media version may also appear in this print version.

For additional information, please contact EDP Sciences – Web of Conferences
at the address below.

EDP Sciences – Web of Conferences
17, Avenue du Hoggar
Parc d'Activité de Courtabœuf
BP 112
F-91944 Les Ulis Cedex A
France

Phone: +33 (0) 1 69 18 75 75
Fax: +33 (0) 1 69 28 84 91

contact-edps@webofconferences.org

Additional copies of this publication are available from:

Curran Associates, Inc.
57 Morehouse Lane
Red Hook, NY 12571 USA
Phone: 845-758-0400
Fax: 845-758-2633
Email: curran@proceedings.com
Web: www.proceedings.com

TABLE OF CONTENTS

Sekolah Vokasi
College of Vocational Studies

E3S Web of Conferences

INVITATION

We cordially invite you to participate in the events of:

ICAS
2nd International Conference on Applied Sciences 2021

"Toward Sustainable Agro-Maritime 4.0 in New Era: Opportunities and Challenges"

Virtual Conference, 8-9th September 2021

Bogor-Indonesia

web page: http://icas-sv.ipb.ac.id/ • email: icas.sv@apps.ipb.ac.id

---

Keynote Speakers

Dr. Beny Bandanadjaya, ST., MT
Director of Vocational Education

Welcoming Speech

Prof. Dr. Arif Satria, S.P., M.Si.
Rector of IPB University

Keynote Speakers

Prof. Dr. Masakazu Komatsuzaki
Ibaraki University, Japan

Speakers

Dr. Ir. Arief Daryanto, M.Ec
College of Vocational Studies
IPB University

Dr. Risti Permani
Deakin University, Australia

Dr. Guillaume Duteurtre
CIRAD, France

Dr. Tammara Soma
Simon Fraser University, Canada

Dr. Vannessa Robitzch
UACh, Chile

Dwi Sutoro, ST, MM, MBA, IPU
PT Perkebunan Nusantara III (Persero)

Free Registration Fee (limited participant)
Registration Link: **http://ipb.link/icas-registration**

e-certificate fee (optional): **Rp 100.000**

Sponsored by: CHAROEN POKPHAND INDONESIA, PTPN 3

Supported by: BLST Holding Company of IPB, BANK BRI

Statement of Peer review

In submitting conference proceedings to E3S Web of Conferences, I certify to the Publisher that I adhere to the Policy on Publishing Integrity of the journal in order to safeguard good scientific practice in publishing.

1. All articles have been subjected to peer review administered by the proceedings editors.
2. Reviews have been conducted by expert referees, who have been requested to provide unbiased and constructive comments aimed, whenever possible, at improving the work.
3. Proceedings editors have taken all reasonable steps to ensure the quality of the materials they publish and their decision to accept or reject a paper for publication has been based only on the merits of the work and the relevance to the journal.

2nd International Conference on Applied Sciences 2021 (ICAS 2021), 8 - 9 September 2021, Virtual Conference from Bogor - Indonesia

Proceedings editor(s):

Dr. Eng. Obie Farobie; Dr. Tammara Soma; Prof. Dr. Yandra Arkeman; Dr. Arief Daryanto; Prof. Dr. Masakazu Komatsuzaki; Dr. Guillaume Duteurtre; Dr. Irman Hermadi; Dr. Bagus Purwanto; Dr. Aidil Azhar; Ganjar Saefurahman, MPhil, MSc; Suhendi Irawan, MSc; Hendri Wijaya, MS

14 April 2022, Obie Farobie

Specialization and diversification as adaptive strategies for smallholder dairy farming systems providing a formal milk chain in Indonesia

Pria Sembada[1*], *Guillaume* Duteurtre[2], *and Charles-Henri* Moulin[3]

[1]College of Vocational Studies, IPB University (Bogor Agricultural University), Jl. Kumbang No. 14 Kampus IPB Cilibende 16128, Bogor, Indonesia;
[2]CIRAD, UMR Systèmes d'élevage Méditerranéens et Tropicaux (SELMET) Campus International de Baillarguet, 34398, France;
[3]UMR SELMET, INRAE – CIRAD – Montpellier SupAgro – Univ Montpellier, 2 Place Pierre Viala, 34060 Montpellier, France.

Abstract. Smallholder dairy farms encounter challenges in minimal production factors that result in a lack of family income. They react differently to these limits by combining on-farm and off-farm activities (diversifying activity), concentrating exclusively on milk production (specializing activity), or leaving the dairy production to secure family livelihood. As far as adaptive strategies are concerned, they may affect milk production growth at the farm and national levels. We performed an observational case study in the West Java Province (Indonesia) to resolve those problems. We gathered information in two phases: a systematic survey (May to September 2015) to 355 farms and an in-depth interview (January to April 2017) with 20 farms. Our result distinguishes four categories of farms, along with a very small specialized dairy farm (T1), a combination of the dairy farm off-farm activity with very limited land (T2), a small specialized dairy farm (T3), and a mixed crop-dairy farm (T4). The technical-economic value varies depending on the farm type. The six trajectories prevail. The main change was the addition of off-farm activities for poor farmers. Farms in the development trajectory, two strategies coexist between the dairy production system's specialization and the mixed crops-dairy system. In conclusion, this study underscored each farm trajectory's different attributes and drivers. The study also underlined the importance of the initial capital of smallholders to illustrate their future farm trajectory.

1 Introduction

The demand for dairy products has increased enormously over the last few decades. In 2013, the need reached 750 million tons of milk equivalent worldwide, increasing demand occurred, especially in developing nations [1]. This circumstance is explained by three main factors: Massive population growth, rapid economic growth, and consciousness in consuming healthy food [2].

* Corresponding author: priasembada@apps.ipb.ac.id

In several developing nations, dairy industries seeking more milk arrange milk collection to allow smallholders access to the market and facilities (equipment, semen, feed, technical assistance, etc.) for milk production [3]. Classically, larger herds delivering more milk are interesting for farmers' economies of scale and lower collection costs for dairy enterprises. Specialized farms are fascinating to enhance household workers' capabilities, engaged in only one activity. As a result, smallholders generally exhibited in production basins organized through specialized or diversified systems, within a range of sizes depending on their assets.

Smallholder farms are associated with challenges due to minimal production factors (herd size, feed, workers, and access to land), which lead to limited family income [4]. They react in different ways to these limits. Some farmers diversify the activities by combining dairy production with other on-farm [5] and off-farm [5, 6] activities. Some others are increasingly specialized in milk production to benefit from economies of scale and enhanced practices. A few others prefer to abandon the dairy business. These strategies may impact dairy production growth at the farm and national levels. Specialization and diversification could be an adaptive strategy, but we raised some issues. To what extent are diversification and specialization a medium-term strategy for smallholders? Or do they quickly shift from one system to another? Then what are the drivers of these changes? What are the linkages between the starting level of the capital and the farm's ability to capitalize and adjust the strategies? What are the motives of many diversification pathways?

We performed an empirical study to deal with these issues. We worked in Indonesia, where smallholder farms play a crucial role in contributing to rural areas' economic growth. Smallholder dairy production provides income and creates employment opportunities for more than 150 thousand households [7]. Many producers with fewer than four cows could hardly contribute about 64 percent of nationwide milk production [8]. Limited capital might pose a challenge to improving farm-level dairy production to guarantee income and provide for the national market, which is insufficient to meet the milk product's increased consumption.

Diversification and specialization as adaptive strategies to address limited production factors of smallholder farms have received considerable critical attention. However, most studies in Indonesia have only focused on a static condition. This study performed a typology and trajectory approach to explore the effect of farmer strategies on developing milk production to secure household income and supply milk to the national market in Indonesia. Using those approaches, we would better understand the static situation and the dynamic one.

2 Method

2.1 Study Design

Study site. A study has been carried out in the North Bandung Milk Cooperative (KPSBU), involving Bandung Barat and Subang districts. Due to comfortable environmental conditions, these two districts have high farm potential (mostly dairy farms).

Sampling. We surveyed 355 out of 4000 dairy farmers, members of KPSBU. Farmers from all sub-district milk collection points were selected randomly. To better grasp the farm dynamics, We conducted in-depth interviews, including 20 out of 355 randomly chosen farms for each farm type.

Data collection. Data were obtained by a formal survey from May to September 2015 to describe the characteristics of the farms in the area. In-depth interviews with dairy producers were also carried from January to April 2017 to better understand each type of farm's trajectory. Additional data were obtained from the coop and the public authorities.

Information is taken into account. We gathered data from farmers & the coop on the characteristics and the diversity of the farm, such as the number of workers, herd size, number of lactating cows, access to land, land for forage cultivation, net income. Net dairy income was obtained from the justification of the cooperative and the farmer. Also, annual net income from on-farm or off-farm activities was collected from the rationale of the farmer.
Essential performance and productivity were also identified: total net income per farm, total net income per family worker, net agricultural income per land unit, and net dairy income per LU. To better comprehend the calculation of net dairy income per LU, we also calculated daily milk yield per LU, milk income per LU, cost of feed concentrate per LU, additional milk income per LU, and other costs per LU.

At present, we described the trajectories of 20 farms that carry out the method suggested by [9] examined the various strategies of maintaining dairy production among farm types. We collected details, including income-generating activities, herd size, and access to land at the starting of the dairy activity and the current situation. As a study reported by [10], we developed qualitative variables to analyze pathways of employees' work on dairy farms, with modalities indicating the dynamics between the starting of the dairy business and nowadays.

2.2 Data Analysis

This study performed typology and trajectory analysis. To describe farm type (typology analysis), this study conducted a multivariate analysis based on seven variables included: a combination of activities (contribution of dairy, crop, and off-farm to total income), the balance between the herd and the land dedicated to foraging (forage per LU), farm size and ratio among factors of production (total land, total workers, herd size). We performed ANOVA, the Fisher test (alpha = 0.05), and the descriptive study for 355 farms to classify the farm types and essential performance using Minitab 17.

This study conducted a visual analysis to define trajectories, as suggested by [11] and [12] in related animal science research. A cross-table was generated with 20 farms in lines and separate variables in rows. A variety of color shades defined the modalities. Moving lines and rows allowed the identification of farms with the same visually profiled modalities.

3 Results

3.1 Farm typology

Four types of farms have been identified (Table 1). Type 1 (T1) was a very small specialized dairy farm (almost 40 percent of the sample). Farmers had minimal land access. Type 2 (T2) still had very restricted land access; however, farmers mixed dairy production with off-farm activity. Type 3 (T3) was a small specialized dairy farm. Type 4 (T4) was a mixed crop-dairy activity.

Based on production factors, T1 and T2 had the same features, with minimal land access and a small herd (3.5 LU). T3 and T4 had higher capital. T1 and T3 had the same high dairy contribution (more than 85 percent) to the total income. It means that they focused more on dairy activity. On the other hand, T2 and T4 farms have less dairy income contribution (less than 60 percent) to the total income. Crops (T4) or off-farm activities (T2) also played an essential role in generating income. On average, T2 had a smaller herd size and limited land access, while T4 had better production factors.

Table 1. Characteristics of the four farm types (mean value and standard deviation)

Farm types	T1 (n=138)	T2 (n=62)	T3 (n=94)	T4 (n=61)
Dairy contribution to total income (%)	96.3 ± 6.6 [A]	53.4 ± 25.8 [C]	85.9 ± 24.7 [B]	42,5 ± 27.2 [D]
Crop contribution to total income (%)	1.2 ± 4.1 [B]	1.5 ± 4.4 [B]	3.6 ± 12.4 [B]	47.6 ± 29.9 [A]
Off-farm contribution to total income (%)	2.4 ± 5.7 [C]	45.1 ± 26.8 [A]	10.5 ± 21.4 [B]	9.9 ± 22.2 [B]
Forage land per herd (m2/LU)	292 ± 294 [C]	725 ± 703 [B]	1907 ± 1824 [A]	541 ± 682 [BC]
Total land access (m2)	1328 ± 1115 [C]	1791 ± 1723 [BC]	7735 ± 5586 [A]	3294 ± 2795 [B]
Total workers (person)	1.9 ± 0.6 [B]	1.8 ± 0.6 [B]	2.4 ± 0.8 [A]	2.0 ± 0.7 [B]
Herd size (LU)	3.5 ± 1.6 [B]	2.2 ± 1.0 [C]	5.5 ± 4.8 [A]	3.9 ± 1.8 [B]

Note: Means in the same line with a different superscript differ significantly ($P<0.05$); NS: not significant. Source: a survey (2015)

3.2 Performances Based on Farm Types

Farms T3 and T4, which had better land access and herd size (Table 2), reached the highest total net income per farm. Labor productivity in both types was the highest. The annual net total income on average was above the average salary. The average labor productivity for T1 and T2, with less land access, was below the threshold. Land productivity was the highest in both types. Farmers with narrow land access remain in a position to produce milk and earn cash by using, for example, natural grass nearby and paddy straw. On the other hand, T3 and T4 farms had lower land productivity.

Table 2. Economic performances and productivity of the production factor (labor, land, and herd) of the four farm types

Farm types	T1 (n=138)	T2 (n=62)	T3 (n=94)	T4 (n=61)
Total net annual income per farm (million IDR)	28 ± 20.8[C]	32.6 ± 20.8 [BC]	51.2 ± 52.9 [A]	44.6 ± 48.2 [AB]
Total net annual income per family worker (million IDR)	16.6 ± 13.7 [B]	20.7 ± 15.6 [AB]	28.2 ± 34.9 [A]	24.9 ± 27.8 [A]
Net annual agriculture income per hectare (million IDR)	532.2 ± 790 [A]	576 ± 1416.7 [A]	136.5 ± 326.6 [B]	379 ± 965.7 [AB]
The net annual dairy income per LU (million IDR) [1]	9.1 ± 5.8 [B]	13.1 ± 18.2 [A]	9.9 ± 6.9 [B]	5.9 ± 5.6 [C]

Note: Means in the same line with a different superscript differ significantly ($P<0.05$); NS not significant (Fisher, LSD). [1] = milk income coop + additional product (cows, calves, manure sales) – cattle purchase – additional feed cost (tofu and cassava waste). Source: a survey (2015)

Two characteristics were apparent given herd productivity (Net Dairy Income per LU, Table 2). Herd productivity was very low in T4 and was the highest but high variability in T2 (Table 2). The milk yield was highest in T2, but it was very variable, which can be seen from three times higher standard deviation than other types. The milk income from the cooperative showed the same trends. In T2 and T4, the proportion of the concentrate ration was different. T2 used an amount of feed concentrates from the cooperative and additional feed (cassava and tofu waste). It might explain the higher milk yield in T2 and with high variability. On the other side, the T4 farms used many cassavas and tofu waste, although far less concentrate feeds from the cooperative. In the survey year, Farms T4 produced just 2.6 million (IDR) additional income per LU from the herd's sales and its products, half of the amount of T2 farms. This was also demonstrated by the disparity in herd productivity between T2 and T4. Specialized dairy farms with more (T3) or less (T1) capital endowment had a similar feeding activity (concentrate) and performance. Thereby, the level of net dairy income per LU was the same and intermediate between the levels of income for T4 (minimum) and T2 (maximum).

Table 3. The dairy activity's technical and economic results are according to the four farm types.

Variables	T1 (n=138)	T2 (n=62)	T3 (n=94)	T4 (n=61)
Daily milk yield per LU (liters)	8.4 ± 3.3^{B}	11.5 ± 12.3^{A}	8.8 ± 3.2^{B}	7.4 ± 2.9^{B}
Annual milk income from coop[1] per LU (million IDR)	9.1 ± 4.5^{B}	13.4 ± 17.1^{A}	9.9 ± 4.7^{B}	8.8 ± 4.3^{B}
Annual concentrate feed cost[2] per LU (million IDR)	4.32 ± 2.21^{B}	5.07 ± 3.52^{A}	4.27 ± 2.01^{B}	3.12 ± 1.86^{C}
Annual additional dairy income[3] per LU (million IDR)	3.41 ± 4.06^{B}	5.62 ± 7.24^{A}	3.78 ± 4.95^{B}	2.57 ± 3.56^{B}
Annual additional feed cost[4] and purchase cattle per LU (million IDR)	3.48 ± 3.90^{B}	5.88 ± 8.73^{A}	3.73 ± 3.59^{B}	5.49 ± 4.53^{A}

Note: Means in the same line with a different superscript differ significantly ($P<0.05$); NS not significant. [1] = milk sold – concentrate feed cost; [2] = concentrate feed cost from coop; [3] = cows, calves, and manure sales; [4] = tofu and cassava waste. Source: a survey (2015)

3.3 Farm Trajectories

In the study of 20 farms, dairy activity started between 1979 and 2013 (Table 4). In the beginning, most farmers engaged in the activity only with one or two cows (n=9) or even only with calves or heifers (n=12). Just one farmer began with five cows. Some farmers (n=9) were able to capitalize on natural growth and the purchase of animals, with a high annual growth rate (more than 10 percent), and reached a herd size of more than 3 LU in 2017. Some have recently begun their business (5 years), while some have been in the dairy business for a long time (30 years). In contrast, with a low annual growth rate (less than 10%), six farmers did not improve the herd above 3 LU, with less than three cows in the herd, after at least ten years of activity (n=6). In the intermediate group (n=4), farmers starting with just one or two heifers could reach 3 LU with 2 or 3 cows in 2017, after 5 or 10 years.

We defined six farm trajectories (Figure 1). These six trajectories could be clustered into three main groups. First, farmers with minimal access to land (types T1 and T2) could stay at the same type or move from T1 to T2 and vice versa, depending on the practice of off-farm activities. Even if they have been in dairy farming for a long time, they could not expand their land access and maintained a limited cow.

Among such types (T1 and T2), it was challenging for most farmers to raise herd size. For the remaining dairy farm-off-farm activity (T2), one farmer profits from off-farm activity (tofu waste trading) than dairy activity. His child operated the dairy business, and the income was also shared equitably. On the other hand, another farmer obtained more income from the dairy activity than from off-farm activity (wage worker). If farmers wanted funds for tuition, wedding ceremonies, health care, or undisclosed regular expenditures, they sold cows. As a result, the herd's scale stayed constant, becoming more considered saving.

Farmers attempted to increase the size of the herds in the trajectory remaining in T1. They added more cows through the cooperative credit scheme and kept the calves born as replacement stocks. Nevertheless, due to the barn's limited space and the need for cash to cover living costs, they had difficulty adding more cows. Farmers might also have credit facilities from the coop, although this amount depends on their resources, such as financial and physical capital. It describes the potential reason for less access to credit for T1 and T2 farms than other types. As a result, some calves or even all of them were sold to get cash. In T1 and T2, the total annual income was smaller than the unskilled worker's salaries. Dairy activities did not entirely support the household income.

Table 4. Development of herd size since the beginning of dairy farming activity until 2017

		Beginning year		2017			
Farmers ID	Beginning dairy business (year)	Cows	Heifers	Cows	Heifers	LU[1]	Annual herd growth
Low increase and do not reach more than 3 LU							
RF074	2006	0	1	1	0	1	0.05
RF063	2007	1	0	2	0	2	0.10
TP045	2004	2	0	2	1	2.5	0.04
RF004	2000	2	0	2	1	2.5	0.03
PS010	1995	1	1	2	1	2.5	0.05
GD003	2007	0	3	2	1	2.5	0.10
High increase but do not reach more than 3 LU							
PS015	2001	0	1	2	2	3	0.16
GD078	2003	0	1	3	0	3	0.18
RF015	2005	0	1	3	0	3	0.21
IN047	2013	0	2	2	2	3	0.50
High increase and reach more than 3 LU							
RF050	1993	0	1	3	1	3.5	0.13
RF055	2008	1	0	3	1	3.5	0.28
RF018	2012	0	1	2	3	3.5	0.60
RF073	2012	1	1	3	2	4	0.50
IN040	1992	0	2	2	6	5	0.16
RF005	2007	0	3	4	4	6	0.45
RF054	1979	0	2	5	2	6	0.13
PS057	2007	0	3	5	2	6	0.45
RF011	2000	1	2	14	10	19	1.00
More than 3 LU from the beginning with no increase							
IN038	2000	5	0	5	0	5	0.00

Note: LU= Livestock Unit (1 cow = 1 LU; 1 heifer = 0.5 LU)

Farmers in T3 had high and fast growth in herd size (Figure 1). Discussions with farmers confirmed that the dairy income could cover living costs. They had better assets than other types, too. Under these conditions, some calves could be kept as replacement stocks. It reflected stable natural growth as they had more heifers than culled cows. This helped them to increase their herd size. Farmers had another chance to obtain cows via credit schemes or grants from a dairy development project. It might increase the herd size. As a result, more forage is needed to feed the cows. Both farmers accessed more land to grow forage. Some of them configured workers to cut and carry grass, both family and hired workers.

Most farmers in the mixed crop-dairy model (T4) did not abandon crops or improve their herd size. Some farmers expanded access to land; however, more was required to produce cropping activity, not dairy activity. This was due to the low availability of workers, access to land, and barn capacity. Another factor was the uncertainty of the dairy business, such as animal health issues. The farmers' income to increase the herd size was affected by this issue.

Farmers ID	Farm traject.	Current farm type	Off-farm activities	Herd growth	Land access	Crops
RF015	1-2	2				
RF063		2				
IN040	2	2				
TP045		2				
IN047	2-1	1				
RF055	1	1				
RF004		1				
IN038		1				
RF018		1				
RF073	3	3				
RF050		3				
PS057		3				
RF054		3				
RF005		3				
RF011		3				
PS015	4	4				
GD078		4				
PS010		4				
GD003		4				
RF074		4				

Figure 1. Six farm trajectories based on activities, land access, and herd growth

Note: ***Off-farm activities.*** *White: no; Soft grey: abandoning; Dark grey: starting; Black: keeping.* ***Herd growth.*** *Soft grey: low increase and do not reach more than 3 LU; Dark grey: high increase but do not get more than 3 LU; Black: high increase and reach more than 3 LU.* ***Land access.*** *White: low land access without increase; Soft grey: low land access with increase; Dark grey: high land access without increase; Black: high land access with increase.* ***Crops.*** *White: low contribution to income / Dark: high contribution to income*

4 Discussion

4.1 Technical and Economic Performances of the Dairy Activity

It is surprising that low milk yields in mixed crops-dairy farms (T4). This is mainly caused by the low nutritional value of feed provided by farmers. Farmers gave only a small amount of concentrate feed, as referred to in feed costs. Farmers preferred tofu waste and cassava waste. However, those feeds have low energy values [13] and nitrogen [14]. It is shown by the high additional feed costs used by the T4 farms.

Our findings are also somewhat surprising since there is high variability in T2 farm performance. Consequently, these differences are linked to the quality of feed and the buy-sale of livestock in the year of data collection. This, in principle, can give rise to a low quantity of feed due to low land access in a very dense population area. Farmers replaced high-quality feed such as Napier grass with low-nutrient feed such as paddy straw [15]. They also provided a high amount of concentrates that could lead to digestive issues and poor performance [16].

4.2 The Future of the Development of Dairy Production based on the Farm Types

The most intriguing finding is that the significant pathways of development of the farms' types are the rise in herd size, access to land, and diversification of the activity generating income. These alterations are the reaction of the capital endowment. It affects milk produced at the farmer's level to guarantee income and supply for the Indonesian market.

Farms with very small land access. These farms are essential in the national market. They provide milk to the dairy industries and the national dairy market. They are multiple and provide significant contributions in milk supply to the coop: T1 and T2 account for 56% of the farms – out of 355 farm samples, and contribute 47% of the milk sold to the cooperative. Small farmers with two cows selling milk to the dairy industry (through cooperative) are ordinary in Indonesia [17].

The smallholders are in a challenging position. Farms with little capital might be more vulnerable to the continuation of the dairy business. Dairy activities may not always cover regular expenditures due to low economic conditions. This is because of the limitations of enhancing other production factors, particularly access to land and herd size. In strategies to adapt to the shortage of land access, some farmers utilize non-cultivated forage by buying, cutting, and transporting natural grass. Even so, the continuity of the future remains unknown, given the availability of fodder and land in the nearby area. Farmers could have credit access from the cooperative to increase the herd, but the total is based on their assets. It means that they only have access to limited cash.

Off-farm activities in very small farms. In their trajectories, poor farmers (T2 farms) seem more changeable. Adding off-farm activity becomes a reasonable reaction if a farm has available workers but limited access to land. There are two explanations for this, incomplete market and risk reduction [6]. Because of incomplete land and labor markets, family workers would be assigned to other activities to optimize production factors. In the case of incomplete financial markets (credit and insurance), off-farm activity plays a vital role in addressing purchasing production factors [6]. Reduction of risk is another reason why farmers broaden their activities [6, 18]. This study showed that poor farmers (especially in T2) add off-farm activities to mitigate the risk of dairy farming. Studies performed by [19] confirm that diversifying activity can minimize the dairy business's risk.

On the other side, the reduction of off-farm activity was mainly due to the lack of availability of family workers. It may be associated with a missing labor market [6]. Reduced

family workers while growing the number of cows allow farmers to quit their off-farm activity and focus solely on dairy farming. Following [20], highlight the significance of labor force and capital on the farm development pathway.
The strategy for adding off-farm activities is a wise decision to ensure family incomes and mitigate production factors' limitations. Also, improving herd by a credit scheme may not be sufficient to encourage these farms. It might be successful if it is accompanied by an improvement in other factors of production. This research confirms findings from previous observations reported by [4].

Farms with more access to land and mixed crop-dairy system. The attention of the crop-dairy system might be seen from both the technical-economic performance. We did not identify any dynamics for dairy production development in the hybrid system (T4 farms). However, the family may have decided to leave the dairy activity and focused on crop production (but not seen based on our sample).

T4 farms can maintain dairy activity. Regarding dairy production growth to guarantee income and meet national demand, these farms appear to be steady or slightly improved. The number of farms is small (17 percent of farms), and they support only 15 percent of the milk supply to the cooperative. The size of the farm (herd, access to land, workers) is slowly growing. It may relate to their capital (especially family workers and land), which should be optimally kept for two activities (crop-dairy) to have higher household incomes. Also, workers' productivity in crops could be better than in dairy farming. Therefore, there is less motivation for the family to improve more dairy than crops, enlarge the land to cultivate forage at the expense of crops. Most of them also rely heavily on the land of Perhutani or PTPN for foraging. Given the complexity of land conversion, it could be fragile in the future. It highlights the importance of self-sufficiency in production factors, such as land [21].

Farms with better access to land and dairy specialization. Bigger dairy farms (T3) have the main reasons for choosing a dairy business specialization rather than a mixed farm. First, there is a lack of family workers. The second reason was limited land for crops nearby. Allowing only one activity is the easiest way to manage capital efficiently.

These farms (T3) are in the development pathway. They were able to capitalize (herd and land access). They had higher capital than other farms' types. Higher capital allows farmers to obtain more credit to overcome limited production factors. Farmers have a complete market under this condition. [6] demonstrated that farmers use a small part of their income and save the rest if a complete financial market exists. It enables farmers to have steadily natural growth of their animals.

The total of T3 farms is around 26% of the total farms, but they can deliver 38% of the total milk supply to the cooperative. On average, other farm types (T1, T2, and T3) provide ten kiloliters of milk to the coop annually. The farm of the T3 produces 17 kiloliters. These farms may play a significant role in developing the dairy sector in the area and at the national level. However, in the context of a strategy to supply the national market, focusing only on T3 farms seems irrelevant. It is because there is not too much difference in size among producers.

5 Conclusion

This study reveals a classical circumstance with a gradient among smallholders in difficulties to secure their families' livelihoods and those in the development trajectory of their farms through dairy and commercial crops. The study also underlines the significance of the smallholders' initial assets to illustrate their further farm trajectory. For those in the development trajectory, two strategies coexist between the dairy business's specialization and the system of mixed crops-dairy. The decision of one or the other is more related to pathway dependency. The mixed system could generate a similar net income with the specialized dairy system but with more levers of protection, combining two activities to secure livelihoods. The dairy activity provides income, and livestock plays a vital role in saving. Even so, this activity on a tiny scale is highly responsive to risks. As they are multiple, these very small producers provide a significant portion of the milk supply to the cooperative. Given the current growth of milk consumption and its significance to food security in developing nations, these very small producers are also crucial in participating in food security at the national level.

Acknowledgment

The authors gratefully acknowledge Danone Ecosysteme, who funded this study, together with CIRAD. We are also thankful to Sahabat Cipta Foundation, the KPSBU coop, local communities, and producers for their cooperation during the field survey.

References

1. FAO, FAOSTAT: Data of Dairy Product. FAO. Rome (2018)
2. J. Moran, Business Management for Tropical Dairy Farmers. Landlinks Press (2009)
3. J. Bernard, P. Y. Le Gal, P. Triomphe, N. Houstiou, & C. H. Moulin, Involvement of small-scale dairy farms in an industrial supply chain: When production standards meet farm diversity. Animal. 5-6: 961-971 (2011)
4. P. Sembada, G. Duteurtre, B. P. Purwanto, & Suryahadi, Improved milk production performance of smallholder farms in West Java (Indonesia). Trop. Anim. Health Prod. 48: 793-799 (2016)
5. A. -I. García-Arias, I. Vázquez-González, F. Sineiro-García, & M. Pérez-Fra, Farm diversification strategies in northwestern spain: factors affecting transitional pathways. Land Use Policy. 49: 413–425 (2015)
6. C. Barrett, T. Reardon, & P. Webb, Nonfarm income diversification and household livelihood strategies in rural Africa: concepts, dynamics, and policy implications. Food Policy. 26: 315–331 (2001)
7. Direktorat Jenderal Peternakan dan Kesehatan Hewan, Statistik Peternakan dan Kesehatan Hewan. Kementerian Pertanian Republik Indonesia, Jakarta (2016)
8. M. Muzayyanah, S. Syahlani, & Y. Suranindyah, Profiles of smallholder dairy farmers groups after volcanic eruption damage in Indonesia: A case study of Sleman Regency, D.I. Yogyakarta. J. Int Dev and Coo. 19(4): 121–129 (2013)
9. C. -H. Moulin, S. Ingrand, J. Lasseur, S. Madelrieux, M. Napoléone, J. Pluvinage, & V. Thénard, Comprendre et analyser les changements d'organisation et de conduite de l'élevage dans un ensemble d'exploitations: propositions méthodologiques. In: Dedieu et al. L'élevage en Mouvement: Flexibilité et Adaptation des Exploitations d'Herbivores. 181–196 (2008)
10. P. D. Malanski, N. Hostiou, & S. Ingrand, Evolution pathways of employees' work on dairy farms according to task content, specialization, and autonomy. Cah. Agric. 26: 65005 (2017)
11. J. Bertin, La graphique et le traitement graphique de l'information. Flammarion, Paris, France (1977)
12. N. Hostiou, & B. Dedieu, A method for assessing work productivity and flexibility in livestock farms. Animal. 6: 852–862 (2012)

13. A. Anggraeni, S. Hasibuan, B. Malik, & R. Wijaya, Improving the quality of tofu waste as a source of feed through fermentation using the bacillus amyloliquefaciens culture. Int. J. Adv. Sci. Eng. Inf. Technol. 3: 285 (2013)
14. R. Antari, & U. Umiyasih, Optimizing the use of cassava plant and its byproduct as ruminant feed. War. Indonesia. Bull. Anim. Vet. Sci. 19: 191–200 (2009)
15. I. Prihartini, S. Soebarinoto, S. Chuzaemi, & M. Winugroho, Nutrient characteristics and fermented rice straw degradation by Lignolitic TLiD and BopR Inoculums. Anim. Prod. 11: 1–7 (2011)
16. H. Dong, S. Wang, Y. Jia, Y. Ni, Y. Zhang, S. Zhuang, X. Shen, & R. Zhao, Long-term effects of subacute ruminal acidosis (SARA) on milk quality and hepatic gene expression in lactating goats fed a high-concentrate diet. PLoS ONE. 8: e82850 (2013)
17. P. Sembada, G. Duteurtre, & C.-H., Moulin, The essential role of farm capital in the sustainability of smallholder farms in West Java (Indonesia). Cah. Agric. 28: 15 (2019)
18. H. Hansson, R. Ferguson, C. Olofsson, & L. Rantamäki-Lahtinen, Farmers' motives for diversifying their farm business – The influence of family. J. Rural Stud. 32: 240–250 (2013)
19. S. Madelrieux, M. Terrier, D. Borg, & L. Dobremez. Family dairy farms in the northern french alps: persistence and adaptation in a changing world. Mt. Res. Dev. 35: 49–56 (2015)
20. M. Baccar, A. Bouaziz, P. Dugué, P. –Y. Le Gal, Shared environment, diversity of pathways: dynamics of family farming in the Saïs Plain (Morocco). Reg. Environ. Change. 17: 739-751 (2016)
21. T. Lebacq, P. V. Baret, & D. Stilmant, Role of input self-sufficiency in the economic and environmental sustainability of specialised dairy farms. Animal. 9: 544–552 (2015)

Treatment of Suspect Feline Panleukopenia in Cat at Healthy Pet Animal Clinic in Madiun

Henny Endah Anggraeni[1*], *Muhammad* Nurhudayanto [1], and *Winantika* Aprillia Mutiara Fitri [2]

[1]Paramedic Veteriner Study Program, College of Vocational Studies, IPB University, Indonesia
[2]Healthy Pet Animal Clinic, Madiun, East Java, Indonesia

Abstract. Feline panleukopenia is a highly contagious, often fatal, viral disease of cats. This scientific report aims to describe information about treatment of suspect feline panleukopenia in cat at Healthy Pet in Madiun East Java, Indonesia. Daily observations on a cat (Moni) hospitalized for dehydration, vomiting and diarrhoea, WBC count 1,400 cell/mm³. The treatment had 3 components of monitoring, medication, and biosecurity application. Monitoring was 4 times a day (temperature, faeces consistency, vomiting and appetite) for 8 days. Medication was antibiotics Sulfadiazine® to treat secondary bacterial infections, Ondansetron® to control of vomiting and Loperamid® to control of diarrhoea. Fluid therapy RL was performed to correct dehydration and electrolyte abnormalities. Biosecurity application was applied as breaking the chain of transmission (isolation room, sanitation 2 times a day and physical distancing). The cat was declared healthy on 8th day based on WBC count 15,900 cell/mm³, good appetite and normal defecation.

Key words: Feline panleukopenia, Healthy Pet Madiun, Treatment

1 INTRODUCTION

1.1 Background

Feline panleukopenia is a viral disease that attacks young cats or kittens, with the primary agent being the Feline panleukopenia virus which is the prototype of parvovirus. panleukopenia means a decrease in the number of all of the white blood cells in the body. White blood cells play a major role in immunity and are important in defending against infections and diseases. Clinical symptoms caused by this disease are leukopenia, vomiting, depression, diarrhoea, and dehydration [1]. The transmission route of this virus is through faeces, vomit, saliva, and urine of infected cats. The treatment for this suspect is to isolate the infected cat to break the chain of transmission.

Feline panleukopenia is most likely to occur in cats younger than 1 year of age, but it can occur in unvaccinated or improperly vaccinated cats of all ages [7]. However, kitten deaths have been reported in households of fully vaccinated kittens, possibly because of exposure to large amounts of virus in the environment [8]. Cats with panleukopenia that survive the first 5 days of treatment usually recover, although recovery is often more

* Corresponding author: hennyendahanggraeni@apps.ipb.ac.id

prolonged than it is for dogs with parvoviral enteritis. In 244 cats with feline panleukopenia from Europe, the survival rate was 51.1% [7].

Healthy Pet Animal Clinic is an Animal Clinic that handles several pets and exotic animal diseases. Intensive infectious disease treatment carried out at the Clinic will affect the recovery rate of the affected cat.

1.2 Purpose

This study aims to describe and provide information about the treatment of suspect feline panleukopenia in cats.

2 CASE ANALYSIS

Methods of collecting data was using primary and supporting data collection. Primary data was carried out by observation of FPV suspects at the Healthy Pet Madiun clinic. Primary data was observations (anamnesis, clinical examination, feeding and drinking, sanitation, daily observation, treatment) of suspect feline panleukopenia at Healthy Pet Madiun clinic, East Java, Indonesia. Supporting data was obtained from the literature related to FPV. Observations were conducted on one cat (her name is Moni) with suspect feline panleukopenia.

3 DISCUSSION

Anamnesis is not appetite and diarrhoea. According to Mahendra et al. [2], common symptoms in cats with FPV infection are vomiting, diarrhoea and decreased appetite. Signalement of suspect feline panleukopenia at healthy pet clinic Madiun is presented in Table 1.

Table 1. Signalement of cats suspect feline panleukopenia

1.	Name	Moni
2.	species	cat
3.	Race/breed	Mix domestic
4.	Fur colour	Red tabby
5.	Gender	female
6.	Age	>1 years old
7.	Special sign	There is an orange line on the left chest

Moni had a weight of 2.8 kg and body temperature of 39.2 °C. The cat was dehydrated and the skin turgor was more than 3 seconds. Results of haematology blood examination are described in Table 2.

Table 2. The results of the first Moni's hematology examination

Inspection	Result	Normal Range	Unit
White Blood Cell (WBC)	1.4	5.5-19.5	$10^3/\mu L$
Red Blood Cell (RBC)	7.4	4.60-10.00	$10^6/\mu L$
Hemoglobin (Hb)	9.2	9.3-15.3	g/dL
Hematocrit (HCT)	39.5	28.0-49.0	%
Platelets (PLT)	75	100-514	$10^3/\mu L$

Based on the physical examination and haematology results, veterinarian diagnosed that it was suspect feline panleukopenia. White blood cell or WBC count reduced from the normal range. According to Ishida [3], the normal number of white blood cells in cats is 5,500 – 19,500 cells / mm^3, Moni's WBC level was 1,400 cells / mm^3. According to Purnamaningsih [4] in his research, the number of white blood cells in cats infected by FPV is generally 1,700-4,900 cells / mm^3.

Monitoring was carried out 4 times a day including examination of activity and appetite, examination of temperature, faeces and vomit. According to Suartha [5], to prevent bad conditions, monitoring is performed every 4 h in 24 h and patient monitoring depends on the patient's condition. Monitoring the development of the cat's condition can be reviewed from the consistency of faeces, and vomit checking was also very important in monitoring the development of the cat's condition. Vomiting is one of the clinical signs of FPV, so vomiting becomes one of the indicators of a cat's condition. During the treatment, the cat vomited on the 1st day, and the 2nd day.

Medication was given using antibiotics Sulfadiazine® to treat secondary bacterial infections that commonly develop due to the lack of white blood cells and the resulting reduced immunity. Ondansetron® was given to control of vomiting and Loperamid® to control of diarrhoea. Vitamin B injections every day was extremely helpful. Fluid therapy RL was applied to correct dehydration and electrolyte abnormalities. Prednisone® at a dose of 1 mg / kg PO [6] is a drug that serves as an anti-inflammatory. The drug was mixed into 10 capsules for 5 days consisting of Sulfadiazine® 50mg, Neurobion® 10mg, Elkana® 35mg and Predison® 1.5mg.

The body temperature of Moni fluctuated until the 4th day. The body temperature stabilized on the 5th day and Moni was declared healthy on 8th day based on WBC count 15.900 cell/mm^3 (Table 3), good appetite and normal defecation. The normal range of white blood cell count was 5.5-19.5x103/µL.

Table 3. The result of Second Haematology Examination

Inspection	Result	Normal Range	Unit
White Blood Cell (WBC)	15,9	5,5-19,5	10^3/µL
Red Blood Cell (RBC)	8,56	4,60-10,00	10^6/µL
Hemoglobin (Hb)	10,5	9,3-15,3	g/dL
Hematocrit (HCT)	346,2	28,0-49,0	%
Platelets (PLT)	223	100-514	10^3/µL

The treatment of FPV disease is to apply strict hygiene and biosecurity in the clinic. Strict hygiene can reduce the spread of FPV disease. Washing hands before and after handling patients will reduce the rate of disease transmission. Infected animals are placed in infectious hospitalizations using glass isolation cages equipped with air ducts located separate from non-infectious hospitalizations with air conditioning. Sanitation is carried out 2 times a day using disinfectants containing active hypochlorite.

4 CONCLUSION

In this study, the treatment consisted of 3 components, i.e., monitoring, medication and biosecurity application. Monitoring was performed 4 times a day (temperature, faeces consistency, vomiting and appetite) for 8 days. Medication was using antibiotics

Sulfadiazine® to treat secondary bacterial infections, Ondansetron® to control of vomiting and Loperamid® to control of diarrhoea. Fluid therapy RL was performed to correct dehydration and electrolyte abnormalities. Biosecurity application was applied as breaking the chain of transmission (isolation room, sanitation 2 times a day and physical distancing). The cat was declared healthy on 8th day based on WBC count 15,900 cell/mm^3, good appetite and normal defecation.

Acknowledgement

The case study was conducted at the Healthy Pet Animal Clinic, Madiun, East Java, Indonesia

References

1. Tilley, L.P. and Smith, F.W.K. The 5-Minute Veterinary Consult: Canine and Feline, 3e, lviii, 1487 p.Baltimore, MD: Lippincott Williams & Wilkins (2004).
2. Mahendra, Y. N., Yuliani, M. G. A., Widodo, A., Diyantoro, D., & Sofyan, M. S. A Case Study of Feline Panleukopenia in Cats at The Educational Animal Hospital of Universitas Airlangga. Journal of Applied Veterinary Science And Technology, 1(1), (2020).
3. Ishida, T, How to Get Maximum Information Out of Feline Hematology. Prosiding WSAVA World Congress Proceedings (2011).
4. H. Purnamaningsih, S.Indarjulianto, Y.Yanuartono, A. Nururrozi, I. Widiyono, R. Hayati, Jurnal Sain Veteriner. Vol 38, No 2 (2020)
5. Suartha IN. Terapi Cairan pada Anjing dan Kucing. Laboratorium Penyakit Dalam Fakultas Kedokteran Hewan Universitas Udayana, Denpasar. Buletin Veteriner Udayana. 2(2):69-83. (2010).
6. Plumb DC. Veterinary Drug Handbook Sixth Edition. PharmaVet Inc. Stockholm, Wisconsin. ISBN: 978-0-8138-1097-3 (2008).
7. Jane E. Sykes. Elsevier Public Health Emergency Collection. Canine and Feline Infectious Diseases. 2014 : 187–194.
8. Kruse B.D., Unterer S., Horlacher K. Prognostic factors in cats with feline panleukopenia. J Vet Intern Med. 2010;24:1271–1276.

Towards Sustainable Fisheries Through Marine Ecological Carrying Capacity Index

A Susandi[1,2], *A* Wijaya[1], *W S* Kuntoro[1], *I* Faisal[1], *F G* Kertabudi[1], and *I* Nurdin[1]

[1]Bandung Institute of Technology
[2]Sekolah Tinggi Intelijen Negara (STIN)

Abstract. Overexploitation and unsustainable fisheries activities will have an impact on environmental quality degradation. In the 4.0 era, technology can assist in monitoring and managing sustainable fisheries resources. Research on sustainable fisheries management activities guided by the marine ecological carrying capacity (MECC) evaluation index has been carried out. The purpose of this study was to determine changes in fishery activity to MECC as an evaluation material for sustainable development in the fisheries sector in the form of a website. This can provide convenience and effectiveness and can be used as a basis by stakeholders in monitoring and making policies in the sustainable development of the fisheries sector. MECC is composed of two components, namely carrying object (OI) and carrier resilience (RI). OI is obtained from changes in the value of changes in human activity (RI) and socio-economic development (SI). Fishery activities are a component of RI in the Nunukan Regency, Indonesia. The HI value decreased during 2015 - 2019 and affected the decline in the MECC index as a whole. This decrease was caused by the increase in fish catches, the number of fishing households, and boat activity in the waters of the Nunukan Regency. Based on the evaluation of the decline in the value of HI, it can be used as an evaluation material in developing integrated and sustainable development in the fisheries sector by mitigating bycatch by regulating fishing gear and methodologies, as well as establishing protected area management and fishing. To make this happen, it is necessary to build partnerships between public authorities, business, civil society, academia, and the financial sector within the framework of the Penta-helix model.

Key Words: Fisheries Management, Sustainability, Marine Ecological Carrying Capacity (MECC), Information System

1 INTRODUCTION

Background

Fishery products make important contributions to the livelihoods, food and nutrition security, and well-being of international coastal groups. Fishery activities are vulnerable due to declining environmental health and climate change. Fisheries sustainability is challenged by several elements and techniques including overfishing and unprofitable fishing practices, improper shore repair, and land-based pollutants [1]. All of these can threaten the productivity of the useful resource base, poor fish handling, and abandoned harvesting practices, in addition to non-existent, or underdeveloped infrastructure, leading to high wastage, and loss of yields. Fisheries sustainability is also low with the cumulative impact of climate change, resulting in shifts in species abundance and variety, increased acidity and deoxygenation of the oceans, in addition to increasing climate turbulence, making sport fishing less certain and more unstable [1].

If climate change which decrease the biodiversity and increase the number of pollutants continue to emerge in the future, it will lead to more critical consequences. Consequently, to reduce the adverse impacts that arise, sustainable marine monetary improvement efforts are needed. Therefore, the idea of Integrated Ocean Management (IOM) techniques is vital. IOM is a sustainable concept that aims to improve the ability to use and sustainably manage marine resources and ensure that the health, productivity, and resilience of marine ecosystems can provide multiple benefits without reducing environmental quality due to human activities [2]. Consequently, it is imperative to effectively measure, monitor, and adaptively manipulate ocean space as a complex adaptive engine [3]. The Marine Ecological Carrying Capacity (MECC) can be used as a means to monitor IOM implementation. The regularly occurring characteristics of the affected MECC and the need for answers to address human activities and socio-economic improvements for marine and coastal management in Nunukan Regency is one of the unique contexts including local knowledge, environmental conditions, scale-up of the surrounding environment. movement, and the need for percentage facts and potential builds.

Strong resource control has been thwarted by the loss of information about how humans impact the ocean. There was an explosion in statistics and a new era for the ocean in the meantime and with it the vast potential for advances in knowledge and management of marine resources. Coordinated efforts by industry, researchers, and governments can create sophisticated sensor networks that provide anyone who wants them with a redundant, real-time record of the oceans, an "Internet of Things" for the oceans [4]. Era can take advantage of important improvements in control. Real-time statistics and automation can enable powerful and agile variations to change situations and create new accountability in authorities and across companies. An urgent priority is to ensure that the new capabilities are shared by all marine stakeholders. In the 4.0 era, technology can assist in monitoring and handling sustainable fisheries resources.

This research conducted a combination of MECC concept with the “4.0” approach. Calculation of MECC parameters was conducted inside a system with algorithm then it is depicted as simple as possible with cloud system. The simplicity of the information is important because it will be used by some of stakeholders including local government or even fisherman. Naturally, not all part of society could easily understand the MECC information, therefore, the visualization method plays an important role in order to transfer the information from MECC system to the stakeholders. The aim of this study is to map the potential of the coastal environment especially in Nunukan Regency with the approach of MECC concept that already integrated with cloud system.

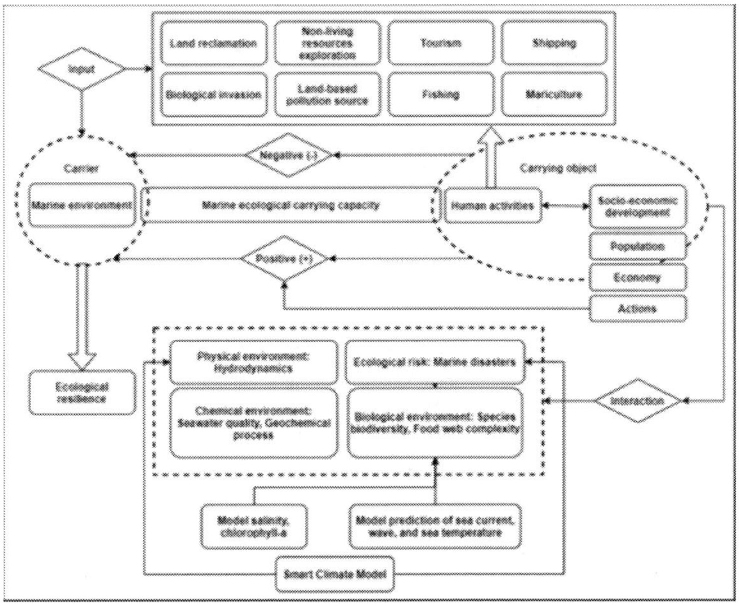

Figure 1. Marine Ecological Carrying Capacity (MECC) Modeling Concept Chart

2 METHODS

Marine Ecological Carrying Capacity (MECC) Monitoring System

Carrying capacity could be described as an ecological principle which states that a certain consumption value will accommodate a limited number of individuals as long as the surrounding ecosystem is not degraded. This concept is to describe the relationship between the community as a supporting entity (carrying object) and the environment as support (carrier) to ensure sustainability. The population carrying capacity is theoretically followed by the carrying capacity of resources, the carrying capacity of the environment, and finally the ecological carrying capacity (ECC). Many studies have been conducted with a focus on comprehensive regional carrying capacity because sustainable development policies have become the main guideline in regional socio-economic development. Compared to other ecosystem assessments used in sustainability research such as those used in health and climate risk studies, the results of this study are more promising [7], the ECC assessment looks at ocean-atmosphere climate, living things, and their relationships from a more holistic perspective. This metric is easier for the public and policymakers to understand because it offers a holistic understanding of the environmental impacts of sustainable economic and social change, and reveals capacity deficits and surpluses for specific ecosystem components [8]. The ECC has become a widely used index to create a sustainable regional ecological environment [8]. Human activities (such as fishing), coastal growth, and pollution have all changed and disrupted the capabilities of marine ecosystems as the marine economy developed [5].

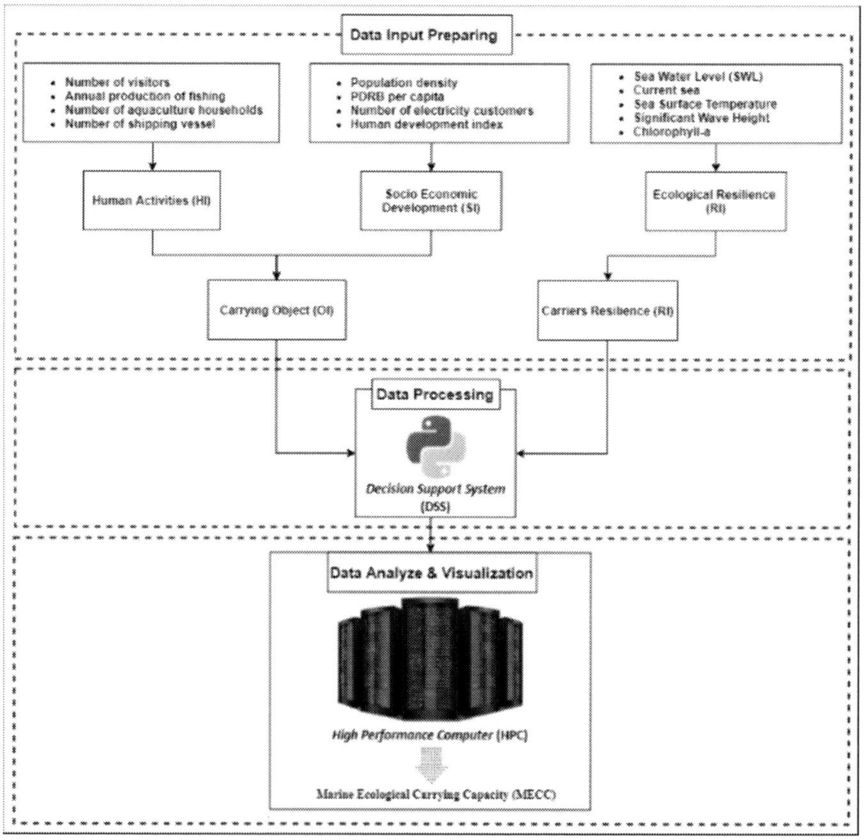

Figure 2. Design of the Marine Ecological Carrying Capacity (MECC) prediction calculation system.

There are three stages on forming the design system, namely data preparation, data processing, and data visualization (Figure 2). The MECC calculation to become An MPI consists of two components, namely the carrying object component (human activity and socio-economic growth) and the carrier resistance component (ecological resilience). Human activities include coastal and marine activities that directly suppress ecosystems (carriers), while socio-economic growth includes elements that represent the population and coastal economy, as well as actions to protect coastal areas. The carrier resilience component consists of elements that maintain or damage the structure and function of coastal and marine ecosystems.

Figure 3. Value of MECC indicators in Nunukan regency during 2015-2019

3 RESULTS AND DISCUSSION

Nunukan Regency is one of the Marine and Capture Fisheries Centers (SKPT) that has been established by the Ministry of Maritime Affairs and Fisheries (KKP) in Indonesia. MECC indicator data for Nunukan Regency for 2015-2019 was obtained from the Central Statistics Agency (BPS) of the Nunukan Regency, European Center for Medium-Range Weather Forecasts (ECMWF), Moderate Resolution Imaging Spectroradiometer (MODIS), and Visible Infrared Imaging Radiometer (VIIRS) as shown in Figure 3.

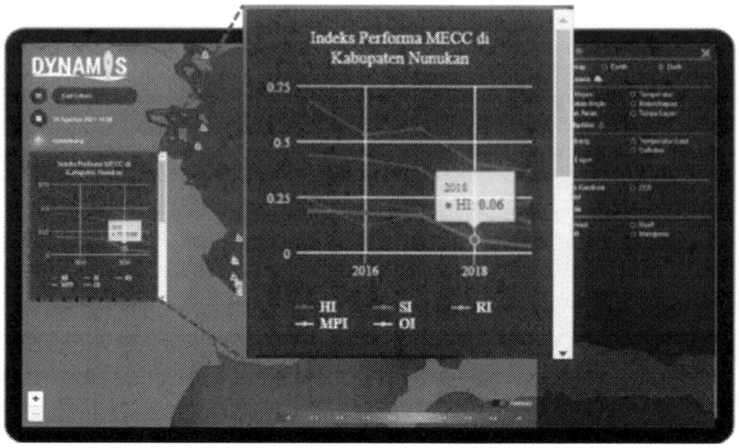

Figure 4. Graph of human activity carrying index (HI) in Nunukan Regency 2015-2019

The graph of the value of the innate human activity index (HI) in Nunukan Regency in 2015-2019 is shown in Figure 4 showing a decrease from 2015 to 2019. The value of HI has decreased by 82% from 0.19 in 2015 to 0.03 in 2019, which shows the effect of increasing human activities that continue to increase. This shows that the increase in the development of socio-economic activities is not proportional to the increase in the human development index in Nunukan Regency.

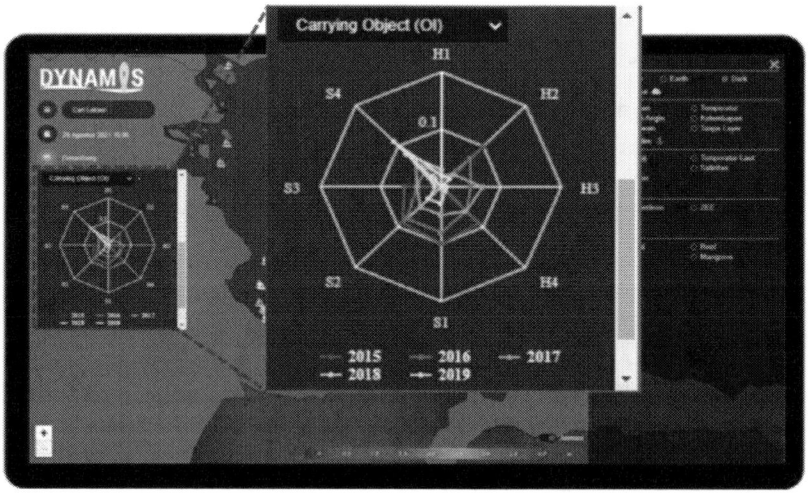

Figure 5. The value of 8 carrying indicators in Nunukan Regency during 2015- 2019

Figure 5 depicts the value of the bearer object indicators (HI and SI) during 2015-2019. In indicators of human activity (H1-H4), the production value of fish catches (H2) and the number of vessels (H4) in the waters of the Nunukan Regency was increased from 2015 to 2019. This shows that the fishing industry, both capture fisheries, and fisheries aquaculture and marine transportation in Nunukan Regency continued to increase from 2015 to 2019. This can be seen from the increase in the number of capture fisheries products (H2, average rate of increase of 6%) and the number of fishermen households (H3, 1% increase rate on average). In addition, sea transportation activities also experienced an increase as seen from the number of ships operating in Nunukan waters (H4, the average rate of increase increased by 4%) while the number of tourists decreased (H1, the average rate of increase decreased by 2%). The increase in activity in this area shows that there has been excessive exploitation in the coastal area so that which can lead to disruption of the stability of the coastal ecosystem in Nunukan Regency.

Based on the evaluation of the decline in the value of HI, it can be used as an evaluation material in developing integrated and sustainable development in the fisheries sector through by catch mitigation by regulating fishing tools and methodologies, as well as determining the management of protected areas and fishing. To achieve this, it is necessary to build partnerships between public authorities, business, civil society, academia, and the financial sector within the framework of the Penta-helix model.

4 CONCLUSION

Based on the MECC analysis, the value of HI has decreased during 2015 – 2019 and has an impact on the decline in the MECC index as a whole. This decrease was caused by an increase in fish catches, the number of fishing households, and boat activities in the waters of the Nunukan Regency. Based on the evaluation of the decline in the value of HI, it can be used as an evaluation material in developing integrated and sustainable development in the fisheries sector through bycatch mitigation by regulating fishing tools and methodologies, as well as determining the management of protected areas and fishing.

References

[1] Stacey, N., Gibson, E., Loneragan, N. R., Warren, C., Wiryawan, B., Adhuri, D. S., ... & Fitriana, R. (2021). Developing sustainable small-scale fisheries livelihoods in Indonesia: Trends, enabling and constraining factors, and future opportunities. Marine Policy, 132, 104654.

[2] Winther JG, Dai M, Rist T, Hoel AH, Li Y, Trice A, Morrissey K, Juinio-Menez MA, Fernandes L, Unger S, and Scarano FR 2020 Nat. Ecol. Evol. 4(11) 1451-1458.

[3] Lubchenco, J., E.B. Cerny-Chipman, J.N. Reimer and S.A. Levin 2016 Proc. Natl. Acad. Sci. U.S.A. 113(51) 14507–14.

[4] Leape, J., Abbott, M., & Sakaguchi, H. (2020). Technology, Data and New Models for Sustainably Managing Ocean Resources. World Resources Institute

[5] Ma P, Ye G, Peng X, Liu J, Qi J, and Jia S 2017 Ocean Coast Manag **144** 23-30

[6] Martire S, Castellani V, and Sala S 2015 Resour Conserv Recy **94** 11-20

[7] Gaillard J 2010 J. Int. Dev **22(2)** 218-232

[8] Wang S, Xu L, Yang F, and Wang H 2014 Science of The Total Environment **472** 1070-1081

The Effect of Audit Experience and Audit Risk on Audit Judgment with Auditor's Perceptions of the Code of Ethics of Public Accountants as Moderating Variables

Eka Merdekawati[1*]

[1]Accountancy, Sekolah Vokasi, Institut Pertanian Bogor, Indonesia

Abstract. This research purposes to empirically analyze the effect of audit experience on audit judgment, to determine auditor's perception of the code of ethics of public accountants as moderating variable that affect audit experience on audit judgment, to assess the effect of audit risk on audit judgment, to determine auditor's perception of code of ethics of public accountants as moderating variable that effect of audit risk on audit judgment. This research used a sample of 140 external auditors who work in the KAP at DKI Jakarta period 2016-2020. Based on method convenience sampling. Data were obtained through a mail survey and personal survey. The analytical method used in this study is Moderated Regression Analysis (MRA). The result of this research indicates that; audit experience not effect significantly on the audit judgment, auditor's perception of the code of ethics for public accountants is not moderating variables that affect of audit experience on audit judgment, audit risk not affect significantly on the audit judgment, auditor's perception of code of ethics for public accountants is moderating variable that effect of audit risk on audit judgment.
Keywords: Audit Experience, Audit Risk, perception of code of ethics for public accountants, and Audit Judgment

1 Introduction

In recent decades, the profession of external auditor (public accountant) has been under sharp scrutiny from the general public because auditors are seen as acting instead to serve or become advocacy for clients. The disclosure of the Jiwasraya case in 2018 caused a loss of up to Jiwasraya's capital minus, and the State is estimated to have suffered a loss of up to Rp. 13.7 billion (beritasatu.com). In this case, it can see that the auditors who audited Jiwasraya have failed to detect or disclose fraud in their financial statements. Such failures can occur because *professional judgment is* not implemented in auditing financial statements [1].

The auditors in making *judgments* must use their professionalism as accountants. According to the ISA 200, *professional judgment* is the application of relevant knowledge and experience, in the context of auditing, *accounting,* and ethical standards, to reach

*eka.merdekawati@apps.ipb.ac.id

appropriate decisions in situations or circumstances during the running of audit assignments and personal qualities, meaning that *judgment* differs among experienced auditors (but training and experience are intended to encourage consistency in *judgment).*

Some experts conduct various studies on factors that influence audit *judgment.* Jayanti [3] conduct research that affects Independence, Experience, Complexity of Tasks, Pressures of Obedience, and Auditor Ethics on Audit *Judgment.* Auditor experience, locus of control, and knowledge of detecting errors positively affect audit judgment [4]. In addition, *audit judgment* can also be affected by audit risk. Audit risk occurs if the auditor, unwittingly, does not modify their opinion as appropriate on a financial statement containing material misstatements. The existence of standard audit reports in the examination of financial statements aims to obtain adequate (not absolute) confidence that the audited financial statements have been free from material misstatements. Rachmawati et al. [5] indicate that audit risk affects the auditor's judgment.

These audit and experience risk factors can also be influenced by one more element, namely the auditor's perception of the public accountant's code of conduct. This is in line with the fact that an external auditor is inseparable from ethics about the *judgment audit* process, dilemmas. The dilemma arises when an auditor is confronted with choices between values as opposed to the client, which asks the auditor to take actions that violate the standards of examination. It found that ethical perceptions affect audit *judgment* [6]. It is argued that the role of the code of ethics is essential in determining audit *judgment* [7].

From the background above, the author is interested in researching whether the auditor's perception of the public accountant's code of conduct can strengthen the effect of the audit experience and audit risk on audit *judgment.*

2 Literature Theory and Hypothesis Development

2.1 Effect of Audit Experience on Audit Judgment

Some research shows that the auditor's experience contributes significantly to the *judgments* made by the auditor. Experience can also affect the auditor's predictive and detection ability against fraud to affect the *judgment* taken by the auditor. Experience is an important attribute that the auditor has.

Learning orientation has a more substantial effect on audit judgment performance than the objective orientation of performance approaches and performance avoidance. Self-efficacy mediates the effect of goal orientation when audit tasks are less complex than when tasks are more complex, so the audit experience affects decision making in audit judgment [8]. Pfulgrath et al. found that experience affects audit *judgment* [7]. Suraida [9] suggests that experience affects audit *judgment.* As for the research, Septyarini [14] and Tampubolon [15] say that the auditor's experience does not affect audit *judgment.* Based on the above understanding, it can be concluded the following research hypotheses:

H1: Experience has a significant effect on *audit judgment*.

2.2 Auditor's Perception of Accountant Code of Ethics Can Moderate The Effect of Audit Experience On *Audit Judgment*

The auditor's understanding of the code of ethics affects audit *judgment.* Perceptions between auditors and clients are sometimes different, which leads to conflict. Conflict situations occur when the auditor meets the client's demands meaning they are passing the rules or vice versa, and if the auditor does not complete the client's needs request, then the auditor will receive sanctions from the client in the form of possible termination of the assignment. Therefore, each ethical and moral perception plays an important role in the auditor's decision making. Fitriani and Daljono [6] Found that ethical perceptions affect audit *judgment*. It is argued that part of the code of ethics is essential in determining audit *judgment* [7]. An auditor who understands the code of ethics will tend to make the right *judgment* compared to an auditor who does not understand the code of ethics. It also reveals auditors with good ethics in obtaining information about client financial statements must be in accordance with established standards [10]. Based on the above understanding, it can be concluded the following research hypotheses:

H2: Auditor's Perception of Accountant Code of Ethics can Moderate the effect of Audit Experience On *Audit Judgment*.

2.3 Effect of Audit Risk on *Audit Judgment*

Audit risk occurs if the auditor, unwittingly, does not modify their opinion as appropriate on a financial statement containing material misstatements. Dutta [11] demonstrates that underestimating audit risk can occur if the audit risk model is used without explicitly considering the risks associated with fraud committed by management. It is revealed that audit risk affects the *skepticism of the auditor's professional* [9]. Rachmawati [5] indicate that audit risk affects the auditor's judgment. Based on the above understanding, it can be concluded the following research hypotheses:

H3: Audit risk has a significant effect on *audit judgment*

2.4 Auditor's Perception of Accountant Code of Ethics can moderate the Effect of Audit Risk on *Audit Judgment*

Audit risk is the risk of providing an improper audit opinion on financial statements that are materially misserved. The purpose of an audit is to reduce the risk of this audit to a low level that the auditor can accept. Ethics is important because the auditor must perform his duties as an opinion giver on financial statements. High ethics will be reflected in the auditor's attitudes, actions and behavior. Auditors with good ethics in obtaining information about the client's financial statements must be in accordance with established standards [10].

The research results [9] found that audit and ethical risks strongly affect the professional *skepticism* of auditors and the accuracy of the provision of auditor opinions by public accountants. It found the same thing that ethics affect professional *skepticism*. It was also found that ethical orientation is significant with *audit judgment* [12]. Based on the above understanding, it can be concluded the following research hypotheses:

H4: Auditor's Perception of Accountant Code of Ethics can moderate the Effect of Audit Risk On *Audit Judgment*.

The thinking model can present this research in Figure 1 below based on the theoretical study outlined earlier.

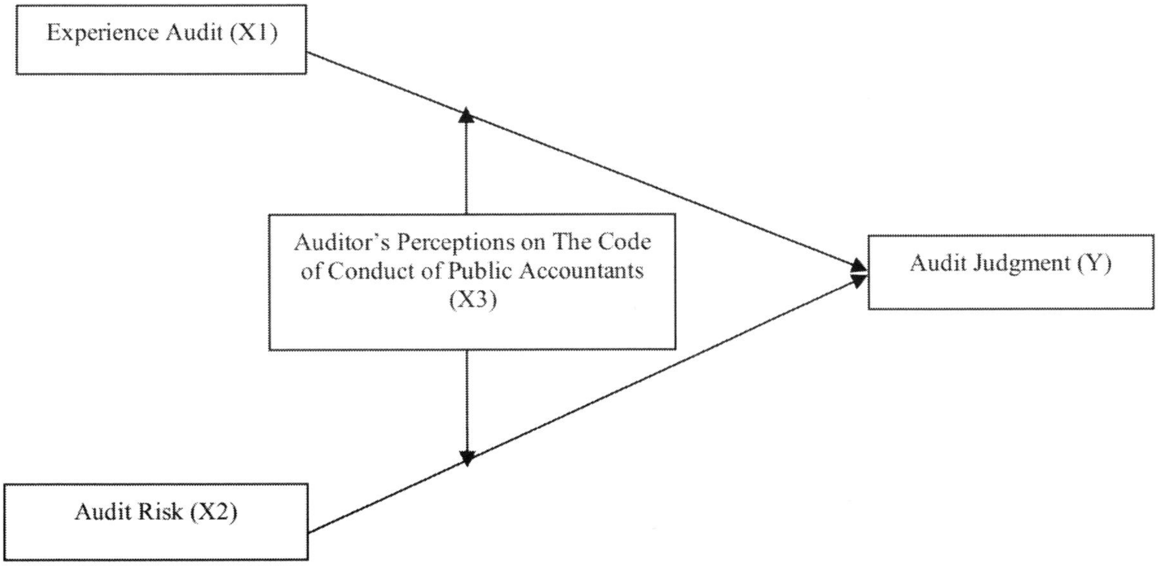

Fig. 1. Research Framework Model

3 Research Method

This study used a population of external auditors working at the Public Accounting Firm (KAP) in Jakarta registered with the Indonesian Akuntan Association (IAI) Directory for 2016-2020. As for the sample in this study, there were 140 External Auditors in Jakarta. The sampling technique used in this study was the convenience sampling method. The respondent criteria were external auditors who work at the Public Accounting Firm (KAP) in Jakarta in accordance with the Director of the Indonesian Akuntan Association (IAI) in 2016-2020 with work experience of at least one known. The data collection method used was the *mail survey* method and *personal survey*. The analytical method used in this study was *Moderated Regression Analysis* (MRA). MRA is a specialized application of multiple linear regression in which it contains elements of regression equation interaction (multiplication of two or more independent variables) [13].

4 Results and Discussion

4.1 Research Instrument Test Result

Testing of research instruments both in terms of validity and reliability of 137 respondents obtained that the results of the research instruments used were valid at a significance level of 5%, and the coefficient of reliability (Cronbach Alpha) was more significant than 0.6 [13].
The results of the normality test using the Kolmogorov-Smirnov test amounted to 0.909. Based on the testing criteria, because the probability value was > 0.05, it can conclude that the regression error (residue) data was a normal distribution. The heteroskedasticity test using the Glejser test showed that all variables have a Sig value more significant than the level of $\alpha = 0.05\%$. This indicates that residual (error) that arises from the regression equation has the same variant or does not occur heteroskedasticity.

4.2 Hypothesis Test Result

The results of the statistical t test can be seen in Table 2 below.

Table 2. Statistical Test Results

Model		Unstandardized Coefficients		Standardized Coefficients	T	Sig.
		B	Std. Error	Beta		
1	(Constant)	9,188	3,859		2,381	,019
	PA	,055	,893	,053	,061	,951
	RA	-1,430	,926	-1,377	-1,543	,125
	KEA	-1,795	,905	-1,749	-1,984	,049
	PA* KEA	,045	,205	,313	,222	,825
	RA*KEA	,410	,219	2,888	1,876	,063
Dependent Variable: AJ						

Source: SPSS 20.0

The results of the equation using coefficients are as follows:

$$AJ = 9{,}188 - 0{,}55\ PA - 1{,}430\ RA - 1{,}795\ KEA + 0{,}045\ PA{*}KEA + 0{,}410\ RA{*}KEA + \varepsilon$$

1. Audit Experience on *Audit Judgment*
   From the results of the test t, it is known that the value of t-calculated Audit Experience to *Audit Judgment* is 0.061 with a significance level of 0.951($\geq$0.1). It can then conclude that hypothesis 1 (H_1) is rejected, which means the audit experience does not affect the *audit judgment.*
2. Experience auditing *judgment with* auditors' perception of the code of ethics of public accountants as a moderation variable.
   From the results of the test t, it is known that the value of t-calculate the audit experience of *audit judgment* with the auditor's perception of the public accountant's code of ethics as a moderation variable of 0.222 with a significance level of 0.825 ($\geq$0.1). It can then conclude that hypothesis 2(H2) is rejected, which means that the audit experience with auditor perception variables about the public accountant's code of ethics as moderation does not significantly affect the Audit *Judgment* so that the auditor's perception of the public accountant's code of ethics is not a moderating variable but is an intervening, exogen, antecedent or predictor.
3. Risks of Audit Audit to *Audit Judgment*
   From the results of the test t, it is known that the value of t-calculated Audit Risk to *Audit Judgment* amounted to -1.543 with a significance level of 0.125 (>0.1). It can then conclude that hypothesis 3 (H3) is rejected, which means that audit risk has no effect on *audit judgment.*
4. Audit Risk to *Audit Judgment* with auditor's Perception of Public Accountant Code of Ethics as Moderation Variable.
   From the results of the test t, it is known that the value of t-calculate the risk of the audit against *the audit judgment* with the auditor's perception of the public accountant's code of ethics as a moderation variable of 1.876 with a significance level of 0.063 ($\leq$0.1). It can then conclude that hypothesis 4 (H_4) is accepted, which means there is a positive influence of audit risk on *audit judgment* with the auditor's perception of the public accountant's code of ethics as a moderation variable. The risk of auditing the *audit judgment* with the auditor's perception of the public accountant's code of ethics as a

moderation variable is included in the type of quadrant moderator variable 3, namely *quasi moderator* (pseudo moderator), where the auditor's perception of the public accountant's code of ethics relates to *the audit judgment* and the auditor's perception of the public accountant's code of ethics as a moderation variable when interacting with audit risk has a significant effect on audit *judgment.*

4.3 Discussion of Research Result

4.3.1 Effect of Audit Experience on Audit Judgment

From the results of data analysis and hypothesis testing above, it is known that the audit experience does not affect *audit judgment,* so we can conclude that the results of this study do not support the hypothesis that the audit experience affects audit *judgment.* That is because the sampling procedure is not perfect. After all, the number of respondents is not too large, and work experience as an auditor is dominated by 78 (56.9%) experienced 0-2 years, 37 (27.0%) experienced 2-5 years, 9 (6.6%) experienced 5-10 years and 10 (9.5%) experience >10 years. In addition, the positions resulting from respondents were uneven, especially among managers and partners, but who answered the most to junior auditors by 72 (52.6%), seniors by 51 (37.2%), managers by 8 (5.8%), and partners by 6 (4.4%).

So that the auditor in making an audit consideration (audit judgment) has no effect because the respondent is generally a junior and senior auditor who in his duties as a member in a team only while who will make the *audit judgment* is the manager and partner. The results of this study are in line with Septyarini [14] and Tampubolon [15], which say that the auditor's experience does not affect *audit judgment.* However, it does not support the results of Pfulgrath [7] and Suraida [9], which suggest that experience has an important role in determining *audit judgment.*

Based on cognitive theory, practices in the field of auditing as independent auditors can be a means of learning and experience for auditors. The auditor will integrate the experience and knowledge in carrying out future tasks. So that the auditor's expertise and knowledge will continuously develop and support the auditor to make *professional* judgments.

4.3.2 Auditor's Perception of Accountant Code of Ethics can Moderate the Effect of Audit Experience on Audit Judgment

From the results of data analysis and hypothesis testing above, it is known that the auditor's variable of the public accountant's code of ethics is not a moderating variable but an intervening, exogen, antecedent, or predictor. Thus, the auditor's association with a high accountant's code of conduct does not strengthen the relationship between the audit experience and the *Audit Judgment.* So, the results of this study do not support hypothesis 2.

The code of ethics established by the Indonesian Accountants Association (IAI) is an indicator of the auditor's perception variables about the public accountant's code of ethics, namely professional responsibility, public interest, integrity, objectivity, competence, professional ingenuity and prudence, confidentiality, and professional behavior. The auditor's perception of the public accountant's code of conduct is not a moderating variable that can affect the experience of audit *judgment,* this can happen because accountants are often faced with a situation of dilemmas that cause and allow accountants cannot be independent. Conflict situations occur if the auditor meets the client's demands meaning he is violating the rules or vice versa. If the auditor does not meet the client's requests, then the auditor will receive sanctions from the client in the form of a possible termination of the assignment. This audit conflict will develop into an ethical dilemma when the auditor must make judgments regardless of the auditor having longer or shorter experience.

It is stated that the role of experience and code of ethics is very important in determining *audit* judgment [7]. An auditor who understands the code of ethics will tend to make the proper *judgment* compared to an auditor who does not understand the code of ethics. However, the auditor's perception of the public accountant's code of conduct is not determined by the magnitude or length of time a person becomes an auditor. So more experienced auditors don't necessarily have a better perception of an accountant's code of conduct than an inexperienced auditor. Thus, the auditor's perception of the public accountant's code of conduct is not a moderating variable to not strengthen the relationship between the audit experience and the audit *judgment.*

4.3.3 Effect of Audit Risk on Audit Judgment

From the results of data analysis and hypothesis testing above, it is known that audit risk has no effects *on audit judgment,* so it can conclude that the results of this study do not support hypothesis three that states that audit risk affects *audit judgment.* That is because the sampling procedure is not perfect. After all, the number of respondents is not too large. The positions resulting from respondents are uneven, especially among managers and partners but who answer the most to junior auditors by 72 (52.6%), seniors by 51 (37.2%), managers by 8 (5.8%), and partners 6 (4.4%).

Managers or partners make audit risk before the audit process is carried out, so the audit risk does not affect *audit judgment* because respondents in this study are generally junior and senior auditors who are in their duties as members in a team only while those who will make audit *judgments* are managers and partners. The results of this study do not support the results of Suraida [9] and Rachmawati et al. [5] indicate that audit risk affects the auditor's judgment. Dutta [11] demonstrates that underestimating audit risk can occur if the audit risk model is used without explicitly considering the risks associated with fraud committed by management.

Inherent risk, control risk, and detection risk are indicators of audit risk variables. According to ISA 200.A34, the risk of material misstatements can occur at two levels: at the level of the overall financial statements and the level of assertion for the types of reports, account balances, and disclosures. ISA 200 A says that auditors are not expected to, and cannot, reduce the audit risk to zero and therefore cannot obtain absolute insurance that financial statements are free from material misstatements caused by fraud or misconduct. That is due to the innate constraints in an audit that cause most audit evidence (which is the basis for the auditor's conclusions and opinions) to be *persuasive* and not *conclusive.* The inherent constraints in an audit arise from: the nature of financial reporting, the nature of audit procedures, and the need for the audit to be carried out within a decent period and at a decent cost [2].

4.3.4 Auditor's Perception of Accountant Code of Ethics can Moderate the Effect of Audit Risk on Audit Judgment

From the results of data analysis and hypothesis testing above, it is known that the auditor's perception variable about the public accountant's code of ethics is a moderating variable because of the interaction of audit risk to audit *judgment* with the auditor's perception of the public accountant's code of ethics as a moderation variable, including the type of quadrant moderation variable 3, namely *quasi moderator* (pseudo moderator). Thus, the auditor's association with a high accountant's code of conduct strengthens the relationship between audit risk to *Audit Judgment.* So, the results of this study support hypothesis 4.

The role of audit risk and code of ethics is significant in determining *audit* judgment [9]. An auditor who understands the code of ethics will tend to make the proper *judgment* compared to an auditor who does not understand the code of ethics. Can determine the auditor's perception of the public accountant's code of ethics by the auditor's ability to detect audit risks. So, an auditor who can minimize the risk of auditing by not committing violations that have been implemented in conducting audits and complying with the code of ethics as an accountant in carrying out his duties will affect the *audit judgment.* Thus, the auditor's perception of the public accountant's code of ethics is a moderating variable.

The code of ethics established by the Indonesian Accountants Association (IAI) is an indicator of the auditor's perception variables about the public accountant's code of ethics, namely professional responsibility, public interest, integrity, objectivity, competence, professional discernment and prudence, confidentiality, and *professional* behaviour. The auditor's perception of the public accountant's code of ethics is a moderating variable that can affect audit risk to audit judgment. However, auditors are often faced with situations of dilemmas that cause and allow auditors to be independent. Still, the auditor's perception of a high public accountant's code of conduct can strengthen the relationship between audit risk to audit *judgment.*

5 Conclusions, Limitations and Suggestions

5.1 Conclusions

Based on the results of research and improvement outlined in the previous chapter, the following conclusions are: Audit experience does not affect *audit judgment.* This hypothesis one is rejected. This indicates that the audit experience does not affect the *judgments* made by the auditor. Auditors' perceptions of the public accountant's code of conduct cannot moderate the influence of experience on *audit judgment.* This hypothesis two is rejected which means that the auditor's perception of the public accountant's code of ethics is not a moderating variable. Audit risk does not affect *audit judgment.* This hypothesis three is rejected, this indicates that the audit risk does not affect *judgment* made by the auditor. Auditors' perception of the public accountant's code of ethics can moderate the effect of audit risk on *audit judgment.* This hypothesis four is accepted, which means that the auditor's perception of the public accountant's code of ethics is a variable moderating.

5.2 Limitations of Research

This research is basically in accordance with research procedures and meets the methodological principles of analysis. However, admittedly the results of this study, there are still shortcomings that can affect the results of this study. The author has made efforts and efforts to avoid things that can reduce the results of this study, but because of the lack of maximum collection of sample data, the busyness of the auditors because the process of taking questionnaires along with the activities of auditors is dense so that it could be that the respondents, in this case, external auditors fill it out in accordance with their conclusions and not based on the reality that happened and experienced by the auditor, so it isn't to distinguish the objective and subjective views that respondents may give.

5.3 Suggestions

This research in the future is expected to present more qualified research results with several things including further research is recommended for filling out questionnaires that should be done by managers, supervisors and partners who in their duties provide an audit consideration.

References

1. http://beritasatu.com. Kasus Jiwasraya di Indonesia. Retrieved on 23 Desember 2020. (2018)
2. Tuanakotta, Theodorus M. Audit Berbasis ISA. (Salemba empat, Jakarta, 2013)
3. Jayanti, Dwi. J. Naskah Publikasi. Universitas Muhammadiyah (2017)
4. Puspita C, Made and Dwirandra. J. Akuntansi Univ. Udayana. **18**, 1, 327-357 (2017)
5. Rachmawati, Yuni, Darwanis and Usman Bakar. J. Akuntansi Pascasarjana Universitas Syiah Kuala **911-102** (2013)
6. Seni Fitriani, Daljono. J. Accounting. **1**, 11-12 (2012)
7. Pfulgrath, Gerry, Nonna Martinov-Bennie and Liang Chen. J. Managerial Auditing **24**, 183-203
8. Mohd Sanusi, Z., Iskandar, T.M., Monroe, G.S. and Saleh, N.M., J. Accounting, Auditing & Accountability. **31**, 1, 75-95 (2018)
9. Suraida. J. Penelit. Disertasi UNPAD (2007)
10. Sabrina, RRK dan Indira Januarti. Simposium Nasional Akuntansi XIV. Aceh, Indonesia (2011)
11. Saurav K. Dutta. The Audit Risk Model Under the Risk of Fraud *Applications of Fuzzy Sets & The Theory of Evidence to Accounting* (Jai Press Inc, 1998)
12. Juhaida, Razana Johari dan Zuraidah Mohd Sanusi. J. Accounting Research Universiti Teknologi MARA Shah Alam (2010)
13. Ghozali, Imam. Aplikasi Analisis Multivariate dengan Program SPSS. (Universitas Diponegoro, Semarang, 2013)
14. Septyarini, K. D. J. Naskah Publikasi **1-76**. (2015).
15. Tampubolon, L. D. *Jurnal Keuangan Dan Perbankan*, **14**, 2, 62–70 (2018)

The Effect of the Covid-19 Pandemic on Changes the Eating Habits of the Community in Bogor

Rosyda Dianah[1*], *Eka Merdekawati*[1]

[1]Sekolah Vokasi, Institut Pertanian Bogor, Indonesia

Abstract. The aim of this study was to determine the effect of the COVID-19 pandemic on changes in eating habits is a quantitative study with a cross-sectional design conducted in Bogor with 110 respondents. Changes in eating habits observed include changes in eating frequency, meal portions, frequency of fruit consumption, frequency of vegetable consumption, frequency of snack consumption, frequency of consumption of fast food/frozen food, and consumption of supplements. Respondents were more women (83,6%), adult age group (89,1%), higher education (79,1%), work (61,8%), income > IDR 4,000,000 (75,5%). The results of the univariate analysis related to changes in eating habits consist of: changes in eating frequency (69,1%), changes in food portions (56,4%), changes in fruit consumption frequency (61,8% more often), changes in vegetable consumption frequency (63,6% more often), changes in the frequency of snack consumption (more often 74,5%), changes in the frequency of consumption of fast food/frozen food (more often 70%) and consumption of supplements (90,9%). The results of the bivariate analysis are showed that changes in eating frequency were significantly related to income; changes in meal portions were significantly related to education and expenditure; changes in the frequency of vegetable consumption are significantly related to income; changes in snack consumption frequency were significantly related to age, gender, occupation; changes in the frequency of consumption of fast food are significantly related to gender and consumption of supplements was significantly related to age. Suggestions for the research are needed to consistently educate about good and correct food habits so that more people understand and apply them to increase immunity and data collection will be better with direct interviews and fill out the FFQ of the food consumed.
Keywords: Covid-19 pandemic, changes in eating habits, influence

1 Introduction

The COVID-19 pandemic causes many changes in daily life. Almost all areas of life are affected by the Covid 19 pandemic. According to the Indonesian Ministry of Health (2020), Coronavirus is a large family of viruses that cause disease in humans and animals. It usually causes respiratory tract infections in humans, ranging from the common cold to serious illnesses such as Middle East Respiratory Syndrome (MERS) and Severe Acute Respiratory Syndrome (SARS). A new type of coronavirus found in humans since an extraordinary event appeared in Wuhan, China in December 2019, was later named Severe Acute Respiratory Syndrome Coronavirus 2 (SARS-COV2), causing Coronavirus Disease-2019 (COVID-19). The spread of the virus was so fast that it caused this outbreak to be designated a pandemic by the World Health Organization (WHO) on March 11, 2020 [7].

*Corresponding author : rosydadianah@apps.ipb.ac.id

Everyone has the opportunity to be infected with COVID-19. This disease can spread through small droplets from the nose or mouth when coughing or sneezing. A person can become infected with COVID-19 when accidentally touching and inhaling droplets from an infected person. This is why it is crucial to maintain a distance of at least one meter from sick people. Experts are still conducting investigations to determine the source of the virus, the type of exposure, and the mode of transmission [7].

The Covid-19 (Coronavirus) pandemic has caused many changes in daily life including the establishment of various policies to reduce the spread of Covid-19. Some of these policies include implementation physical distancing, hand washing as health protocols, or carrying out several other strict policies such as limiting an activity such as work or school [6]. Some countries even lock down an area, so that the spread of the virus can stop [19].

One of the effects is the eating habits of a person or society and affects the amount of household spending, especially related to food. Eating habits are a human behaviour or a group of humans in meeting their food needs, include attitudes, beliefs and food choices. Attitudes towards food can be either negative or positive depending on the values in which a person or group of people grew up. Food quality will be related to good or bad, attractive or unattractive food. These attitudes and beliefs will ultimately affect the choice of food [11]. Ramania, *et..al* (2021) stated that 58.2% of students consumed more food and 82.5% of them consumed snacks that were low in nutritional content [17].

The COVID-19 pandemic has greatly affected people's attitudes and beliefs about the function of food, thus affecting the choice of food to be consumed. The household's food security condition is also disrupted during this Covid 19 period. People who are economically capable will give more attention to the kind of food consumed related to nutritional fulfillment to increase body resistance. Nutrition is the link between food and health. Food is related to every aspect of human life, including culture, gender, physical work, education, and the environment. In the end, it will affect the health of individuals and populations. Maintaining a healthy diet is very important during the Covid-19 pandemic. Although there is no food or dietary supplement that can prevent Covid-19 infection, maintaining a healthy balanced nutritional diet is very important in boosting the immune system [10].

Changes in eating habits related to the COVID-19 pandemic include changes in the number of meals in a day, the amount of food consumed, the variety of food consumed, breakfast habits, the habit of cooking their own food and trying new recipes, the habit of consuming organic food, the habit of consuming fruit and vegetables, habits consuming herbal drinks (empon-empon), and the habit of choosing to eat at home. Changes in eating habits were followed by other changes, including changes in the location and method of purchasing groceries, the priority of the food purchased and the amount of expenditure for groceries.

The results of research [18] with 200 respondents in the study show knew that 62.5% had changed their eating habits and experienced an increase in the diversity of food consumption by 59%, as many as 76% of respondents tended to make empon-empon (spices) as drinks during the pandemic Covid-19. The most widely used type of spice was ginger, as much as 44%, followed by orange/lemon and turmeric. Respondents who experienced an increase in the frequency of eating were 54,5% and the amount of food consumption increased by 51%. Respondents experienced an increase in body weight of 54.5%, and respondents were also not concerned about lack of food, which was 54.5% higher than those who were worried about food shortages as much as 44.5%.

Saragih [18] also showed the relationship between several factors related to eating habits there are changes in eating habits with age, diversity of food with the type of work, breakfast habits with type of work, variety of consumption with changes in eating habits and frequency of eating during the COVID-19 pandemic with respondents weight gain. The unrelated factors include the diversity of consumption with fear of lack of food, breakfast habits with fear of lack of food, drinking empon-empon with eating frequency and types of empon-empon with weight gain.

2 Research Methodology

This research is quantitative research using a cross-sectional research design. The research variables were observed simultaneously and taken when the research was in progress, including the respondents's characteristics (independent variables) and changes in eating habits (dependent variables) [8].

Characteristics of respondents include age, gender, education, occupation, number of family members, income, and expenses. Changes in eating habits include changes in eating frequency; change in meal portions; changes in the frequency of consumption of fruit, vegetables, snacks/snacks, fast food/fast food and consumption of supplements. The criteria for the respondents selected were living in Bogor, aged >17 years, already working, not having physical limitations and providing complete data related to the study. Based on these criteria, the Bogor community who became respondents in this study were 110 people.

The stages of the research were proposal preparation, research preparation, questionnaire creation, questionnaire distribution, data collection, data processing and research report preparation. The research location in Bogor (Bogor City and Bogor Regency). Research respondents spread across Bogor are West Bogor, Central Bogor, North Bogor and South Bogor. The purpose of this study was to produce informative data in the form of an overview of changes in people's eating habits in Bogor during the Covid-19 pandemic.

The data in this study were obtained from filling out the questionnaire. The questionnaire used a google form, and respondents filled it from the link that is informed to them. The questionnaire consisted of questions related to the identity of the respondents, anthropometric data and eating habits. Questions related to respondent data include name, gender, age, address, cellphone number, number of family members, education and occupation. Questions related to anthropometric data were weight, height and changes in body weight. This anthropometric data can give information about respondents nutritional status [20].

Questions related to eating habits before and during the Covid-19 pandemic included eating frequency, menu patterns, meal portions, fruit consumption, vegetable consumption, snack consumption, consumption of fast food/frozen food, consumption of snacks and supplements. This questionnaire is filled out directly by the respondent without direct monitoring when they fill it so that each main question will be followed by other questions to support the answer to the main question. For example, questions related to eating frequency have one main question, there are questions related to changes in eating frequency during the covid-19 pandemic. To support the answer for this question, other questions arise about the frequency of eating before the Covid-19 pandemic, the form of changes that occur in addition or subtraction frequency of eating, the number of frequencies before and during the Covid-19 pandemic. The same was done for the other questions.

The data obtained from the results of filling out online questionnaires (google forms) by respondents will be processed and analyzed so it can provide information related to changes in people's eating habits in Bogor during the Covid-19 pandemic. Data analysis carried out in this study was univariate analysis and bivariate analysis. Data analysis using statistical processing software.

Bivariate analysis was used to determine the relationship between the independent and dependent variables. This research used chi square test (X^2) because each variable was a categorical type of data [8]. The chi square test (x^2) in this study used a significance degree of 95% ($\alpha = 5\%$). If the statistical test results get a p value < 0.05 that means there is a significant relationship between the two variables. If the p value > 0.05 that means there is no significant relationship between the two variables.

The research hypothesis of this research is that the Covid-19 pandemic affect changes in eating habits based on the relationship between the factors observed with the results that age, gender, education, occupation, number of family members, income and outcome have a relationship with changes in respondents eating habits during the Covid-19 pandemic.

3 Results and Discussion

The results of the study consist of the description of the respondents, the results of the univariate analysis (the description of changes in eating habits) and the results of the bivariate analysis. The description of respondents in this study includes age, gender, education, occupation and income, which can be seen in Table 1.

In data processing, age is divided into two categories: adults and the elderly [3]; work is divided into two categories there are not working and working; income is divided into 2 categories there are < 2,000,000 and > 2,000,000, while gender and education data use the same category.

Table 1. Frequency distribution of age, gender, education, occupation and income of the people in Bogor during the Covid-19 Pandemic

Research Variables	n	(%)
Age		
Early adulthood (26 – 35 years old)	42	38.2
Adult (36 – 45 years old)	56	50.9
Elderly (46 – 55 years old)	12	10.9
Gender		
Male	18	16.4
Female	92	83.6
Education		
Basic Education	2	1.8
Middle Education	21	19.1
High Education	87	79.1
Occupation		
Housewife	40	36.4
Businessman	9	8.2
Private employees	14	12.7
civil servant	14	12.7
Etc.	31	28.2
unemployment	2	2
Income (IDR)		
0 – 2.000.000	12	10.9
>2.000.000 – 4.000.000	15	13.6
> 4.000.000 – 6.000.000	20	18.2
> 6.000.000 – 10.000.000	30	27.3
> 10.000.000	33	30.0

Univariate analysis for data on changes in food habits during the Covid-19 pandemic consists of changes in eating frequency, changes in meal portions, changes in the frequency of fruit consumption, changes in the frequency of vegetable consumption, changes in the frequency of snack consumption, changes in the frequency of fast food/frozen consumption and consumption of supplements which can be seen in Table 2.

Table 2. Results of Univariate Analysis of Changes in People's Eating Habits in Bogor during the Covid-19 Pandemic

Research Variables	n	(%)
Changes in eating frequency		
Increased Frequency of Eating	58	52.7
Reduction in Eating Frequency	18	16.4
Not changed	34	30.9
Changes in meal portions		
Changed	62	56.4
Not Changed	48	43.6
Changes in the frequency of fruit consumption		
Increased Frequency	68	61.8
Not Changed	42	38.2
Changes in the frequency of vegetable consumption		
Increased Frequency	70	63.6
Not Changed	40	36.4
Changes in the frequency of snack consumption		
Increased Frequency	82	74.5
Not Changed	28	25.5
Changes in the frequency of fast food/frozen consumption		
Increased Frequency	77	70.0
Not Changed	33	30.0
Consumption of Supplements		
Yes	100	90.9
No	10	9.09

Based on Table 2, it is known that most of the respondents experienced a change in the frequency of eating, which consists of adding and reducing the frequency of eating that the percentage of increasing the frequency of eating was higher than the others. This is because the COVID-19 pandemic gives people more time to consume delicious and nutritious food at home to increase immunity [15]. The increase in the number of servings of food consumed is also very high. Increase in food portions was also followed by an increase in food consumption with a complete menu consisting of staple foods, side dishes (animal and vegetable), and vegetable dishes complemented by fruit. A small number of respondents only consume staple foods, side dishes (animal and vegetable) and vegetable dishes without fruit. These results show that respondents know the importance of balanced nutrition to meet nutritional needs during the Covid-19 pandemic. The same results were stated by Jumalda, *et.al.*, (2021) that there was a change in eating habits between before and during the Covid-19 pandemic, which was shown by the majority of respondents experiencing a change in eating habits become more regular and balanced diet, especially those with Covid-19 cases in their surroundings [9].

The change in eating habits that also occurred during the Covid-19 pandemic was increase of fruit and vegetables consumption frequency around 60-70% of respondents more often during the Covid-19 pandemic than before the Covid-19 pandemic. Increase of fruit and vegetables consumption frequency about 3-9 times/day for vegetables and 3-9 times/week for fruit. Fruits and vegetables contain many vitamins and minerals that are very useful for the body's resistance, so respondents increase the frequency of fruits and vegetables. The same results were stated by Dieny, *et.al.*, (2021) that subjects experienced changes in eating habits, namely an increase in the frequency, variety, and portion of consumption of animal protein, vegetable protein, vegetables and fruit between before and during the COVID-19 pandemic ($p < 0{,}001$) [5].

The results of this study are also in line with research conducted by Ruiz-Roso, *et.al.* (2020) there are 820 adolescents in several countries. The study showed significant differences in vegetable and fruit consumption between before and during the Covid-19 pandemic [16]. The results of another study in China also showed that 40% of research subjects experienced increased consumption of vegetables and fruit [15]. However, a study in France showed the opposite result, where there was a decrease in consumption of fresh food products, especially fruit and fish, during the pandemic [4].

Changes in the frequency of snack consumption also occurred in respondents, and the snack consumption frequency became more often during the Covid-19 pandemic. The types of snacks consumed are very diverse and include dry foods, sweet and salty cakes (cake, sponge cake, cireng, spring rolls), chocolate, side dishes (meatballs, dumplings). The application of the work from home system causes respondents to have a lot of time at home so that respondents have a lot of time to provide snacks or buy online. The same results were stated by Ramania, *et.al.*, (2021) that 82.7% subject know that they consume snacks that are considered low in nutritional value [17].

Fast food/frozen food consumption frequency also becomes more often in a week. Fast food/frozen food is consumed by respondents as a companion to staple food and as a snack. Fast food/frozen food is considered more practical, especially for respondents who rarely cook their food. The results show that 47% of people now have the habit of eating processed foods that are considered more practical. Meanwhile, 36% of other respondents admitted that they consume food from crops less frequently, and 26 % of people rarely eat meat or poultry [21]. The research result by Ammar, *et.al.*, (2020) there is an increase in eating habits during the pandemic and tends to lead to an increase consumption of snacks and fast food, which of course is very dangerous for the health of the human body [1].

The majority of respondents (90.9%) carried out the habit of consuming supplements to increasing immunity in the face of the Covid-19 pandemic. Supplements consumed include multivitamins, herbs, empon-empon drinks and so on. Saragih [18] showed that 76% of respondents tend to make empon-empon (spices) as drinks during the COVID-19 pandemic. The most widely used type of spice was ginger, as much as 44%, followed by orange/lemon and turmeric. The habit of consuming fluids, supplements, and spices also experienced a significant increase, while the pattern of snacking decreased between before and during the pandemic ($p < 0{,}001$) [5]

The COVID-19 pandemic has caused various changes, including changes in eating habits. Respondents experienced changes in eating habits as many as 125 people (62,5%), and the remaining 75 people (37,5%) did not experience changes in eating habits. Changes in eating habits can be caused by respondents working from home, so they are more regular in eating, and there is an increase in food diversity [18].

Bivariate analysis was carried out between the variables of changes in eating habits with age, gender, number of family members, education, occupation and income during the Covid-19 pandemic. The results of the bivariate analysis between the dependent and independent variables can be seen in Table 3.

Table 3. Results of Bivariate Analysis for Interrelated Variables

No	Research Variables		Chi-Square Test (P-Value)
	Dependent	Independent	
1	Changes in eating frequency	Income	0,021
2	Changes in meal portions	Education	0,038
3	Changes in the frequency of vegetable consumption	Income	0,045
4	Changes in the frequency of snack consumption	Age	0,01
5	Changes in the frequency of snack consumption	Gender	0,009
6	Changes in the frequency of snack consumption	Occupation	0,039
7	Changes in the frequency of fast food/frozen food consumption	Gender	0,043
8	Consumption of Supplements	Age	0,009

Description: *p-value $\leq$ 0.05 there is a significant relationship

Based on Table 3, it is known that income has a significant relationship with changes in eating frequency and changes in vegetable consumption frequency. This can happen because the amount of income will affect the power/ability to buy food for daily consumption, including buying vegetables which are one of the components in the main/complete dish.

Education has a significant relationship with changes in food portions. This can happen because most of the respondents are highly educated and quite exposed to efforts to prevent Covid-19. One of the efforts made is the consumption of food in sufficient types and quantities so that respondents who usually do not eat at regular times and portions will change to eating at times with more portions than usual for increasing body resistance to prevent being exposed by Coronavirus.

During the Covid-19 pandemic, the portion of food consumed by a person is mostly increases from the previous condition. This is because most activities are carried out at home, so boredom will arise, especially for people who are used to their time for activities outside the home. Food is one way of diverting boredom. The portion of the meal means main meal or side dish or snack.

Age has a significant relationship with changes in the frequency of snack consumption and changes in supplement consumption. Based on the identity data of the respondents, it is known that most of the respondents are working adults. During this Covid-19 pandemic, many activities are carried out at home, and all family members gather so that complete meals and snacks will be more than normal days. They do that to avoid purchasing food from outside the home. Adults and the elderly are the group that is quite vulnerable to the coronavirus; therefore most of the respondents take supplements (multivitamins, herbs, empon-empon) as an effort to increase endurance.

During the Covid-19 pandemic, the portion of food consumed by a person is mostly increases from the previous condition. This is because most activities are carried out at home, so boredom will arise, especially for people who are used to their time for activities outside the home. Food is one way of diverting boredom. The portion of the meal means main meal or side dish or snack.

Age has a significant relationship with changes in the frequency of snack consumption and changes in supplement consumption. Based on the identity data of the respondents, it is known that most of the respondents are working adults. During this Covid-19 pandemic, many activities are carried out at home, and all family members gather so that complete meals and snacks will be more than normal days. They do that to avoid purchasing food from outside the home. Adults and the elderly are the group that is quite vulnerable to the coronavirus. Therefore, most of the respondents take supplements (multivitamins, herbs, empon-empon) as an effort to increase endurance. Research on people in Pontianak showed different results there was a relationship between gender and the habit of drinking empon-empon. Women tend to consume empon-empon more during the Covid19 pandemic (by 1,442 times) [9].

Empon-empon is a herbal drink made from mix of various spices. Herbs has been shown to increase immunity [18]. Some of these herbs include Brotowali (*Tinospora cordifolia*), Aloe Vera (*Aloe vera*), Mulberry (*Morus alba*), Lime (*Citrus aurantifolia*), Garlic (*Allium sativum*), Turmeric (*Curcuma longa*), Red Ginger (*Quercetin*), Temulawak (*Curcuminoid*), Gotu Kola (*Centella asiatica*), ginseng (*Panax ginseng*), and so on [2, 12, 14]. Consumption of spices/herbs may be able to improve the immune system and protect the body from COVID19.

Gender has a significant relationship with changes in snack consumption frequency and changes in fast food/frozen food consumption. Most of the respondents in this study were women who preferred to consume snacks. The Covid-19 pandemic has caused many people to avoid buying ready-made food from outside. For respondents who rarely cook at home, buying fast food and frozen food is practical and easy [15]. The results of research by Noviasty and Susani (2020) showed different results that the increasing frequency of cooking at home causing a decrease in the frequency of fast food consumption [13]. There are 65.63% of respondents reduce consumption of fast food during the Covid-19 pandemic [13].

Work has a significant relationship with changes in the frequency of snack consumption. Most of the respondents in this study worked. The COVID-19 pandemic has led to the implementation of work from home, which opens up opportunities to work more casually and consume snacks that they like. Especially the respondents in this study were many women who really liked to consume snacks.

4 Conclusions

Based on research on the effect of the Covid-19 pandemic on changes in eating habits, it can be concluded that:

1. Most of the respondents belong to the adult age group (89,1%), female (83,6%), highly educated (79,1%), most of the work (61,8%) and with income > IDR 4 million/month.
2. Changes in eating habits observed include changes in eating frequency (69,1% change); changes in meal portions (56,4% change); changes in the frequency of consumption of fruit (61,8% more often), vegetables (63,6% more often), snacks (74,5% more often), fast food (70,0% more often); and consumption of supplements (90,9% consume the supplements)
3. Results of the bivariate analysis showed the relationship between changes in eating frequency with income; changes in food portions with education; changes in the frequency of vegetable consumption with income; changes in the frequency of snack consumption with age, gender and occupation; changes in the frequency of consumption of fast food/frozen food with gender; consumption of supplements with age.

5 Recommendations

Recommendations for this research are:

1. Education on good and correct food habits must be carried out consistently to make more people understand and apply it, and ultimately it increases each person's body resistance.
2. Data collection would be better if completed with direct interviews with respondents and filling out FFQ related to the frequency of food consumed to be more detailed.

6 Acknowledgments

The authors want to thank College of Vocational Studies IPB University and all participants in this study.

References

1. Ammar, A., Brach, M., Trabelsi, K., Chtourou, H., Boukhris, O., Masmoudi, L., Bouaziz, B., Bentlage, E., How, D., Ahmed, M., Müller, P., Müller, N., Aloui, A., Hammouda, O., Paineiras-Domingos, L. L., Braakman-Jansen, A., Wrede, C., Bastoni, S., Pernambuco, C. S., Hoekelmann, A. (2020). Effects of COVID-19 home confinement on eating behaviour and physical activity: Results of the ECLB-COVID19 international online survey. Nutrients, 12(6),

1–14. https://doi.org/10.3390/nu12061583

2. Carrasco, F. R., Schmidt, G., Romero, A. L., Sartoretto, J. L., Caparroz-Assef, S. M., Bersani-Amado, C. A., & Cuman, R. K. N. (2009). Immunomodulatory activity of Zingiber officinale Roscoe, Salvia officinalis L. and Syzygium aromaticum L. essential oils: evidence for humor- and cell-mediated responses. The Journal of Pharmacy and Pharmacology, 61(7), 961–967. https://doi.org/10.1211/jpp/61.07.0017
3. Depkes RI. Klasifikasi umur menurut kategori umur, (Ditjen Yankes, Jakarta, 2009)
4. Deschasaux-Tanguy, M., Druesne-Pecollo, N., Esseddik, Y., Szabo de Edelenyi, F., Alles, B., Andreeva, V. A., Baudry, J., Charreire, H., Deschamps, V., Egnell, M., Fezeu, L. K., Galan, P., Julia, C., Kesse-Guyot, E., Latino- Martel, P., Oppert, J.-M., Peneau, S., Verdot, C., Hercberg, S., & Touvier, M. (2020). Diet and physical activity during the COVID-19 lockdown period (March-May 2020): results from the French NutriNet-Sante cohort study [Preprint]. Nutrition
5. Dieny, F. F., Jauharany, F.F., Tsani, F.A., dan Nissa, C. (2021). Eating behaviour before and during the covid-19 amonng adolescent and adults in Indonesia. http://dx.doi.org/10.30867/action.v6i2.418
6. Dwyer, M. J., Pasini, M., De Dominicis, S., & Righi, E. (2020). Physical activity: Benefits and challenges during the COVID-19 pandemik. Scandinavian Journal of Medicine and Science in Sports, 30(7), 1291–1294. https://doi.org/10.1111/sms.13710
7. Europe World Health Organization. Food Nutrition Tips During Self-Quarantine. Accessed on 11 June 2020. (2020) Accessed at http://www.euro.who.int/en/health-topics/health emergencies/coronavirus-covid-19/technical-guidance/food-and-nutrition-tips-during-self-quarantine
8. Hastono, S.P. *Analisis Data Kesehatan*. (FKM UI, Depok, 2007).
9. Jumalda, A.E., Suwarni, L., Marlenywati, Selviana dan Mawardi. (2021). Pola makan masyarakat di Kota Pontianak selama pandemi Covid-19. Jurnal Kesehatan Masyarakat Indonesia, 16(10, 1-6.
10. Kementerian Kesehatan RI. *Panduan Gizi Seimbang Pada Masa Pandemi COVID-19.* (Kementrian Kesehatan RI, Jakarta, 2020).
11. Kemenkes RI. Pedoman Gizi Seimbang. (Kemenkes RI, Jakarta, 2014)
12. Mathew, D., & Hsu, W.L. (2018). Antiviral potential of curcumin. Journal of Functional Foods, 40, 692–699. https://doi.org/10.1016/j.jff.2017.12.017
13. Noviasty, R. Susani, R. (2020). Perubahan kebiasaan makan mahasiswa peminatan gizi selama masa pandemic Covid-19. Jurna Kesehatan Masyarakat Mulawarman, 2(20), 90 – 99.
14. Panyod, S., Ho, C.T., & Sheen, L.Y. (2020). Dietary therapy and herbal medicine for COVID-19 prevention: A review and perspective. Journal of Traditional and Complementary Medicine, 10(4), 420–427. https://doi.org/10.1016/j.jtcme.2020.05. 004
15. Perdana, A. H. Dampak COVID-19 terhadap Pola Industri Makanan Minuman Asia Tenggara, (2020). https://www.idntimes.com/business/economy/hana-adi-perdana-1/dampak-covid-19-terhadap-pola-industri-makanan-minuman-asia-tenggara/4
16. Ruiz-Roso, M. B., de Carvalho Padilha, P., MantillaEscalante, D. C., Ulloa, N., Brun, P., AcevedoCorrea, D., Arantes Ferreira Peres, W., Martorell, M., Aires, M. T., de Oliveira Cardoso, L., Carrasco-Marín, F., Paternina-Sierra, K., Rodriguez-Meza, J. E., Montero, P. M., Bernabè, G., Pauletto, A., Taci, X., Visioli, F., & Dávalos, A. (2020). Covid-19 Confinement and Changes of Adolescent's Dietary Trends in Italy, Spain, Chile, Colombia and Brazil. Nutrients, 12(6), 1807. https://doi.org/10.3390/nu12061807
17. Ramania, N.S., Syafriani, R., Ganeswara, R., Nugraha, A.B. dan Hasan, M.F. (2021). Efek Pandemik Covid-19 Pada Aktifitas Fisik, Pola Makan, dan Psikologi Pada Mahaiswa Persiapan Bersama di ITB. Jurnal Sains Keolahragaan dan Kesehatan, 6(1), 1 – 11. http://dx.doi.org/10.5614/jskk.2021.6.1.1
18. Saragih, B. Hubungan Berbagai Faktor dengan Kebiasaan Makan pada Masa Pandemi Covid 19, (Universitas Mulawarman. 2020)
19. Sun, T., Guo, L., Tian, F., Dai, T., Xing, X., Zhao, J., & Li, Q. (2020). Rehabilitation of patients with COVID-19. Expert Review of Respiratory Medicine, 14(12), 1249–1256. https://doi.org/10.1080/17476348.2020.1811687

20. Supariasa, I.D.N. Penilaian Status Gizi Edisi 2, (Penerbit Buku Kedokteran EGC, Jakarta, 2014)
21. Shutterstock. Riset: 7 Perubahan Kebiasaan Belanja di Swalayan Selama Pandemi Virus Corona. (2020) https://kumparan.com/kumparanfood/riset-7-perubahan-kebiasaan-belanja-di-swalayan-selama-pandemi-virus-corona-1u3wOPNY5io. 24 Agustus 2020 9:58
22. Wang, X., Lei, S. M., Le, S., Yang, Y., Zhang, B., Yao, W., Gao, Z., & Cheng, S. (2020). Bidirectional Influence of the COVID-19 Pandemic Lockdowns on Health Behaviors and Quality of Life among Chinese Adults. International Journal of Environmental Research and Public Health, 17(15), 5575. https://doi.org/10.3390/ijerph17155575

Noise Variation Between Days and Hours in an Urban Areas

Yudith Vega Paramitadevi[1], *Shakilah Faradiba Basalamah*[1], *Okta Viani*[1], *and Milgamas Ratna Prigel*[1]

[1]The Vocational Studies of IPB University, Environmental Engineering and Management Program, Kumbang 14th Bogor, West Java, Indonesia

Abstract. Noise disturbance quality standards which refer to Indonesian Government Decree number 22 of 2021 concerning Environmental Management in the future will be drawn up. This possibility has prompted several local governments to measure and map noise in their respective areas. The commonly used pattern in noise studies is taking grab samples and processing the test sample data using extrapolation techniques in order to obtain long-term noise levels. The approach requires a sufficient literature base. This study analyzed noise based on daily data at three different locations in the City of Jakarta, Indonesia. The three locations were construction site, industrial area and along inner-city toll road. Analysis of ANOVA variance at daily and hourly levels for 42 LA eq was presented in this study. Statistical results showed significant differences between hourly and daily periods, so further attention is needed on the extrapolation approach that represents long-term noise levels. The longer the noise analysis period in an area, the more accurate the characterization of noise in the study area will be.

1 Introduction

The Indonesian government has various regulations that conduct noise threshold levels in every specific area or zone. One of the regulations is the Decree of the Minister of the Environment Number 48 of 1996 concerning Quality Standard of Noise Level (The Decree Number 48/1996). It administers the maximum noise level allowed to be discharged into the environment from activities. Furthermore, the noise disturbances to human health and surroundings will be reduced.

The Decree Number 48/1996 divides eight categories of measurement area and three activities such as housing area, a trading area, office area, green open space, industrial area, government areas and public facilities, recreation, and special area (airport, train station, port, cultural heritage). Those three activities consist of hospitals, schools, and places of worship. The noise threshold standard in this regulation is 55 dBA to 70 dBA depending on area category. The newest regulation, Indonesian Government Decree number 22 of 2021 concerning Environmental Protection and Management, has defined noise as a form of disturbance that must be controlled and the quality standard must be set. The quality standard itself was revised from the previous regulation. Simultaneous noise measurement and grab sampling are useful methods for determining the noise level then noise mapping can be carried out [1]. If the noise measurement is done by grab sampling, then prediction efforts can be made to complete the mapping in an area [2].

Massive noise measurement efforts have been carried out in developed countries in the European Union [3,4]. Noise mapping for the entire region in Indonesia itself is not yet mandatory unless the area is subject to environmental impact analysis. The obstacles that arise are the unavailability of fixed stations for noise in cities, the unavailability of policy regarding reliable mapping methods, and limited resources in mapping techniques. Noise measurement methods in Indonesia have used the methods that stated in the Decree of the Minister of the Environment Number 48 of 1996 or SNI (Indonesian National Standard) Number 7231:2009 entitled Measuring Noise Intensity Method in Workplace. The stated method was to analyze the measurement results using the sound level meter, so the value can be compared with recent regulations. In Indonesia, the measurement data obtained are generally by grab sampling and then analyzed using the interpolation method then the mapping points are complete [5,6]. Based on Geraghty and O'Mahony [1], the noise measurement gap occurs if the measurement is based on grab sampling only or the deviation is large if the measurements are not carried out simultaneously.

Based on the problems above, this study was made to describe the differences in daily and hourly measurement variations on specific sources of industry, highways, and construction areas in Jakarta, Bogor, and Bekasi, Indonesia. This study focused on weekly and monthly measurements that can be applied in environmental impact analysis studies where the results will be better than grab sampling. This study hypothesized that there were variations in measurements between weekdays and holidays, there were also variations in measurement between day and night. If the hypothesis was not proven, simultaneous monthly and annual noise measurements are recommended to obtain background data for environmental monitoring purposes.

2 Methodology

2.1 Noise Level Measurement

The measurement area consists of three areas, i.e. industrial area, ring road, and construction area (Fig. 1). The reasons for choosing the areas are currently under construction or development, so noise monitoring needs to be carried out as a condition for compliance. Those data measurements were conducted for 24 hours in the same month using several instruments of sound level meters. The name of the instrument is Lutron brand type SL-4001, GM1351, and IEC 651 TYPE 2. They have specifications measurement distance divided into three range selectors, namely 30-80 dB, 50-100dB, and 80-130 dB.

Measurements were also assisted with a tripod and a GPS to determine the coordinates of the sampling point. Position of the tool was at a height 120 cm to 150 cm from the ground, with a minimum of 3.5 meters far from buildings or trees. Noise level measurement was done in two ways (SNI Number 7231:2009):

1) Noise mapping method
   The method must be done by 2 people as shown in Fig. 2, one person observed the time and gave the clue instantaneously to read noise level per 5 seconds in 10 minutes. The other recorded the instantaneous noise level reading from a sound level meter. Noise measurements were made at several points (7 points for industrial area, 4 points for ring road, and 4 points for construction area) from the noise source on each location using a sound level meter equipped with a windscreen and not in rainy conditions.

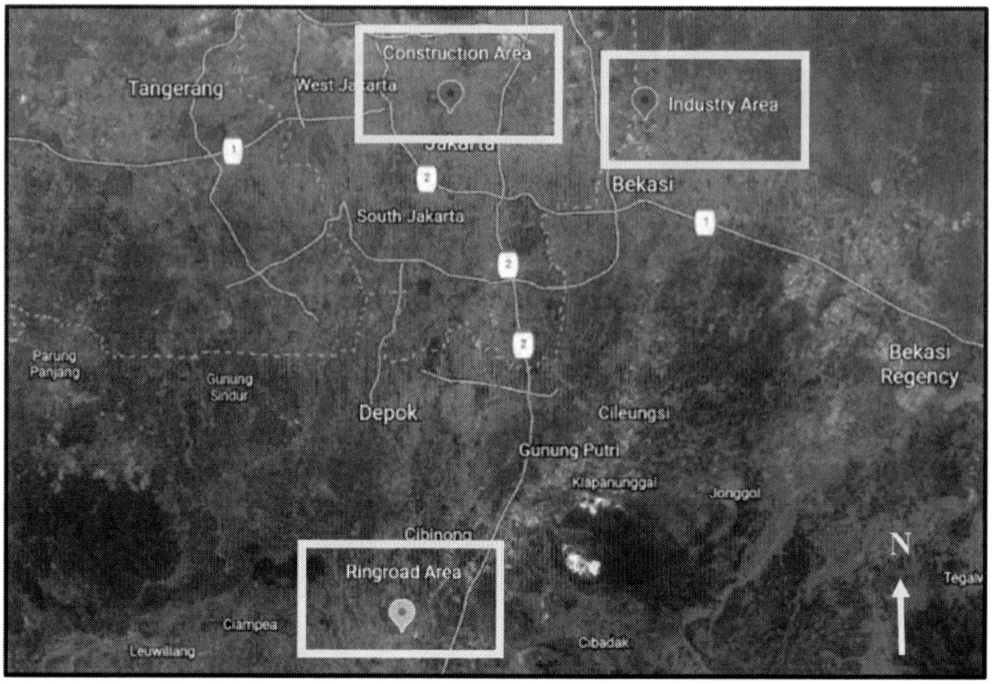

Fig. 1 Three location of measurement named as Site 1-3

The sound level meter is used to measure the pressure level of transitory sound dB(A) for 10 minutes for each measurement. As a result, the surveyor obtains up to 200 pieces of data which represent certain time intervals.

Fig. 2. The first method of measurement

2) Noise variance collection method
The method was carried out with the integrating sound level meter having L_{TM5} measurement facilities, namely Leq with measuring time every 5 seconds, recorded automatically for 10 (ten) minutes. By the instrument, Leq and T have been obtained and read autonomously on the sound level meter. Noise measurements were made at one point which represents the entire area.

2.2 Analysis

The variance was obtained by analysis of variance (ANOVA) and the results were tabulated in post hoc form for one month of observation. Temporal analysis was performed based on a daily rate and hourly rate. If a significance value of <0.05 is obtained and homogeneous, then post hoc visualization is carried out based on the Tukey method. If the homogeneity of variance is not proven, then post hoc visualization is carried out based on the Games Howell method. After the analysis of variance is carried out, the data obtained from noise mapping are calculated to obtain constant levels based on the formula (Minister of Environment Decree No. 48/1996):

$$Leq = 10 \log \{f_1 \times 10^{0.1L1} + f_2 \times 10^{0.1L2} + \ldots f_n \times 10^{0.1Ln}\} \text{ dBA} \quad (1)$$

Leq is the equivalent of constant levels, f are the fractions of time that occur when the shift is 8/24, L is the calculated levels in the time interval of each shift (n). The obtained Leq values were then mapped and visualized using Surfer GIS version 10.

3 Result and Discussion

3.1 Daily Variance

The post hoc ANOVA test (Table 1) was compared to obtain significant results and the results were similar to those of [1,7]. Site 1 is a ring road, site 2 is a construction area and site 3 is an industrial area. Based on Table 1, site 2 and site 3 variations on holidays and weekdays have the same pattern. Site 1 variations are not visible on holidays and weekdays. This is possible because, on the ring road, noise sources come from various sources. The ring road also includes open areas [3] so those daily variations cannot be interpreted directly. Longer observations such as monthly in the span of one year are needed then daily variations are more reliable.

3.2 Hourly Variance

The variation between day and night for site 2 and site 3 has similarities between industrial and construction areas. Site 2 is significant for working hours outside the shift change hours of employees, this also happens for site 3. Both have 3 times shift changes with different times, 05.00-06.00 am; at noon; and 06.00-07.00 pm for site 2 and 07.00; 3 pm; 11.00 pm for site 3. The effect of employee rotation and length of working hours is similar to research [4].

The hourly pattern for site 1 is different from the others (Table 2), this is because the location is a project with an open-source noise. The hourly pattern at this location is the same as the daily pattern, where the rush hour pattern occurs starting at 05.00-10.00 pm. The noise started to decrease in the early hours of the morning, even though the project implemented a 2 shift employee system. Noise level cannot be determined using analysis of variance, but temporally, this analysis of variance can be used to make policy [1].

Table 1. Post hoc statistical significance indication for between day comparisons of Leq for Site 1-3

Site 1							
	Mon	Tues	Wed	Thurs	Fri	Sat	Sun
Mon							
Tues							
Wed							
Thurs							
Fri							
Sat							
Sun							

Site 2							
	Mon	Tues	Wed	Thurs	Fri	Sat	Sun
Mon							
Tues							
Wed							
Thurs							
Fri							
Sat							
Sun							

Site 3							
	Mon	Tues	Wed	Thurs	Fri	Sat	Sun
Mon							
Tues							
Wed							
Thurs							
Fri							
Sat							
Sun							

Based on the analysis of daily and hourly variations, noise measurements with significant variance can be applied for at least one day, but it is recommended that there are previous observation data for one year. Especially in areas with open noise sources, the measurement time may be longer because the noise level is more difficult to estimate. To fulfill environmental compliance and reduce errors, it is better if noise measurements are based on direct measurements.

Table 2. Post hoc statistical significance indication for between hours comparisons of Leq for Site 1-3

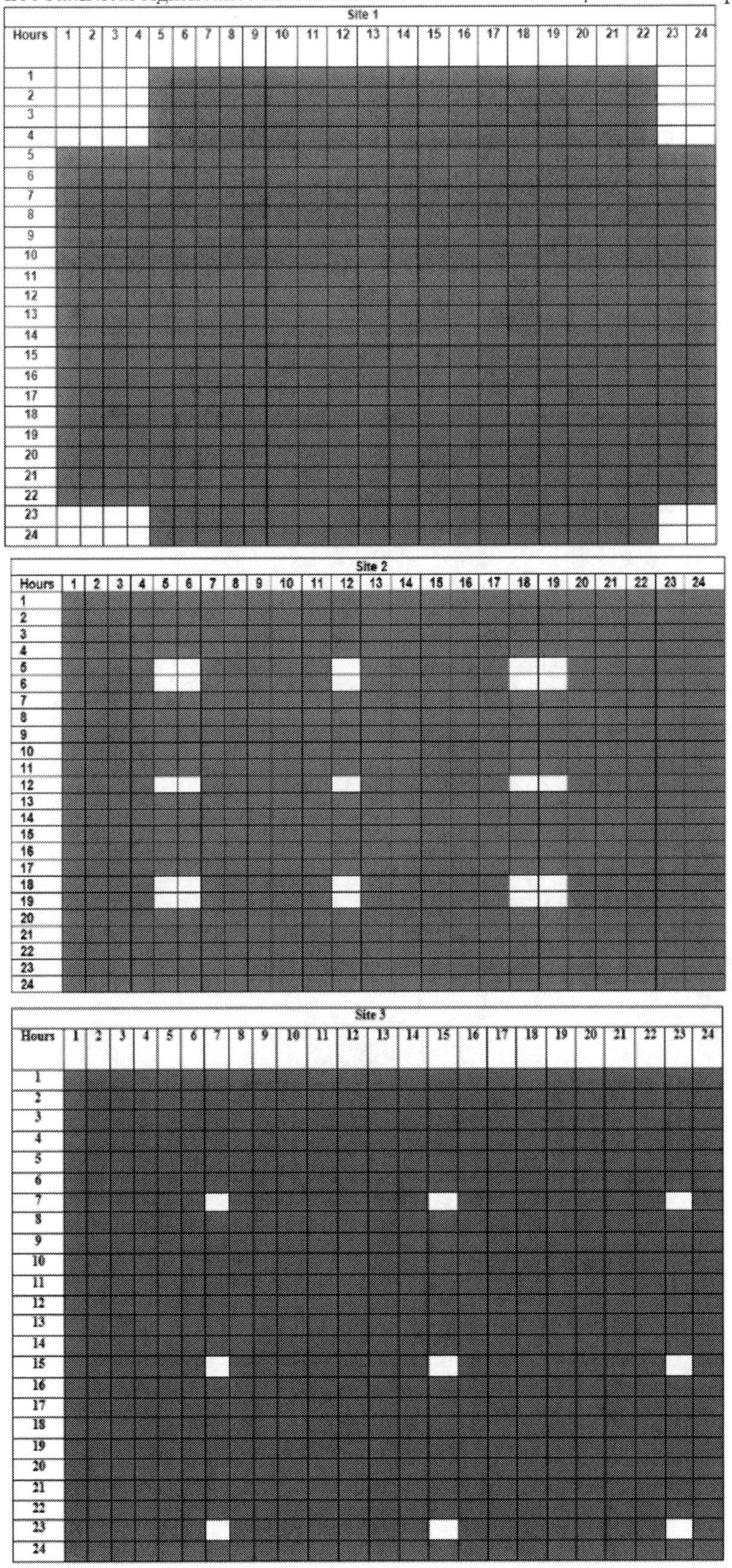

3.3 Noise Mapping

The noise level of each site (Fig. 3) was visualized using a noise map. The red contour represents the high noise area, the yellow contour represents the medium noise area and the green contour represents the low noise area. The noise measurement refers to direct measurements, sampling points represent construction sites and nearby residential areas. Spatial noise analysis needs to be complemented by temporal noise analysis, this is in line with the statement of [1,4,8].

According to the quality standard for noise levels in the work area and the environment (Regulation of the Minister of Manpower and Transmigration Number 13/Men/X/2011 concerning Threshold Values for Physical and Chemical Factors in the Workplace), noise at site 1 is still below the threshold value or below 85 dBA. The noise level at site 2 is dominated by an area with noise levels above the quality standard, which is 86 – 90 dBA. Workers entering the area are required to wear personal protective equipment/ earmuffs. The construction area or site 3 has a high concentration of noise throughout the structural work process. High noise levels are managed by business owners through substitution efforts, engineering control, and administration [6].

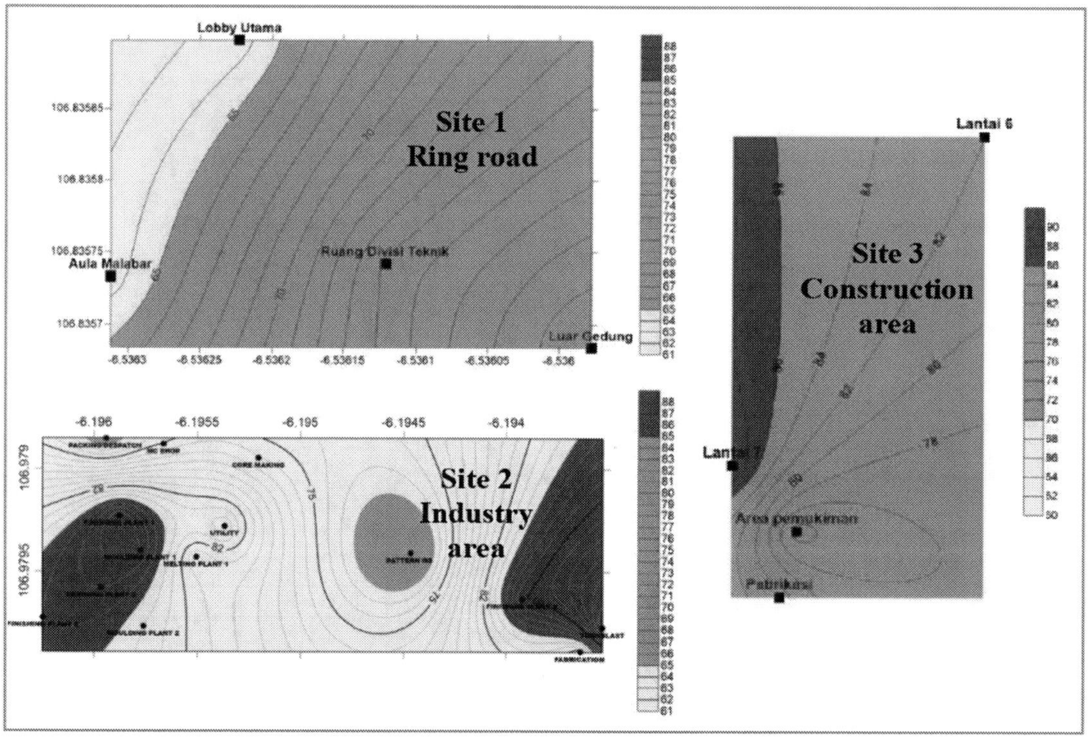

Fig. 3. Noise mapping of site 1 to site

4 Conclusion

Significant measurement variations exist at sites 2 and 3, both daily and hourly. Especially for site 1, the open condition causes the need for a longer noise monitoring effort. Both sites 1, 2, and 3 have noise measurement points that already represent noise mapping. Moreover, to comply with environmental regulations and obtain background data, it is necessary to measure simultaneously with a longer period.

References

[1] D. Geraghty and M. O'Mahony, Int. J. Sustain. Built Environ. **5**, 34 (2016)
[2] W. Wei, T. Van Renterghem, B. De Coensel, and D. Botteldooren, Appl. Acoust. **101**, 127 (2016)
[3] E. Murphy and E. A. King, in Environmental Noise Pollution (Elsevier, 2014), pp. 173–201
[4] P. Aumond, L. Jacquesson, and A. Can, Appl. Acoust. **139**, 34 (2018)
[5] O. P. Rosmala, T. Budi Santoso, and M. Huda, in 2019 International Electronics Symposium (IES) (2019), pp. 148–153
[6] Susanto, D. O. Setyawan, F. Setiabudi, Y. M. Savira, A. Listiarini, E. K. Putro, A. F. Muhamad, J. C. Wilmot, D. Zulfakar, P. Kara, I. Shofwati, S. Sodikin, and M. Tejamaya, Noise Mapping **8**, 1 (2021)
[7] U. Hernandez-Jayo and A. Goñi, Sensors **21**, (2021)
[8] Benocci, Bellucci, Peruzzi, Bisceglie, Angelini, Confalonieri, and Zambon, Environments **6**, 79 (2019)

Waste reducing efforts in the kitchen area of the hotel industry using lean management (a case study of XYZ hotel in Bogor)

Annisa Kartinawati[1*], *Sazli* Tutur Risyahadi[1], *Firman* Muhammad Bashar[1]

[1]College of Vocational Studies, IPB University, Kampus IPB Cilibende Jl. Kumbang No. 14, Bogor, 16151, Indonesia

Abstract. In the hotel kitchen, the production activities, from preparing raw materials to serving dishes, potentially produce waste. This research investigated the application of lean management in the kitchen area of the hotel industry and to find out its suitability and the possibility to optimize resources. Lean Assessment, Value Stream Mapping (VSM) preparation, and priority selection of lean tools with Analytical Hierarchy Process (AHP) were used as the methods. Results revealed that the lean assessment on the production aspect, namely the multi-skill worker and 5S, have been implemented. Based on the VSM, breakfast dish products are processed in semi-finished form during the day and then left for a particular time to be re-processed before breakfast time. There are three lean tools priorities calculated using AHP, (1) the application of a pull system both in production and at suppliers, (2) forming a Kaizen team, (3) visual management, and four criteria weighted by pairwise comparisons, (1) Implementation costs, (2) Implementation Time, (3) Relating to Hotel KPI, and (4) Sustainability of results. The final values of the total alternative priorities show that the highest weight values are Kaizen team (0.643), visual management (0.283), and implementation of Kanban (0.074), consecutively.

1 Introduction

1.1 Background

Indonesia is a famous country destined for foreign and domestic tourists due to many attractive tourist places. Along with Indonesian tourism, the hotel industry is expanding and playing an essential role in the economy and society. Such progress can create a chain and distribution of income for the region. However, tourism and hospitality businesses face future challenges, significantly reducing costs and restructuring internal processes to be more efficient. This action is done because there is a lot of competition in the hotel industry, which tends to compete in price and service

* Corresponding author: annisakartinawati@apps.ipb.ac.id

quality. Moreover, lately, the hotel industry has been negatively affected by the COVID-19 pandemic, which has affected its revenue.

The hospitality industry must adopt strategies or methods to improve service efficiency, such as reducing costs and time to win the competition. These challenges have been successfully met in other industrial sectors, especially the manufacturing industry, with a lean management approach. Unfortunately, this lean approach is not very popular in the hospitality industry, although some have implemented it [1]. A very dynamic/fluctuated customer demand and a partial implementation of the lean approach in this sector might influence this.

According to the lean service concept reported in the research of Andrés-López et al. [2], it explains that waste in the industry can be categorized into 7 types, namely excess, delay, unnecessary transportation and displacement, quality levels that are less than standardization, disruption in customer demand, and resources with low utility. Since these types of waste can also occur in the kitchen area of the hotel industry, implementing the lean approach in this part would benefit the hotel industry. Studies of the application of Lean in the hotel industry have begun with various tools and specific area objects of study, for example on warehouse hotel management using 5S [1], supply chain hotel using SIPOC Approach [3] and working process on hotel room using Eliminate, Combine, Rearrange, and Simplify (ECRS) tools [4]. However, not all of the selected tools of lean are suitable at all for the hospitality area. The influence of the managerial implications of lean on medium and small hotel scales has been conducted by Vlachos et al. [5]. However, not many are enhanced with prioritizing lean tools, especially by using the Analytical Hierarchy Process (AHP). The objective of this work is to investigate the potential application of Lean Management tools in the hotel area and to make prioritization using the AHP Methods. a case study of XYZ Hotel in Bogor, and to determine their suitability and the possibility to optimize resources, especially in the kitchen area.

1.2 Literature review

In the context of lean management, lean is essentially defined as being flexible, agile, and concise. Lean is a package of principles, methods, and actions for the productive configuration of the processing activity. The main goal of lean management is to create value without generating waste. Value is any action or process that customers are willing to pay. Furthermore, lean tools help to identify and eliminate waste so that quality is better, reduces costs and processes [6].

According to a book of Womack and Jones [7], the philosophy of lean management is based on the following 5 principles:

1. Identify specifically the value from the customer's point of view
2. Identify and map a value stream process by reviewing all stages of the value stream for each product or service to eliminate any processes that are not generating value
3. Create flow by designing sequential production processes to flow smoothly to customers.
4. Respond to customers with a pull system
5. Encourage perfection; the previous four philosophies are repeated, then continue until perfection is achieved.

Value stream describes the process from the beginning to finished product in one industry, either providing value (value-added) or not providing value (non-value added). Beforehand, lean management has shown its application to the service industry, including the tourism industry. However, there is still limited literature on the examples of companies implementing lean management in the hospitality industry. A research by Rauch et al. [1] applied lean management at Hotel Yukai Resort. The results indicated that it is possible to provide the same standard of service to hotel guests, but at a cost that is half either high or low in hotel occupancy. Some of the things that have been done are (1) sharing all information regarding work assignments to all workers, (2) using buffet style for dinner and breakfast activities to reduce employee needs, (3) involving staff who are always active and involved in lean management or lean Kaizen, (4) implementing cross skill training so that in his spare time an employee will go to areas that need assistance, such as the kitchen or laundry.

A research of Rauch et al. [1] conducted at Hotel Apex applies lean management to find process activities that do not provide value by eliminating these activities and carrying out the process and organizational restructuring and workforce re-planning. The lean team has been able to save 5,728 man-hours per year. The application of lean methods saves money and significantly improves customer service and working conditions for hotel staff. A research of Pearlman and Chacko [8] conducted at Starwood Hotels has implemented Lean Six Sigma. As a result, the company received financial benefits in the form of an increase in revenue of up to 19% and an increase in the number of customers up to 12%. The food procurement process in the hotels follows procedures in the thriving food industry using Lean and Six Sigma principles.

Currently, there are many lean management methods and tools. Based on the fact that lean was born in the manufacturing industry, most tools came from the manufacturing sector. Through adaptation over time, lean can now be applied in various fields and not just for one particular type of industry. However. Not all available methods are suitable for every process. So there must be a qualified evaluation procedure to rank the most suitable lean methods in the hospitality industry [1]. Lean tools and methods are a set of technical options in lean that have been proven successful in improving performance. According to Rauch et al. [1], forty-seven lean tools may be applied in the hotel industry, and they can be classified into five clusters of methods, namely (1) Machinery and Equipment, (2) Organization and Staff, (3) Material flow and Layout, (4) Production Planning and Control, and (5) Quality.

However, it is more suitable for the Organization and Staff and Material flow and Layout clusters for the hotel industry. The Machinery and Equipment cluster is unsuitable due to the high cost and lengthy implementation time. In finding the most suitable lean method in the hotel industry, it is necessary to have a clear definition and weighting of the criteria. Four criteria have been defined and weighted to each criterion, namely (1) implementation costs, (2) implementation time, (3) impact on KPI (Key Performance Index), and (4) outcome sustainability and its implementation [1]. A criterion can have different weights compared to other criteria. In the case of the hospitality industry, the weighting is determined jointly by involving the hotel management. There are 4 best tools selected from the four criteria, namely Kaizen workshop, 5S implementation, visual management, and Poka-Yoke. This research refers to the latest journals or literature related to applying lean management in the hospitality industry. Research of Rauch et al. [1] and Kuaites et al. [9] become the primary reference.

A research of Kuaites et al. [4] used the flow process chart method to analyse work processes and the flow of materials, staff and equipment from the beginning to the end of the process. The technique used is ECRS (Eliminate, Combine, Rearrange and Simplify), which means that rearrangements are carried out in carrying out work or process efficiency as much as possible. Still, it would be better to be simplified, combined or eliminated. However, the study only focuses on housekeeping activities. Rauch et al. [1] describes the criteria for selecting the proper lean method for the hotel industry and weights these criteria to make it easier for hotel practitioners to apply it. However, the procedure for weighing and quantitatively selecting criteria is not explained; the author directly determines the weights qualitatively based on his experience. The tools chosen to be applied to the case study were immediately selected, namely CIP (Continuous Improvement Process)-Workshops, FIFO Warehouse Management, implementation of 5S, and reduction of housekeeping movements.

A research of Ahmed [10] adopted a quantitative approach to obtain the objective of identifying lean methods that hotels have applied. However, it does not apply significantly to improve the hotel's performance. The research investigated using a series of questionnaires for what hotel managers have done concerning lean in Egypt. A research of Indrawati et al. [11] adopted lean management and applied it to fast food restaurants with an integrated approach between Six Sigma and the TRIZ (Theory of Inventive Problem-Solving) method to get creative ideas to overcome the problems faced. The result of the study is the identification of waste that occurs, namely the delay in the flavouring process from the total time. The improvement strategy is to reduce the setup time for changing flavours and implementing 5S.

Al-Aomar et al. [3] assessed the application of lean management techniques in hotel supply chains that have identified potential waste in hotel management with the SIPOC (Supplier, Input, Process, Output and Customer) approach. Besides determining the type of waste, it has identified lean techniques that have been carried out by the hotel and methods that may be used to improve hotel performance in the future. Al-Aomar et al. [12] developed a framework for identifying, categorizing, and prioritizing the choice of lean techniques to be adopted by the hospitality industry. The methodology used is a survey of hotels in the United Arab Emirates with a SIPOC based design to find lean techniques that have been used by hotel management. Then it used the AHP method to prioritize the existing techniques by considering various criteria. The results show that there are 6 main categories of lean techniques and 19 lean techniques relevant to the hospitality industry. The main categories are JIT, Kaizen, Quality, Inventory, Maintenance and Standardization.

2 Method

This study was conducted at Hotel XYZ through observation for one month in October 2020. The observations consisted of three kinds: direct observation on the hotel kitchen floor by looking at the production process, breakfast and a *la carte*, how the operator works, work procedures, environmental conditions, and production floor layouts. The two interviews were with several workers in the procurement department and Food and Beverage Manager to operators in hotel kitchens, such as cook helpers and head chefs. It needs to do documentation of the process and the required data. The data can be in the form of photos, production data, and processing time. The research flow chart can be seen in Figure 1. The diagram includes the activities performed in sequential research steps.

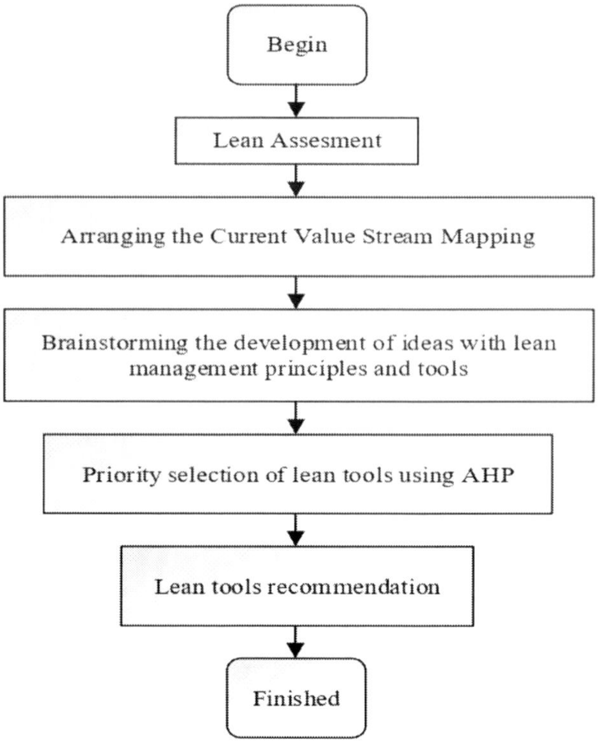

Fig. 1. Research flow chart

2.1 Lean assessment

The research begins by conducting a lean assessment that assesses how lean a hotel kitchen is based on principles and indicators. Assessment is carried out to determine the lean implementation plan to be more appropriate to the company's conditions. The procedure for conducting a lean assessment using a checklist [13]. The people who are the focus of lean assessment are a Food and Beverage manager and the head chef of Hotel XYZ.

2.2 Value Stream Mapping (VSM)

The value stream is used to describe the process from the beginning (raw materials) to finished products in the form of dishes in a hotel kitchen, either providing value (value-added) or not providing value (non-value added). VSM is divided into two, namely current and future VSM. The initial stage of VSM current is carried out by direct observation of the dish production process in the kitchen area for one week.

VSM future is structured to map improvement ideas based on the lean method that has been selected, while the way of compiling is the same as VSM current, but includes ideas such as supermarket, pull, flow, heijunka, and so on. The VSM future will be a reference for gradual implementation. How to prepare VSM Future is done by sharing data from the previous stages, namely lean assessment, VSM current, brainstorming improvement ideas. The parties involved are the research team, FBM, and chef to the waitress for breakfast preparation.

2.3 Choosing the right lean method

Of the many proposed tools, prioritization of the methods to be used by the hotel is carried out. The initial stage is the weighting of the criteria in choosing the right lean tools priority. The criteria refer to the research of Rauch et al. [1], namely implementation costs, implementation time, impact on Hotel KPI and the sustainability of their application. The data obtained from the questionnaire were processed using pairwise comparison and then calculated using the AHP.

3 Results and discussion

3.1 Lean assessment

Lean assessment refers to the research of Laoha and Sukto [14] for production and Tortorella et al. [15] for procurement. It is rated from 1 to 5 for each aspect of the lean. The lean production aspects are presented in Table 1.

Table 1. Lean assessment on production aspect

Aspect	Implemented?	Assessment
Kanban/Pull System	Yes	2
Visual management	Yes	2
5S (Concise, Neat, Clean, Care, Diligent)	Yes	4
Operation base layout	Yes	3
Quick changeover	Yes	3
Multi skill is workers	Yes	5
Standard operation procedure	Yes	3
Poka-Yoke	Yes	3
Total preventive maintenance	Yes	2
Awareness of 7 waste	Yes	3
Kaizen Production	Yes	1

The Kanban and pull system aspects are given a value of 2 to identify that the number of occupied rooms is known and is updated every hour as a reference Kanban signal for the production of when and how many breakfast dishes for the following day. However, this information is sometimes not conveyed to the kitchen. In addition, some dishes have been prepared from noon D-1 breakfast with a predetermined amount based on experience. For example, fried chicken that has been provided ranges from 25 kg on weekends and 10-15 kg on weekdays. This system shows that there is a tendency to apply a push rather than a pull system so that the impact is that there is an excess or less potential that can cause the waste or reduce the value of satisfaction desired by customers.

However, for the activity of providing meals for meetings and weddings, a checker is assigned to update the condition and number of remaining dishes at any time to the Food and Beverage Manager (FBM) so that an FBM can determine what, when, and how many more dishes to cook properly. For example, meeting activities have been booked for 40 guests; 40 are made indirectly, but with smaller lots and gradually increasing according to data from the checker. This action aligns with the principle of lean management, namely producing small lot sizes or small batches [16]. The hotel has implemented it in the visual management aspect but is given a score of 2. Signs/pictures and indicators indicate this action to make it easier for management to make decisions, for example, via smartphones capable of sharing the latest data on room occupancy to all managers. However, there is no display in the kitchen regarding the number of rooms. In addition, based on observations regarding inventory, it is not visible to see the current stock condition to determine the types and quantities available. Chefs must check and open their refrigerators, which can result in a potential shortage or excess of raw materials.

In applying 5S, cooking utensils and ingredients have been placed neatly in their place. The chef periodically cleans cooking utensils with a chamois and removes water on the floor. Trash cans have been provided with easy-to-reach locations, and the size is also quite large. Dirty dishes and cooking utensils do not stay in the dishwasher for too long because multitasking assistants are provided who can also serve as a stewardess. However, it showed a leaking water pipe from above the ceiling, causing the floor always to be slippery and the chef potentially slipping so that the chance of wasting food would occur when the chef was transporting the dish. The application of 5S in the kitchen refrigerator found that there were no labels for storing ingredients. There was a potential for an error in the location of storing ingredients in the fridge so that the next chef shift needed time to look for it.

The U-shaped layout makes it easier for chefs to pick up and cook with minimal movement. Even in the middle of the room, a plating table is not far away. The size of the room is appropriate, not too big and not too narrow. However, it is necessary to adjust the placement of cooking utensils periodically, considering the hotel occupancy rate is constantly changing. Layout U can already represent lean production [17]. Change over time shows the setup time from one dish to another is quite fast, especially for a *la carte* dishes. For example, the same kettle is used for two different dishes, but the chef can quickly arrange and clean the kettle. This action is because the frying area is provided with a water faucet to clean the kettle soon. The best change over achieves SMED (Single Minute Exchange Dies), reducing downtime and increasing output quantity [18]. Multi-skill workers have been implemented. Very multi-skill workers, a service or waitress can do several chef activities. A stewardess, who is in charge of washing dishes, can work as a cashier and checker. And a chef can cook a variety of Chinese, European, and Indonesian dishes. However, cross-training has not been seen to be implemented by the hotel. Lean production tends to apply cross-training to achieve Just in Time to eliminate waste [19].

The implementation of the Standard Operation Procedure (SOP) has been carried out well by the Hotel, but this has not been seen because it has been used by experience. However, Poka-Yoke is recommended to minimize errors made by the operator [20], based on observations that the microwave has been equipped with a countdown timer that will sound when it is finished cooking. Based on the observations, Total Preventive Maintenance has not been carried out properly. It can be seen that some cooking utensils are damaged but are still stored in the kitchen for a long time. The refrigerator is used as a temperature indicator to identify if the fridge is damaged. There needs to be a maintenance schedule for cooking utensils to prevent sudden damage when it is needed.

The Food and Beverage Manager (FBM) concerns for the 7 types of waste is good; this is indicated by the efforts to reduce waste that has been achieved, such as the target of raw material costs of 20% of sales. Besides, FBM is efficient in determining the number of the kitchen staff. Inventory is constantly monitored every month, and there are almost no food defects. Overproduction of dishes has been minimized by deep-frying. However, other potential wastes such as delay and lead time need further improvement. Kaizen Production in this hotel has not yet formed a formal Kaizen team. However, continuous improvement activities are always there, as can be seen from the changes in the breakfast banquet system for new regular changes.

Aspects of lean implementation in the hotel supply chain are reviewed with an assessment of 1-5 in detail, which are presented in Table 2.

Table 2. Lean assessment on supply chain aspects

No	Aspect	Assessment
1	Collaborative relationship based on mutual trust and commitment	4
2	Not many main suppliers	3
3	Small batch size and high delivery frequency	3
4	Kanban with supplier/pull	2
5	Supplier selection criteria based on capacity, added value and historical background	4
6	Build communication and information	4
7	Supplier adopts quality techniques	4
8	Cooperation with suppliers in solving problems in the hotel kitchen	2

Based on the observations, suppliers are made a strategic part by the hotel, which is indicated by the existence of a long-term cooperation agreement. Hotels are trusted to make payments after 1-month of running, and suppliers are guaranteed to supply regularly to the hotel within a certain time. The hotel is also not easy to change suppliers, so they tend to choose not many suppliers for the same type of goods [21]. For the small batch category with frequent deliveries, the hotel has applied it to delivering raw material for dishes. It shows that deliveries are carried out regularly on a scheduled basis with different quantities depending on the estimated hotel room occupancy rate; however, the hotel needs to request that it be sent more frequently with fewer quantities to ensure the freshness of the raw materials.

The hotel prefers to choose a flexible supplier. For example, if it is known that there are unripe papayas that are still sticky, the supplier can flexibly change the type of papaya that the hotel needs. With the capacity owned by the supplier, hotel demand that wants a fixed price for a particular time can be fulfilled by the supplier, although in terms of quantity, it can vary. To build good communication with suppliers, the hotel always holds regular meetings to discuss the supply of raw materials for dishes. However, it has not made a joint performance measure between hotels and suppliers, and it only evaluated on a 6-month or 1-year basis.

3.2 Value Stream Mapping (VSM)

Current Value Stream Mapping and Future Value Stream Mapping are presented in Figures 2a and 2b. Based on the observations, many breakfast products are still in the semi-finished form, which are processed during the day and then left for a particular time to be reprocessed before breakfast time, which begin in the early morning at 3 a.m. This action is done for labour efficiency so that the daytime shift workers H-1 are not unemployed. However, the impact is that the freshness is reduced. The role of the refrigerator is crucial in this regard. In addition, there are still some raw materials that are stored for quite a long time because the procurement system pushes a scheduled system.

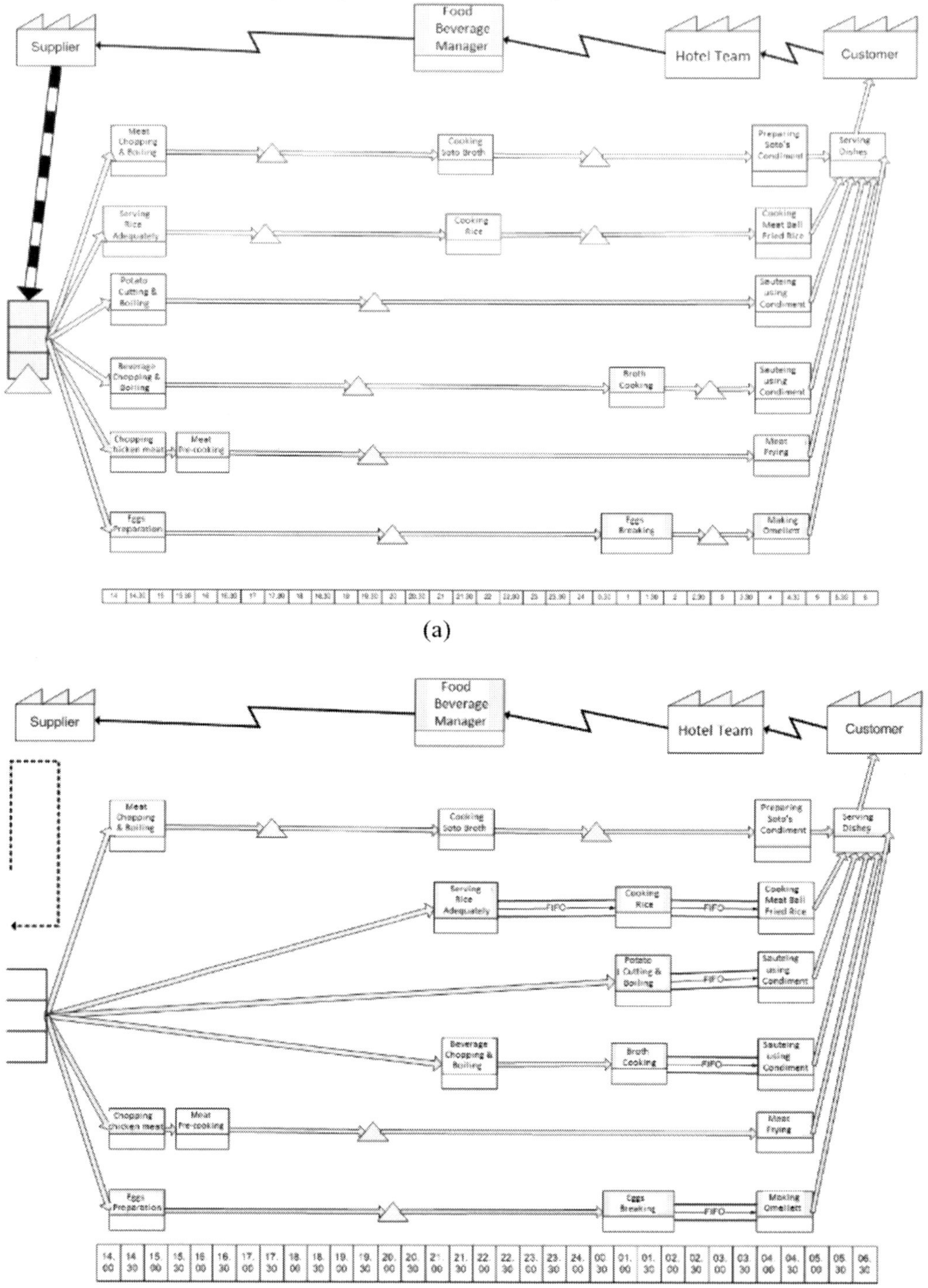

(a)

(b)

Fig.2. Current Value Stream Mapping (a) and Future Value Stream Mapping (b)

3.3 Selection of lean tools with AHP

Based on the results of the lean assessment, the preparation of the VSM current and discussions with several related parties, seven activity plans that need to be prioritized to be implemented are prepared. Table 3 shows these seven activities plans.

Table 3. Lean tool alternatives

No	Aspect	Lean method alternative plan
1	Kanban/Pull System	Application updates the number of rooms filled in the kitchen screen as Kanban
2	Visual management	Using colours that are easy to see immediately as an indicator of a display on the number of rooms and room service orders so that lead times are short. Visual colours describe the number of stocks of various products
3	Poka-Yoke	Increase Poka-Yoke for every process that tends to have human error
4	Total Preventive maintenance	Implementing preventive maintenance, make maintenance schedules and autonomous maintenance
5	Kaizen Production	Creating a specific improvement team that is assessed by top management
6	Small batch with often frequency	Applying heijunka and milk-run process
7	Kanban with supplier/pull	Changing the system for certain raw materials that are still being pushed into a pull system to suppliers

Three plans were chosen to be prioritized for implementation from these seven plans. Based on an in-depth discussion, the plans are (1) implementing a pull system both in production and at suppliers, (2) forming a Kaizen team, and (3) visual management. Determination of the order of the tools can be carried out based on 4 criteria weighted by pairwise comparison, as follows (Figure 3):

1. Implementation costs,
2. Implementation time,
3. Impact on Hotel KPI,
4. Sustainability of outcomes and their implementation.

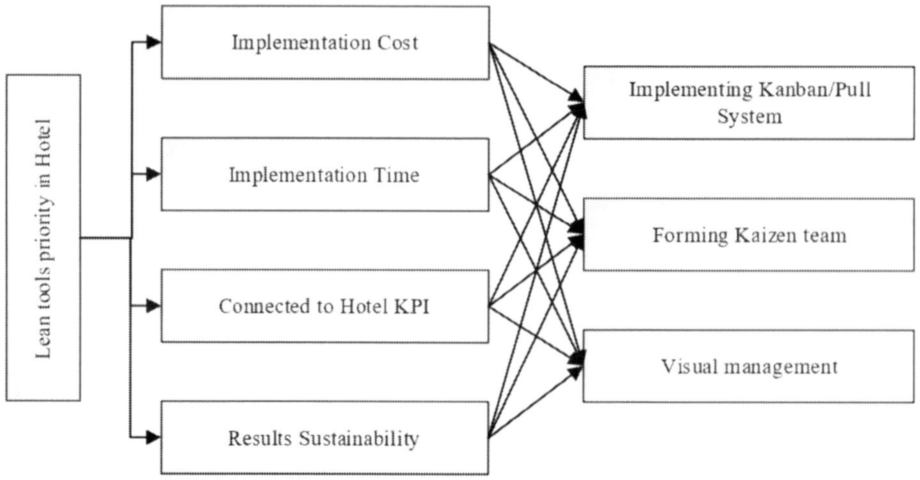

Fig. 3. Lean tools priority in hotel

The cost of investment and the use of resources should be as low as possible to guarantee returns to the shortest possible period [22]. The less time it takes to see the improved results, the better the lean method is applied. However, the failure of implementing lean tools is often caused by the desire to get results in the short term. KPI has a significant influence on choosing the proper lean method by hotel management and help decide on the sustainability, expansion and limitation or even termination of a lean management implementation project. Lean thinking is a long-term benefit for a company. Changes in people's mindsets do not occur daily but require a longer time. Therefore sustainability criteria are essential in choosing the proper method.

Table 4. Pairwise comparison criteria

	Cost	**Time**	**KPI**	**Sustainability**
Cost	1	3	0.25	0.33
Time	0.33	1	0.33	0.5
KPI	4	3	1	1
Sustainability	3	2	1	1

From the pairwise comparison, the consistency was observed, after which the weight was calculated (Table 4).

Table 5. Comparative matrix design

Criteria	**Weight**	
Cost	5.565	0.167
Time	4.110	0.114
KPI	4.013	0.388
Sustainability	3.746	0.311
Total	17.433	1.000

The results of the level 1 weighting show the highest value in the sequence, starting from the criteria of KPI, sustainability, cost, and time (Table 5).

Table 6. Pairwise comparison alternatives

	Implementing Kanban	**Forming Kaizen team**	**Visual Management**
Implementing Kanban	1	1/7	1/5
Forming Kaizen team	7	1	3
Visual Management	5	1/3	1
Total	13.000	1.476	4.200

From the above pairwise comparison alternatives (Table 6), the consistency was observed, after which the weight was calculated.

Table 7. Comparative matrix results

Criteria	**Weight**	
Implementing Kanban	2.877	0.074
Forming Kaizen team	2.849	0.643
Visual Management	3.564	0.283
Total	9.290	1.000

From the comparative matrix results above (Table 7), forming Kaizen team, visual management, and implementing Kanban weights 0.643; 0.283; and 0.074, respectively. Thus, the formation of the Kaizen team has the highest weight value.

Table 8. Results of prioritizing alternatives

Criteria	**Weight**	**Alternative weight for each criteria**		
		Implementing Kanban	**Forming Kaizen team**	**Visual Management**
Cost	0.167	0.074	0.643	0.283
Time	0.114	0.074	0.643	0.283
KPI	0.388	0.074	0.643	0.283
Sustainability	0.331	0.074	0.643	0.283

Determination of each final value for each alternative is calculated by multiplying the weight of each criterion by each alternative weight and then adding up for the entire criteria as many as 4 indicators so that the results can be seen as follows (Table 8 – 9, Eq. 1): [4]

The final value for each alternative

$$= \sum_{i=1}^{4} Criteria\ weight\ i\ x\ Alternative\ value \qquad (1)$$

Table 9. Final value of total alternative priority

No	Alternative	End value
1	Kanban implementation	0.074
2	Forming Kaizen team	0.643
3	Visual Management	0.283

Based on the final value of the total alternative priorities, the highest weight values consecutively are forming a Kaizen team, visual management, and implementing Kanban.

4 Conclusion

Competitiveness in this modern era is essential for a company to survive. It also prevails in the hotel industry. Based on this study, waste elimination is a crucial part of increasing the competitiveness of the hotel industry. The proper method to eliminate waste is to do a lean assessment to describe how lean the XYZ Hotel's kitchen is, which is then visualized through the VSM current and given improvement through the VSM future. Further, it should determine the lean tools and use the AHP method to specify which lean tools would be applied first. Based on the AHP calculations, the competitiveness of Hotel XYZ can be increased by forming a Kaizen team.

References

1. E. Rauch, A. Damian, P. Holzner, D. T. Matt, *Lean hospitality-application of lean management methods in the hotel sector*, Procedia CIRP, vol. **41**, no. April, pp. 614–619, (2016), Available at : doi 10.1016/j.procir.2016.01.019.
2. E. Andrés-López, I. González-Requena, A. Sanz-Lobera, *Lean Service: Reassessment of Lean Manufacturing for Service Activities*, Procedia Eng., vol. **132**, pp. 23–30, (2015), Available at : doi 10.1016/j.proeng.2015.12.463.
3. R. Al-Aomar, M. Hussain, Tour. Manag., vol. **69**, no. June, pp. 553–565, (2018), Available at : doi 10.1016/j.tourman.2018.06.030.
4. T. Kuaites, T. Thueanphae, N. Chanthranapasawat, J. King Mongkut's Univ. Technol. North Bangkok, (2020), Available at : doi 10.14416/j.kmutnb.2020.08.003.
5. Ilias Vlachos, Aleksandra Bogdanovic, Lean thinking in the European hotel industry, Tourism Management, Volume 36, 2013, Pages 354-363, ISSN 0261-5177, https://doi.org/10.1016/j.tourman.2012.10.007.
6. O. Taiichi, *Toyota Production System: Beyond Large-Scale Production*, Reprint. (Boca Raton: CRC Press, 2019).
7. J. P. Womack, D. T. Jones, *Lean Thinking, Banish Waste and Create Wealth in Your Corporation*, Revised, (New York: Simon and Schuster, 2013).
8. D. M. Pearlman, H. Chacko, Int. J. Hosp. Tour. Adm., vol. **13**, no. 1, pp. 48–66, (2012), Available at : doi 10.1080/15256480.2012.640212.
9. T. Kuaites, T. Thueanphae, N. Chanthranapasawat, *The Application of Lean Concept to Waste Reduction of Working Process in Hotel Service Industry*, proceeding of IEEE 7th International Conference on Industrial Engineering and Applications (ICIEA) (2020), Available at : doi 10.1109/ICIEA49774.2020.9101938.

10. M. Ahmed, Int. J. Heritage, Tour. Hosp., vol. **12**, no. 3, pp. 181–194, 2018, Available at : doi 10.21608/ijhth.2018.32119.
11. S. Indrawati, E. A'Azzam, E. Adrianto, S. Miranda, A. D. Prabaswari, IOP Conf. Ser. Mater. Sci. Eng., vol. **722**, no. 1, (2020), Available at : doi 10.1088/1757-899X/722/1/012044.
12. R. Al-Aomar, M. Hussain, Int. J. Lean Six Sigma, vol. **10**, no. 1, pp. 375–396, (2019), Available at : doi 10.1108/IJLSS-10-2017-0119.
13. W. Urban, The lean management maturity self-assessment tool based on organizational culture diagnosis, Proceeding - Soc. Behav. Sci., vol. **213**, pp. 728–733, (2015), Available at : doi 10.1016/j.sbspro.2015.11.527.
14. C. Laoha, S. Sukto, Int. Bus. Manag., vol. **9**, no. 4, pp. 590–595, (2015), Available at : doi 10.3923/ibm.2015.590.595.
15. G. L. Tortorella, M. V. L. L. Rosa, R. Caiado, D. Nascimento, R. Sawhney, *Assessment of lean implementation in hotels' supply chains*, Producao, vol. **29**, (2019), Available at : doi 10.1590/0103-6513.20190044.
16. N. Adlin, H. Nylund, M. Lanz, T. Lehtonen, T. Juuti, Procedia Manuf., vol. **00**, no. 2019, pp. 1371–1378, (2020), Available at : doi 10.1016/j.promfg.2020.10.191.
17. J. Metternich, S. Bechtloff, S. Seifermann, Procedia CIRP, vol. **7**, pp. 592–597, (2013), Available at : doi 10.1016/j.procir.2013.06.038.
18. A. A. Karam, M. Liviu, V. Cristina, H. Radu, Procedia Manuf., vol. **22**, pp. 886–892, (2018), Available at : doi 10.1016/j.promfg.2018.03.125.
19. J. R. Jadhav, S. S. Mantha, S. B. Rane, J. Ind. Eng. Int., vol. **11**, no. 3, pp. 331–352, (2015), Available at : doi 10.1007/s40092-014-0092-4.
20. A. Zhang, Int. J. Six Sigma Compet. Advant., vol. **8**, no. 2, pp. 147–159, (2014), Available at : doi 10.1504/IJSSCA.2014.064260.
21. H. Abdul Zubar, P. Parthiban, Int. J. Logist. Syst. Manag., vol. **18**, no. 1, pp. 72–99, (2014), Available at : doi 10.1504/IJLSM.2014.062122.
22. D. Gracanin, D. Ciric, B. Lalic, J. Curcic, N. Tasic, Procedia Manuf., vol. **38**, no. 2019, pp. 316–323, (2019), Available at : doi 10.1016/j.promfg.2020.01.041.

Reducing the product defects using lean production perspective: a case study at CV Cita Nasional

Hendri Wijaya[1], *Yulita* Rohmasari[1], *Hanif* Asmoro Kaundy[1], *Rizqi* Pratama Kurnia Indah[1], and *Ide* Bagus Nurjaya[1]

[1]Industrial Management Program Study, College of Vocational Studies, IPB University, Jl. Kumbang No 14 Bogor 16151, Indonesia

Abstract. This article discusses the issues of how to reduce pasteurized milk defects as part of waste in a lean production system. Product defects are the most important problem faced by the dairy enterprise daily. This problem makes production costs higher and decreases the profit. A set of seven tools and methods is used to identify and analyse the defect problem, and then make a solution plan. Value Stream Mapping is used to describe the current condition of the production system and make improvement planning. The lean production concept has been successfully used not only to find ways to reduce product defect, but also to plan and eliminate other wastes in the pasteurized-homogenized milk production process. Waste found and eliminated are production lead time, transportation, waiting time, defective products, and downtime machines. Overall, the research results in a reduction of overtime for 41 minutes, and make downtime on the packaging machine to become 0 minutes.

1 Introduction

Defective products are one of the problems faced by companies. The more defective products produced by the company, the more inefficient and effective the company is. Production that produces defects economically will lead to higher production costs. Defective products also cause the company's ability to sell products to decrease, resulting in reduced company revenues. Moreover, higher production costs and lower revenue from product sales resulting in lower profits for the company. In the end, the company's ability in earning lower profits will result in lower competitiveness. Company competitiveness is the key for companies to be able to compete in an increasingly competitive industry due to globalization.

Defective product from a lean production perspective is a waste that must be endeavored to be eliminated [1]. Lean production's ability to eliminate waste is seen by practitioners and researchers as one of the reliable methodologies used to improve company competitiveness. Lean production is a production philosophy that emphasizes minimizing the use of resources in various company activities. Lean production is also defined as a systematic approach to identify and eliminate waste (non-value-added activities) through continuous improvement by flowing products according to consumer needs to achieve perfection. The application of Lean production aims to increase product value by reducing the occurrence of waste. Waste is reduced by sorting out activities that generate value-added (VA) and do not produce added value (non-value-added / NVA) [2].

CV Cita Nasional is a privately owned company in Semarang, Central Java, which processes pure cow's milk into processed milk products. There are two products produced, namely pasteurized-homogenized milk and yogurt. Pasteurized-homogenized milk products are the main product of CV Cita Nasional, which is produced every day and in much higher quantities than yogurt that is not produced every day. Pasteurized-homogenized milk products are sold in five packages, namely 150 ml cup, 250 ml pure pack, 500 ml pure pack, and 125 ml minipack, and 180 ml minipack. The company has problems that occur in every production process carried out, namely there are defective products. CV Cita Nasional has a high commitment to implementing quality management.

The purpose of this study is to find the root cause of the occurrence of defective products and other potential waste and make an improvement plan. The products used as research objects are pasteurized-homogenized milk products packaged in 150 ml cups, 125 ml mini packs, and 180 ml mini packs. Lean production philosophy is used as the basis for achieving the objectives. Quality control tools in the seven tools method are used to help identify and analyze problems. The why-why analysis is used to find the root of the problem.

2 Literature Review

The principle of lean production system developed from Japan, exactly from Toyota. The term lean production came into prominence in the late 1980s when the Massachusetts Institute of Technology (MIT) researched the global automotive industry, focusing on the Japanese Toyota model to derive best-practice industry models. The concept of lean production itself has been implemented after the Second World War ended when Japanese industry began to think about changing its production model. The first industry to apply was the Toyota Company and the system used at that time was called the Toyota Production System [3].

Although lean production systems were originally used for the automotive industry, recently, many researchers have used lean production principles for other sectors, including agriculture and agro-industry. The benefits of lean production in the red meat industry value chain to respond to consumer needs [4]. The same research takes pork products in Catalan [5]. A framework for lean thinking for the agricultural sector [6]. Melin and Barth examined the application of lean thinking in 34 agricultural companies in Sweden and found that management commitment, training of managers and employees, corporate leadership, and the role of change agents are important factors for the success of lean implementation [7]. The use of lean production in the agribusiness sector to produce lean production that can facilitate the decision-making process, increase profits, market share, and customer evaluation [8]. Lean production is the key to the sustainable development of agricultural enterprises [9].

The concept of lean production is not whatever associated with the type of the manufactured products and therefore it applies in the energy sector to the same extent as in any other industry [10]. The lean production approach can be used to reduce waste in the enterprise [11].

The implementation of lean production can also be used in the industrial era 4.0. lean production supported by the use of technology 4.0 from the level of production planning to product distribution to consumers can increase flexibility and customer satisfaction [12].

The application of lean production which is supported by advanced technology (advanced tools) such as big data and smart technology will help companies to find their core competitiveness and achieve excellence [13].

3 Methodology

Data and information were collected in three ways, such as field observations, interviews, and discussions, as well as literature from company records and reports. Identification and analysis of data using checksheet tools, fishbone, root cause analysis using ask why 5 times, and value stream mapping techniques.

Checksheet is a basic tool for well-arranged data collection of quality in processes and simplifies post-processing of collected data [14]. Checksheet in this research is used to present the data, type, and value of defective products. Based on the checksheet, the highest type of defect is determined for further analysis. Fishbone diagrams are used to identify and show possible causes of problems and especially when a team tends to fall into thinking on a routine [15]. Fishbone diagrams are used to describe the causes of the highest types of defective products. The causes of problems are grouped into 4M, namely man, machine, materials, and method. Root cause analysis is used to find the root cause of the highest defective product, which is happening in the company. The simplest technique in finding the root of the problem uses the ask why 5 times technique [16]. Value Stream Mapping (VSM) is a visual method for mapping the production line of a product that includes material and information. VSM is used to describe the comparison of the current condition with the condition after the improvement plan has been carried out [17].

4 Results

4.1 Identification of Number and Type of Defective Product

Check Sheet is used to identify the number and type of defective products. Data collection using a checksheet for 4 months obtained the number of defective products presented in Table 1.

Table 1. Number of defective product

Month	Product		
	150 ml cup	125 ml minipack	180 ml minipack
Dec 2020	11,253	21,658	6,486
Jan 2021	9,888	20,686	6,064
Feb 2021	13,272	23,499	3,062
Mar 2021	10,173	21,114	6,132
Average/ month	11,146	21,739	4,702

Based on the results of the checksheet for the number of defective products, it was found that the 125 ml and 180 ml minipack product had the highest number of defects. Furthermore, a checksheet of the type of damage that occurred on the minipack product was carried out. The checksheet for the type and value of defective products of 125 ml and 180 ml minipack per month is presented in Table 2.

Table 2. Type and value of defective product 125 ml and 180 ml minipack per month

type	Number of defective product		Loss value
	125 ml	180 ml	
Leaky packaging	10,652	2,210	IDR 21,503,000
Lack volume	7,609	1,787	IDR 15,881,000
Expired date not printed	3,478	705	IDR 6,979,500
Total	21,739	4,702	IDR 44,363,500

There are three types of damage for minipack products such as leaky packaging, lack of volume, and expired date not printed. The highest type of damage as presented in Table 2 is leaky packaging. The value due to damage the minipack leaky packaging is IDR 21,503,000 per month. The second type of damage is lack of volume with a value of IDR 15,881,000 per month, and the smallest is an expired date not printed with a value of IDR 6,979,500 per month.

4.2 Analysis of the Causes of Defective Products

The cause of leaky packaging is illustrated using a fishbone diagram. There are 4 possible causes of leaky packaging based on the fishbone diagram, such as man, machine, material, and method. The results of the depiction of the leaky packaging fishbone diagram in 4M are presented in Figure 1.

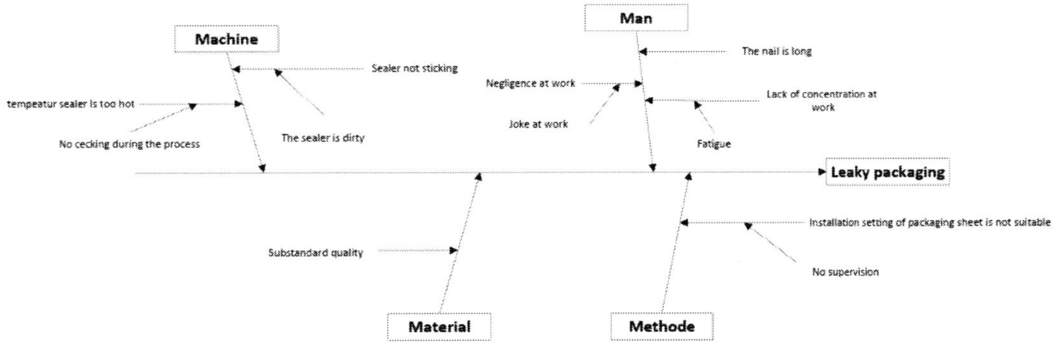

Fig. 1. Leaky packaging fishbone diagram

Root cause analysis by using why 5-time technique was used to find the root cause of the problem that causes the packaging to leak. The root of the problem in the leaky packaging occurs because there were no technical instructions for the use and maintenance of the sealer machine. Technical instructions specifically relate to two things, namely cleaning the sealer after use and replacing the sealer machine element within a certain period. The root cause analysis is presented in Figure 2.

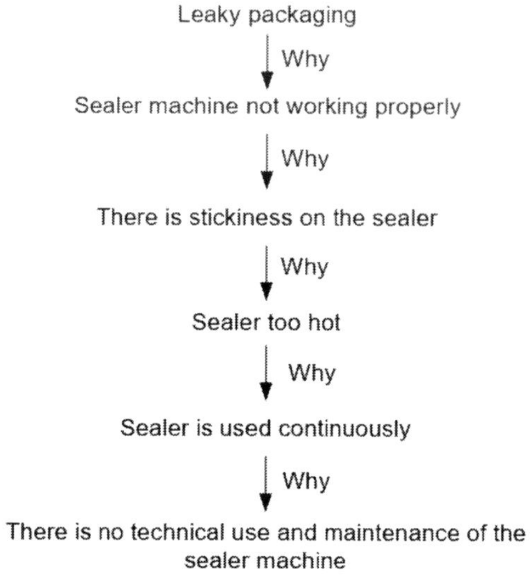

Fig. 2. Root cause analysis of leaky packaging

4.3 Waste Reduction Plan using VSM

The depiction of the VSM milk production process of the pasteurized-homogenized was begun with a description of the current state mapping. Current state mapping was needed as the first step in the process of identifying activities that may cause waste in the production process of pasteurized milk and homogenization of minipacks. There are 6 processes for making pasteurized milk minipacks, laboratory analysis, fresh milk cooling, mixing ingredients, pasteurization and homogenization, cooling and quality control, and packaging. The current state mapping is presented in Figure 3.

 https://doi.org/10.1051/e3sconf/202234800008

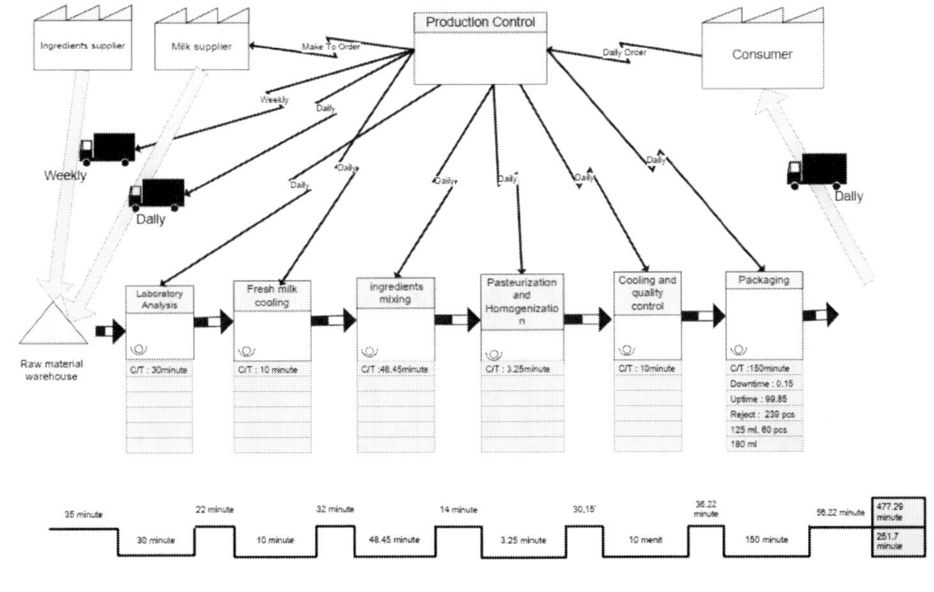

Fig. 3. Current state mapping of minipack product

After observing and analyzing, the VSM can identify the presence of waste other than defective products, such as production leadtime, transportation, waiting time, and downtime. production lead time and transportation waste occurs between the process of cooling fresh milk with the mixing ingredients. Waiting time occurs between the cooling process and mixing raw materials, also in the pasteurization process, cooling, QC, and the packaging process. Another waste found is downtime in the packaging process. The results of the current state mapping analysis are presented in Figure 4.

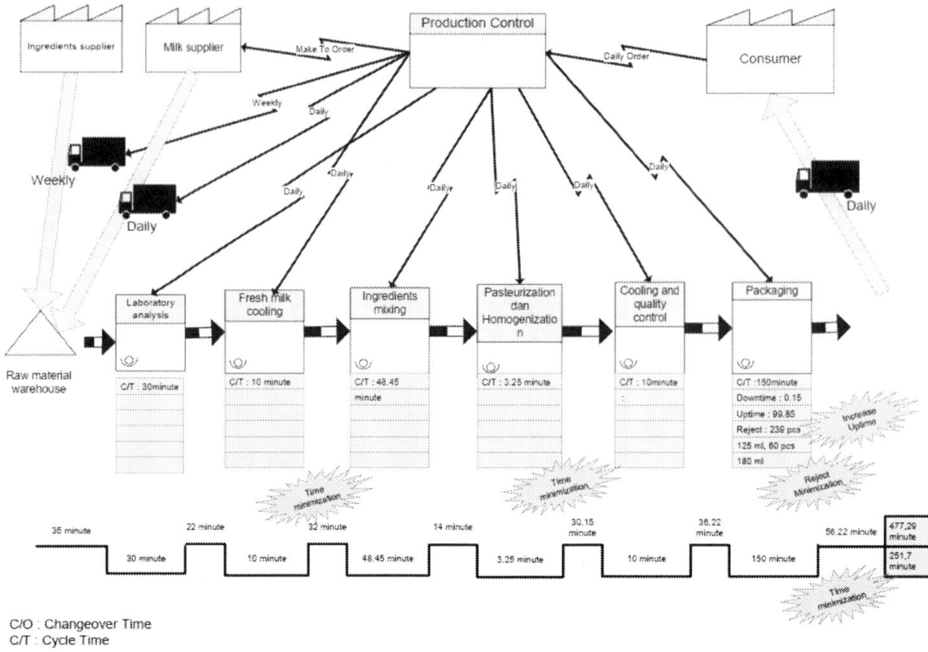

Fig. 4. Current state mapping analysis of minipack products

The way of handling waste based on the current state mapping is done by making improvement plans. The results of the waste analysis that may be improved are presented in Table 3.

Table 3. Improvement plan

Where	Waste	Cause	Improvement
Between the process of cooling fresh milk with the mixing ingredients	Production lead time	Weighing of ingredients the cooling process is complete	Place additional raw materials in the production room and weight them before the cooling process is complete thereby eliminating waiting time and transportation
	Transportation	1. Transport for additional raw materials from laboratory to dairy production room. 2. Transport for additional raw materials over the slippery floor.	
Between pasteurization process with cooling and quality control	Waiting time	controller walks from the QC laboratory room to the cooling room, after the homogenization process is complete	controller is ready in the cooling room, so that after the homogenization process is complete, samples can be taken immediately
Packaging process	Leaky packaging reject products	There is no sealer machine operating manual yet	1. Setting the temperature correctly and checking every 15 minutes. 2. Sealer is cleaned every 2.5 hours.

	Downtime	1. Cutter is not filed regularly and scheduled so that it is blunt 2. Replacement of the heater element wait until it is damaged	1. File cutter once a week. 2. Replace the cutter every 3 months so that it doesn't get dull. 3. Lubricate the machine on a scheduled basis every 30 minutes. 4. Set the sealer temperature properly and check every 1 hour. 5. Replace the heater element every 96 hours of use. 6. Socialization of autonomous maintenance to operators.
	Waiting time	1. The crates are not always ready to use because they are still being washed. 2. The product packaging arrangement works on the truck jokes a lot so it takes longer time. 3. Re-dismantling the packaging that is arranged in the truck because it looks like there is a leaking product.	1. Schedule for washing the crate, so that it is clean before the packaging process begins. 2. Packaging arrangement work on trucks must be more orderly. 3. Clear separation between product crates and reject product crates.

The improvement plan in addition to reducing waste can also reduce changeover time by 41 minutes and eliminate the downtime. The results of the improvement plan are depicted in the future state mapping presented in Figure 5.

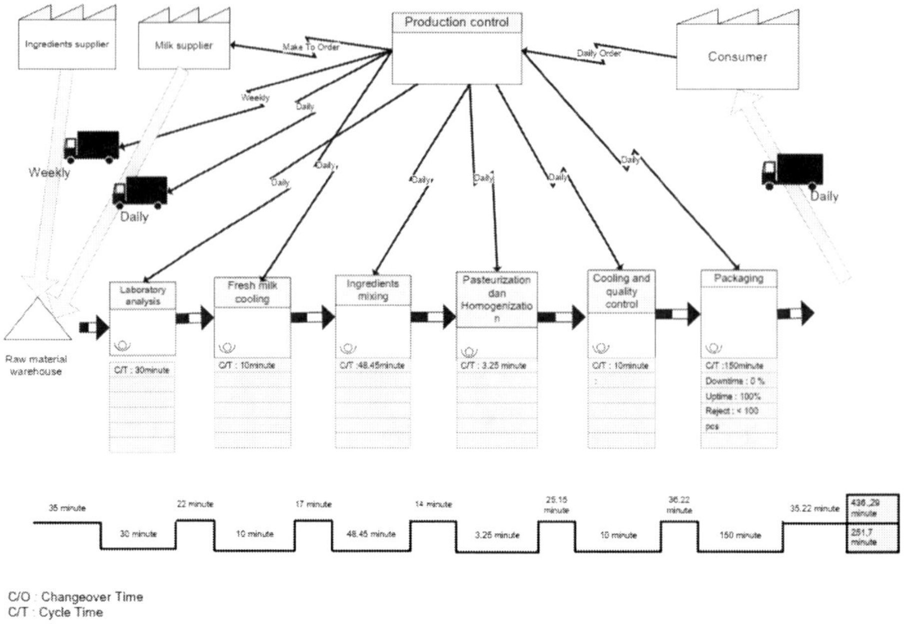

Fig. 5. Future state mapping of minipack product

Future state mapping shows the efficiency of change over time compared to current state mapping, after carrying out plant improvements. The change over time efficiency that occurs between the process of cooling fresh milk with the mixing ingredients is 15 minutes, from 32 minutes to 17 minutes. Efficiency of change over time between the pasteurization process with cooling and QC is 5 minutes, from 30.15 minutes to 25.15 minutes. Efficiency in the packaging process is 21 minutes, from 56.22 minutes to 35.22 minutes.

Efforts that need to be made to support the implementation of the improvement plan were to form a VSM team. The VSM team involves cross-functional members consisting of a combination of laboratory, production, purchasing, and maintenance service divisions. The task of the VSM team was to collect data, analyze, and make improvement plans to eliminate waste.

5 Conclusion

Lean production perspective can be used to make improvement plans to reduce product defects in dairy companies. The advantage of using lean production is that it can find other problems that become waste, such as production leadtime, transportation, waiting time, and downtime in this case. Sustainability of handling waste problems with the concept of lean production requires a team consisting of cross-functional people.

References

1. P. Hines, N. Rich, Inter. J. of Oper. & Prod. Man, 46-64 (1997)
2. R. Sundar, A.N. Balaji, R.M.S. Kumar.Proc. Eng. **97**, 1875-1885 (2014)
3. T. Ohno, Toyota Production System: Beyond Large Scale Production, (Productivity Press, Boca Raton, 1988)

4. D. Simons, K. Zokaei, Brit. Food J., **107**,4, 192–211 (2005)
5. C. Perez, R. De Castro, D. Simons, G. Gimenez., Supply Chain Management, **15,** 1, 55–68 (2010)
6. G. Narayanamurthy, A. Gurumurthy, Inter. J. of Oper. & Prod. Man, **36,** 10, 1115–1160 (2016)
7. M. Melin, H. Barth, Prod. Plan.Con., **29**, 10, 845-855, (2018)
8. E.G. Satolo, L.E.DS. Hiraga, G.A. Goes, W.L. Lourenzani. Inter. J. O. Lean S. Sig., **8**,3, 335-358 (2017)
9. T. Ostapchuk, N.Valinkevych, H.Tkachuk1, K.Orloval, T. Melnyk, E3S Web Conf. **166,** 13008 (2020)
10. N. Ketoeva, N. Soldatova1, S. Ilyashenko2, E3S Web Conf. **124,** 04015 (2019)
11. Z. Nihlah, T. Immawan, E3S Web Conf. 73**,** 07010 (2018)
12. B. Mrugalska, M.K. Wyrwicka. Proc.Eng. **182**, 466 – 473 (2017)
13. X Ji, E3S Web Conf. 253**,** 03037 (2021)
14. P. Ondra, D. Tucek, R. Rajnoha, Polish J. O. Man. Stud, **17**,2, (2018)
15. N. R. Tague, The Quality ToolBox, (Milwaukee, Wiconsin: ASQ Quality Press, 2005).
16. P.M. Williams, BUMC Proc., **14**,2, 154-157 (2017)
17. T. M. Aken, Val. Strm. Mapp: A Rev. and Comprtv. Anlyss. of Rcnt. App, (2010).

Quality of Garlic Bulbs with Irrigation Application according to Plant Needs

Ika Cartika[1], *Suwarni Tri* Rahayu[1*], *Rofik Sinung* Basuki[1], *Darkam* Musaddad[1]

[1]Indonesian Vegetable Research Institute, Tangkuban Perahu Street. 517 Lembang, West Java, Indonesia

Abstract. The irrigation technique that has been done by farmers so far has not been optimal in increasing garlic production. This study aims to evaluate the quality of garlic tubers with irrigation applications according to plant needs. The experiment was carried out in Lembang, West Java, in 2020. The method used in this study was a group design of 6 treatments with four replications. Treatment A (daily ET accumulation within two days), B (daily ET accumulation within three days), C (daily ET accumulation within four days), D (daily ET accumulation within five days), E (daily ET accumulation within six days), F (farmers' irrigation practices). The plant data obtained were analyzed using Analysis of Variance (ANOVA) at the 5% level. Data analysis was performed using the PKBT - STAT 1.01 program. The results showed that the tuber diameter in treatment B (daily ET accumulation within three days) was significantly greater than control. Weight, weight loss, and water content were not significantly different between all treatments. The volatile content in all treatments was significantly higher than control.

1. Introduction

The productivity of garlic in Indonesia still needs to be improved by improving varieties, cultivation techniques, fertilization, and irrigation. Vegetable growers worldwide are continually under increasing pressure to optimize irrigation and nutrient management [1]. The irrigation technique that has been done by farmers so far has not been optimal in increasing garlic production. Irrigation and water management models are needed for the efficient watering of crops on the land. In addition, new advances in remote sensing technology, offering data and management, offer opportunities to optimize decision-making in terms of watering regimes, especially in the absence of water quality [2].

Optimized irrigation scheduling to better match applied water to crop requirements can reduce the wasteful use of water and energy, improve water use efficiency, and increase gross profit margin. The water demand in plants can be estimated by calculating the plant evapotranspiration rate (ET) because the ET value is the value of water demand given to plants as a substitute for water lost through evapotranspiration. ET values are calculated based on the FAO method. This method calculates the plant's water requirements by considering the physical characteristics of the soil and the depth of roots of each phase of plant growth. Within ± 120 days of local garlic growth, farmers generally water it at intervals of 3-5 days, considering the availability of irrigation water and the cost for watering. Thus, the value of ET per day must be accumulated per certain period so that the plant's water needs can be fulfilled maximally but at an economical watering cost.

*Corresponding author: swarnit@yahoo.com

The cultivation of garlic can positively contribute to the local economy of rural areas, but needs tools to help in decision making for more efficient use of water (3). Adaptation of plants to environmental conditions is very important to produce crops under conditions of sustainable climate change. Significant differences were found in the crop water stress index and water consumption. On the other hand, local cultivars show lower yield potential and more stable production in water supply environments [4]. However, given the increased cost of the energy used by the irrigation systems, the reduction of water availability for agriculture, and the low profitability of barley, farmers have shown interest in using regulated deficit irrigation. Thus, improving irrigation water productivity and coping with droughts is a priority objective to improve yield and profitability [5, 6].

Garlic contains bioactive compounds such as organosulfur, phenol with biological activities , namely antioxidant, anticancer, anti obesity, and antifungal [7]. Several quality parameters of garlic are the bulb diameter, bulb weight, number of cloves, the content of nutrients, and active compounds beneficial to the body. In addition to cultivation techniques and varieties, harvest age affects the size and quality of the tuber produced. The nutrition content of garlic on 100 g of dried tuber: 68% water, 3.5 g protein, 0.3 g fat, 1 g ash, Ca 29 mg, P 202 mg, 529 mg K [8]. This research aims to evaluate the quality of garlic tubers with irrigation applications according to plant needs.

2. Materials and Methods

The experiment was carried out in Lembang, West Java, from January 2020 to September 2020. Garlic bulb seeds that have been prepared are separated from the cloves; Bulbs used are class B types with a diameter of 3-4 cm. The beds are watered with sufficient water before the seeds are planted to maintain soil moisture. The spacing of the planting holes used is 12.5 cm x 12.5 cm. Each hole is inserted one clove of source then covers the hole with a little soil. After the seeds are planted, the straw mulch is placed transversely over certain doses and intervals. Watering time is done in the morning when the water evaporates in the soil, and the air temperature is low.

Meanwhile, pesticide spraying for plant pest control is carried out at 4-day intervals starting at 14 days after planting. Each time the application of the active pesticide used is different and adjusted to the type of HPT that is likely to attack. Weeding is done by cleaning the planting area from weeds that grow. Weeding is done every week starting at the age of 14 days after planting. Weeding must be stopped at the period of ± 100 days after planting or has entered the generative phase because it will interfere with the formation and enlargement of tubers. The method used in this study was a 6 treatment randomized block design with four replications. Treatment A (daily ET accumulation within two days), B (daily ET accumulation within three days), C (daily ET accumulation within four days), D (daily ET accumulation within five days), E (daily ET accumulation within six days), F (farmers' irrigation practices). Physically experimental parameters were weight, diameter, weight loss. Chemical parameters were moisture content, ash content, Total Soluble Solute (TSS), and Volatile content. The plant data obtained were analyzed using Analysis of Variance (ANOVA) at the 5% level. Data analysis was carried out using the PKBT–STAT 1.01 program.

3. Results and Discussion

Quality will influence consumer acceptance of a product. Size is one quality parameter that determines market prices. Table 1 shows the weight of garlic bulbs ranging from 21.24 - 34.33 mm. Bulbs with the largest weight in treatment D (34.33 mm), while the smallest bulb weight in treatment A was 21.24 mm, but not significantly different between all treatments. The diameter of the garlic bulbs from all treatments was greater than 4 cm, with the largest diameter in treatment B. (44.89 mm) and significantly different from the control treatment (farmer method). Weight loss during storage showed results that were not quite different between all treatments. The productivity of garlic bulbs is influenced by the amount of water given. The optimum irrigation treatment in the tuber formation phase gave the highest yield. Reducing the amount of water in this phase affects the quality and quantity of tubers [9].

Table 1. Physical Parameters of Garlic Bulbs from Six Treatments

Threatment	Weight (g)	Diametre (mm)	Weight loss (%)
A	$21.24^{a} \pm 4.63$	$42.29^{ab} \pm 2.39$	$13.32^{a} \pm 5.56$
B	$26.88^{a} \pm 3.16$	$44.89^{a} \pm 2.27$	$17.89^{a} \pm 5.83$
C	$22.16^{a} \pm 3.38$	$42.64^{ab} \pm 2.16$	$21.84^{a} \pm 6.84$
D	$34.33^{a} \pm 7.72$	$43.20^{ab} \pm 2.97$	$15.21^{a} \pm 5.83$
E	$24.87^{a} \pm 5.88$	$43.66^{ab} \pm 3.63$	$16.27^{a} \pm 3.12$
F	$22.56^{a} \pm 5.57$	$42.08^{b} \pm 3.50$	$11.83^{a} \pm 5.93$
LSD 5%	13.59	2.69	10.95

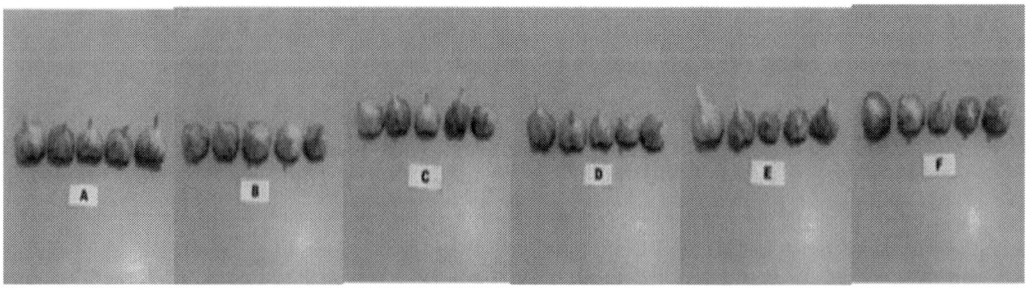

Fig. 1. Garlic Bulb of Six Treatments

The influence of the type of fertilizer such as the content of nitrate and the duration of irradiation affect the growth of plants and the condition of the garlic bulbs produced [10]. The maize plant height, leaf area index, above-ground biomass and yield did not significantly decrease due to the decrease in fertilizer application. Fertilization management increased the nutrients and rainfall use efficiency and net profit [11]. The water content between all treatments was not significantly different, which ranged between (61.96% - 63.49%). Meanwhile, the ash content of garlic bulbs in all treatments was significantly lower than the farmer's technique. The TSS content ranges from 36.80 - 38.55 Brix. The quality of the harvests and the significant differences found in this study (weight of barley and maize grains, and calibres size of bulbs in onion) did not affect the price perceived by the farmer. Due to this, the deficit irrigation reached a higher profitability than full irrigation methods in the scenarios with lower availability of irrigation water (≤5000 m3 ha−1). Therefore, the use of these methods at great scale may increase the profitability of farms located in water scarcity areas through a more efficient use of land and irrigation water [12].

Table 2. Garlic Bulb Chemical Parameters of Six Treatments

Treatment	Water (%)	Ash (%)	TSS (Brix)	Volatile content (ppm)
A	$63.49^{a}\pm 1.67$	$1.10^{b}\pm 0.05$	$37.10^{b}\pm 0.12$	$255.60^{b}\pm 29.69$
B	$62.22^{a}\pm 0.72$	$1.13^{b}\pm 0.02$	$37.85^{ab}\pm 0.10$	$284.33^{a}\pm 7.15$
C	$63.42^{a}\pm 1.32$	$1.23^{b}\pm 0.15$	$36.80^{b}\pm 0.43$	$266.00^{ab}\pm 18.49$
D	$62.09^{a}\pm 1.48$	$1.10^{b}\pm 0.07$	$37.10^{b}\pm 0.12$	$244.69^{bc}\pm 16.76$
E	$63.36^{a}\pm 1.06$	$1.16^{b}\pm 0.07$	$38.50^{a}\pm 1.41$	$269.80^{ab}\pm 13.90$
F	$61.96^{a}\pm 1.41$	$1.39^{a}\pm 0.12$	$38.55^{a}\pm 0.87$	$222.87^{c}\pm 8.43$
LSD 5%	1.92	0.15	1.15	17.95

In table 2, volatile content in treatment B was significantly higher than other treatments, namely 284.33 ppm. Garlic contains sulphur compounds, including a chemical called alliin which makes raw garlic feel bitter. The use of fertilizers affects the sulphur content in garlic. The use of fertilizers containing sulphur at a dose of 30 kg/ha shows a high sulphur content. In comparison, the use of fertilizers that contain lots of Nitrogen is less good for the sulphur content of garlic. TIBBP, the technology from IVEGRI, has been proven to increase garlic bulb quality compared to the farmer's conventional technology. Tawangmangu Baru, cultivated with TIBBP, has a 32.69% bigger diameter and is 11.5 grams heavier than Tawangmangu Baru, which is produced with the farmer's conventional technology [13, 14]. The ET value in the application of garlic irrigation shows a rather low sensitivity, which allows certain formulas to calculate the ET of plants. The other research reported that irrigation at 20 days intervals showed better performance over-irrigation at 10 days intervals and control [15, 16].

4. Conclusion

The diameter of bulb in treatment B (daily ET accumulation within three days) was significantly greater than the control. Weight, weight loss, and water content were not significantly different between all treatments. The volatile content in all treatments was significantly higher than control (farmers' irrigation practices).

References

[1] M. Gallardo, A. Elia, R.B. Thompson, Agric. Water Manag. **240** (2020)

[2] J. Zinkernagel, J.F. Maestre-Valero, S.Y. Seresti, D.S. Intrigliolo, Agric. Water Manag. **242** (2020)

[3] B.C. Léllis, A. Martínez-Romero, R.C. Schwartz, J.J. Pardo, J.M. Tarjuelo, A. Domínguez, Agric. Water Manag. **260** (2022)

[4] Á. Sánchez-Virosta, V. O. Sadras, D. Sánchez-Gómez, Scie. Horti. **285** (2021)

[5] B. Hanson, D. May, R. Voss, M. Cantwell, R. Rice, Agric. Water Manag. **58**, 1(2003)

[6] J.J. Pardo, A. Martínez-Romero, B. C. Léllis, J. M. Tarjuelo, A. Domínguez, Agricultural Water Management. **228** (2020)

[7] L.Tavares, L. Santos, C. Pelayo Z. Noreña, Trends in Food Science & Technology 114 (2021)

[8] Q.P. Meer. and A. H. Permadi. *Allium sativum* L. In Siemonsma, J.S., and K. Piluek (Eds). *Plant Resources of South- East Asia **8** Vegetables*. Prosea Foundation. Bogor, Indonesia (1994)

[9] C. F. Cortés, F. M. de Santa Olalla, R L. Urrea, Agric. Water Manag. **59**, 2 (2003)

[10] A. Sánchez-Virosta, D. Sánchez-Gómez, Scie Horti, **252** (2019)

[11] J. Fan, X. Lu, S. Gu, X. Guo, Agric. Water Manag. **241** (2020)

[12] A. Martínez-Romero, A. Martínez-Navarro, J.J. Pardo, F. Montoya, A. Domínguez, Agric. Water Manag. **192** (2017)

[13] S.T. Rahayu, R. S..Basuki, P.S. Levianny. J. Widya Riset, **6**, 2 (2020)

[14] F.J. Villalobos, L. Testi, R. Rizzalli, F. Agric. Water Manag. **64**, 3 (2004)

[15] Wang, Haiping, X. Li, D. Shen, Y. Oiu, J. Song. Euphytica **198**, 2 (2014)

[16] M.A. K.Bhuiya, M.A. Rahim, M.N.A Chowdhury. Asian J. of Plant Scie. **2,** 8 (2003)

Vertical farming application with several growing media for coffee nurseries

Restu Puji Mumpuni[1*], *Hidayati* Fatchur Rochmah[2], and *Undang*[1]

[1] College of Vocational Studies of IPB University, Bogor, West java, Indonesia

Abstract. Coffee is one of the plantation commodities that play an important role as a source of foreign exchange for the country. One alternative to overcome the land decrease in rural area is to utilize urban farming in the form of a vertical planting system. This study aimed to obtain alternative techniques for planting coffee seedlings using vertical farming and to determine the effect of vertical planting with several growing media on the growth of coffee seedlings. The study was arranged in a non-factorial Randomized Block Design (RBD) with four treatments, namely: 1) conventional nursery; 2) topsoil verticulture nursery; 3) topsoil and compost media verticulture, and 4) topsoil, compost, and husk charcoal media verticulture. The data were analyzed using the Duncan Multiple Range Test (DMRT). The observed variables included observations of the morphology of coffee seedlings and verticulture environmental factors. The results showed that vertical planting on the vegetative growth of coffee seedlings was not significantly different from conventional planting. Topsoil planting media and husk charcoal in vertical planting did not significantly affect the morphological growth of coffee seedlings. The combination of topsoil, husk charcoal and compost gave the best plant height of 12.4 cm at 13 weeks after planting.

1 INTRODUCTION

Coffee is one of the plantation commodities that has a fairly high economic value compared to other plantation crops and plays an important role as a source of foreign exchange for the country. The taste and psychological effect of its freshness make it much in demand by consumers around the world and even become one of the main menus in various official banquets. Indonesia itself is the fourth largest coffee bean producer in the world after Brazil, Vietnam and Colombia [1].

The area of coffee plantations in Indonesia is 1.236 million hectares with the largest exploitation or around 96% by smallholder plantations. Indonesian coffee bean production fluctuates throughout the year. The production of large state and private coffee plantations in 2017 decreased by 4.95% from 2016 and in 2018 decreased again by 7.1%. Likewise, with smallholder plantations, coffee production in 2018 reached 685,790 tons or decreased by 0.002% compared to 2017. The decline in national coffee production was mainly because the

* Corresponding author: restu_puji@apps.ipb.ac.id

area of coffee plantations owned by state and private plantations decreased from 46.82 thousand hectares to 42.17 thousand hectares in 2018 [1]. The decline in production occurred due to replanting land, lack of technical understanding of Good Agricultural Practices (GAP) on smallholder plantations, prolonged drought, and a decrease in land area due to land conversion. If this continues, the national coffee production will continue to decline.

Coffee plantation land is usually located in rural or rural areas far from the city, the land is currently experiencing a lot of land use change. One alternative to overcome the decrease in land area in rural areas is to utilize urban farming. Urban communities in Indonesia are currently promoting urban farming. According to [2] several agricultural methods in urban areas that have been developed and applied in Indonesia are verticulture farming and hydroponics. Verticulture is an agricultural cultivation system that is carried out vertically or stratified [3]. Verticulture is a farming model that can be done in a narrow area such as a yard or roof of a house.

Verticulture farming so far has focused on cultivating horticultural crops, especially leaf and fruit vegetables, these vegetables are generally seasonal crops that are harvested within 1-3 months. In vertical farming plantations, it can be applied in the nursery phase. Sari's research [4] conducted morphological observations of oil palm plants in the early nursery phase vertically and the results on physical observations of plant morphological growth that were planted vertically were better than conventional methods.

The growth of coffee seedlings in the pre-nursery still has short roots and morphological growth that is not too high so that it is in accordance with plant requirements for vertical planting. Vertical planting media must be able to support plant growth and nutrient requirements during the nursery period. Planting media with a combination of soil and compost is a planting medium that gives the highest dry weight yield of a vegetable verticulture system [5]. The combination of the planting media also gave the best yields in Sitawati's [6] research on optimizing planting media to support vertical crop yields.

The potential for vertical coffee breeding in urban areas will support the development of coffee cultivation technology in the upstream sector which is still conventional. The use of appropriate planting media for coffee seedlings needs to be considered to support the availability of superior seeds for the success of vertical coffee cultivation in Indonesia.

The purpose of this study was to obtain alternative techniques for planting coffee seedlings using vertical racks, to determine the effect of vertical planting on the growth of coffee seed morphology, and to determine the effect of several planting media on vertical planting on the growth of coffee seedlings morphology.

2 METHODS

2.1 Time and Place

This research was carried out in the Field Laboratory of the nursery area of the Gunung Gede Campus, College of Vocational Studies IPB University which is at an altitude of 350 meters above sea level and an average temperature of 25 degrees Celsius. The research was carried out from August 15 to December 4, 2020.

2.2 Materials and Tools

The tools used in this research were 5 sets of vertical planting, pump, net pot using 220 ml plastic cup, fiber tube, caliper, analytical balance, EC meter, lux meter, thermohygrometer, pH meter, image-J software. The materials used were catimor coffee seeds from Puslitkoka (Coffee and Cocoa Research Center), NPK fertilizer, topsoil, compost and husk charcoal.

2.3 Research Methods

This study used a randomized block design (RBD) with 4 treatments with 3 replications. Each treatment contained 20 plants so that 12 experimental units and 240 coffee plant seeds were obtained. The treatments used were:
P0 : conventional coffee nursery (control)
P1: verticulture with top soil planting media
P2: verticulture with top soil + husk charcoal
P3: verticulture with top soil + husk charcoal + compost
If the results of the variance show a significant effect on the 5% level of the F test, then a comparison is made between the treatment mean and the DMRT (Duncan Multiple Range Test) test at the 5% level.

2.4 Observations

The variables observed in this study included observations of coffee plant morphology and verticulture environmental factors. The first vegetative observations were observed after the coffee plant had germinated and the cotyledons had separated. Each experimental unit was randomly selected 10 sample plants. Observations were carried out every 2 weeks. Observations of vegetative growth observed were plant height, number of leaves and stem diameter. Observation of environmental factors for vertical growth consists of Electrical Conductivity, pH, temperature and humidity of the vertical environment.

Observation Environmental factors observed were Electrical Conductivity or EC was carried out to determine the amount of fertilizer in water solution, EC observations were carried out using an EC meter carried out every day in the morning from the results of observations the average EC on plant nutrition was 1.5 - 2 ms. The EC is in accordance with the needs of coffee plant seeds. Subsequent observations, namely pH, were carried out to determine the pH in aqueous solutions, pH observations were carried out using a pH meter every day in the morning.

3 RESULTS AND DISCUSSION

3.1 Results

The results of observations of the average pH of 5.8 and still in accordance with the requirements for growing coffee plants. The temperature and humidity of the vertical environment were observed using a thermohygrometer placed in the nursery area and observed every day. The results of observations of the average temperature in the morning are 21°C and during the day 25°C with an average humidity of 83%.

Observations of plant vegetative growth observed were the number of leaves, plant height and stem diameter. The growth of the number of leaves on coffee seedlings in several treatments in Table 1 The growth of the number of leaves did not show significant differences in either conventional or vertical nurseries with several growing media. Number and length of coffee leaves
Prenursery breeding period is more influenced by genetics and less responsive to the treatment given

Table 1. The growth of the number of leaves on coffee seedlings in several treatments

	Seedling age (Week After Plant)				
Treatments	5	7	9	11	13
Conventional	2.1	2.4	2.9	4.1	5.4
Verticulture top soil	2.1	2.5	2.7	3.9	4.8
Verticulture top soil + husk charcoal	2.2	2.4	2.5	3.1	47
Verticulture top soil + husk charcoal + compost	2.1	2.6	2.7	3.2	4.7

The growth of plant height in Table 2 shows that conventional treatment and top soil verticulture and husk charcoal did not show a significant difference. The best and significantly different plant height growth was shown by the verticulture treatment of top soil, husk charcoal and compost at weeks 11 and 13. Verticulture planting media that had high aeration and water absorption could support the high growth of coffee seedlings. Mixed husk and compost media provide high porosity, husk charcoal contains C and SiO_2 which affects P availability, compost provides slowly available macro nutrients [6].

Table 2. Growth of plant height (cm) on coffee seedlings in several treatments

	Seedling age (Week After Plant)				
Treatments	5	7	9	11	13
Conventional	10.5	10.7	11.0	11.1 b	11.5 b
Verticulture top soil	10.4	10.6	11.0	11.1 b	11.4 b
Verticulture top soil + husk charcoal	9.8	10.0	10.1	10.4bc	10.6bc
Verticultuer top soil + husk charcoal + compost	11.0	11.5	11.8	12.1 a	12.4 a

Note : a number followed by the same letter indicates that there is no significant difference in the 5% DMRT test.

Table 3 shows that from all treatments, both vertical and conventional, there was no significant difference in the growth of coffee seedling stem diameter. Annual plants have relatively slow seedling growth, so the resulting stem diameter results are not much different.

Table 3. Growth of stem diameter (mm) in coffee seedlings in several treatments

	Seedling age (Week After Plant)				
Treatments	5	7	9	11	13
Conventional	1.43	1.62	1.60	1.60	1.65
Verticulture top soil	1.55	1.55	1.55	1.57	1.56
Verticulture top soil + husk charcoal	1.46	1.45	1.45	1.52	1.52
Verticultuer top soil + husk charcoal + compost	1.56	1.56	1.56	1.59	1.59

The vegetative growth of coffee seedlings verticulture was only significantly different in plant height after 11 WAP, meaning that plant growth remained normal even though it was planted vertically. The number of plants that can be planted vertically with a size of 1 m^2 is 80 plants, while conventionally only 50 seeds (Fig. 1.)

Fig. 1. Vertical farming set up that is applied to coffee seedlings

3.2 Discussion

Coffee is a type of plantation plant that has long been cultivated and has high economic value. From the results of the study, the temperature in Bogor City, especially in this research environment, in the morning the average temperature is 21°C and during the day 25°C. The coffee grown in this study is Robusta Coffee, depending on the growing conditions for environmental temperature, planting in the city of Bogor is suitable for the requirements for growing Robusta coffee. Robusta coffee is a type of coffee plant that can grow in the middle to lowland areas. Robusta coffee grows in the lowlands with an altitude of 400-700 m above sea level with an average daily temperature of 25-26°C. In general, coffee plants can grow in areas with loose and fertile soil conditions with a pH of around 4.5-6.0 [7]. The results of the measurement of the average pH of the nutrients given to vertical planting have a pH of 5.8 so that it is in accordance with the requirements for growing coffee plants.

Vertical farming is the vertical cultivation of plants, which began to appear in Europe when greenhouse technology was applied in the early 1700s, initially the installation still used wood and soil planting media for planting ornamental plant [8]. Vertical farming then developed for the cultivation of leaf vegetables, fruits and herbs. Further developments in the early 1900s verticulture cultivation in England has used automatic fertigation with pumps that deliver water and nutrients to plants automatically according to plant needs and monitored using a computer. Verticulture comes from the English words, namely vertical and culture, so verticulture is a system of agricultural cultivation that is carried out vertically or stratified, both indoor and outdoor. This vertical or stratified agricultural cultivation system is a greening concept that is suitable for urban areas and limited land. The vertical requirement is that the installation is strong and easy to move around. Plants to be planted should be adapted to the needs and have high economic value, short life, and short roots [3]. Vertical farming is a farming technique that is very prospective in terms of economy and architecture. In times of crisis, this vertical farming method can increase food security in urban areas, if it is integratedly applied it will certainly reduce poverty, unemployment and can meet food needs in a sustainable manner [9].

The results of the study, it can be seen that verticulture planting has no significant effect on the growth of leaf number and stem diameter compared to conventional methods, meaning that plant growth remains normal the same as when conventional planting. The growth of plant height was significantly different in the verticulture treatment of top soil, husk charcoal and compost. The number of plants that can be planted in an area of land with the verticulture method is more than conventional. In this study, a vertical plant with a size of 1 m^2 can be planted with 80 medium plants with the conventional method of only 50 seeds so that it is more land efficient.

In this study, three types of planting media were used, namely soil only, soil and husk and soil charcoal, husk charcoal, compost. The results of the analysis of variance showed a significant difference in the verticulture treatment with top soil, husk charcoal and compost media. Planting media is a place where plants grow to support roots. From this planting medium, plants absorb nutrients in the form of nutrients through their roots. According to [10] The media of soil, husk and manure had a significant effect on the percentage of coffee seed germination and seedling height. In verticulture planting media must also use media that have aeration and high water absorption, the media that are often used in verticulture are top soil, cocopeat, spagnummoss, husk charcoal, compost, rockwool [11]. According to [12] coffee seedlings treated with coffee husk and skin compost had plant height growth, number of leaves, stem diameter, leaf surface area, total wet weight. plants and higher plant total dry weight than soil media alone.

4 Conclusion

Morphological growth of coffee seedlings in vertical cultivation was as good as conventional methods. Top soil planting media and husk charcoal in vertical planting had no significant effect on the morphological growth of coffee seedlings. The combination of topsoil of husk charcoal and compost gave the best plant height. The vertical planting technique can be used as an alternative for land efficient coffee seedlings.

References

1. BPS. *Indonesian Coffee Statistics 2018*. (Badan Pusat Statistik, Jakarta 2019)
2. Juniawati, M. Hayuningtyas. 2017. *Urban Agriculture Development: Strategy to support food security* in Proceeding of the 2nd International Conference on Sustainable Agriculture and Food Security, ICSAFS, Okt 2017, Jakarta, Indonesia (2017)
3. L. Lukman. *Teknologi Budidaya Tanaman Sayuran Secara Vertikultur*. (Balai Penelitian Tanaman Sayuran, Bandung, 2015)
4. V. I. Sari. J. Citra Widya Edu. **10**, 2 (2018)
5. D. S. O. Yosandy, M. Baskara, N. Herlina. J. Prod Tan. **6,** 2 (2018)
6. Sitawati, A. Suryanto, E.E. Nurlalelih. Research J. of Life Sciences. Vol **3** (1): 55-64 (2016)
7. S. N. Andini, R.N. Sesanti. J. Wacana Pertan. **14**,1 (2018)
8. D. Despommier. *Vertical Farms in Horticulture*. (Fordham University, New York, 2015)

9. F. Kalantari, O. M. Tahir, A. M. Lahijani, S. Kalantari. Advanced Eng. Forum. **9**, 24 (2017)
10. T. Suharjanto, T. Wardhani, R. Risfandi. J. ilmu-ilmu pertan. Agrika. **13**,1 (2019)
11. H. Gustia. E-Journal WIDYA Kes. dan Lingk. **1,** 1 (2013)
12. H. Debora, K. P. Wicaksono. J. Prod. Tan. **1**, 8 (2020)

Implementation of technology 4.0 in achieving the effectivity and efficiency of the production process in palm oil plantation

Lili Dahliani[1*], *Sutria* Wirandayu[1], and *M* Dewantara[1]
[1] Technology and Management of Plantation Production Department, College of Vocational Studies, IPB University, 16128 Bogor, Indonesia
[1*]Corresponding author: lilidahliani@apps.ipb.ac.id

Abstract. Palm oil (*Elaeis guineensis* Jacq.) is a plantation commodity that has a role in the Indonesian economy. The area of palm oil plantations in Indonesia reaches 14,300,000 hectares with 43,000,000 tons of *Crude Palm Oil* (CPO) production. The implementation of technology 4.0 on palm oil production is needed not only for technology innovation, but also for effectivity and efficiency on production process. This research aims to study the efficiency of resources use and the effectivity of output achievement on palm oil production. This research was done descriptively and quantitatively using primary data, which was direct interviews with the people who works in palm oil plantations, and the secondary data was obtained from research results, plantation company's yearly report, and other literature review. The research result showed that the implementation of technology 4.0 on the recording of employee attendance register and calculation of harvest density rate (HDR) including field and fruit quality was more efficient 25-50% in use of employee and time. The implementation of technology 4.0 was also more effective in achieving output compared to the manual method. The conclusion of this research is the implementation of technology 4.0 on some production process in palm oil plantation is more effective and efficient compared to the manual method. The starting budget on implementing technology 4.0 is higher by 30% compared to the manual method, but it becomes ten times cheaper on the next maintenance cost.

Keywords : palm oil, plantation commodity, technology 4.0, effectivity, efficiency

1 Introduction

Palm oil (*Elaeis guineensis* Jacq) is one of the mainstay plantation commodities that has a positive and impacts and benefit on the Indonesian economy and creates a multiplier effect. Those benefits are contributing to foreign exchange earnings, providing employment and sources of farmers income, generating regional original income, contributing to the development of the area and producing region, producing gross domestic product, and accelerating of community economic development to alleviate poverty in rural areas [3, 4, 8]. Although palm oil plantations lead to a decrease in the level of interaction and mutual assistance, they contribute greatly to the provision of public facilities and education in the community [5]. Palm oil plantation development activities have resulted in high population mobility, the emergence of centers of economic growth, the circulation of money in rural areas, especially for routine household needs and the need for production facilities, as well as the growth of trade and services [8]. Palm oil is classified as a more effective crop in terms of land use, productivity and marketing, which can help reduce dependence on fossil fuels [1].

Total oil and palm kernel production were 51.443.315 tons in 2019. The national export volume of palm oil in 2019 was 29.135.179 tons with a value of 20.802.708.000 US$. Palm oil plantations in Indonesia are carried out by large state plantations (PBN), large private plantations (PBS), and smallholder plantations (PR). The area of Indonesia's oil palm plantations according to the estimated 2019 data is 14.677.560 ha, of which the PR area is 41% of the total oil palm area. Palm oil production from smallholders' plantations in the same year was 14.846.112 tons, which is 16% of national production [2].

Palm oil development strategy is very necessary considering the increasingly competitive global market competition. One of strategy that is able to increase competitiveness is a program to increase added value through innovation and industry technology 4.0 to increase productivity through innovation and technology plays a role in increasing productivity. The application of technology will strengthen the sustainability of the industry, including the palm oil industry. Digital technology has made many jobs easier in the palm oil industry. The leading cloud service provider in Indonesia, Datacomm Cloud Business together with partners, introduced the SMART AGRICULTURE Integrated Plantation Digital solution for plantations in general and Palm Oil plantations. Smart Agriculture supports Industry 4.0 and the Sustainable Development Goals.

Indonesia has compiled an industry 4.0 roadmap by specifying five manufacturing sectors which will be the main priority in its development. One of them is the agricultural sector that included in the food and beverages sector [7].

2 Method

This research was conducted in a palm oil plantation located in Labuhan Batu and South Tapanuli regency, North Sumatra, Indonesia. The determination of the research location was done intentionally, with the consideration that the location was a place for field work practice for students who majoring in Technology and Management of Plantation Production, College of Vocational Studies, IPB University. The research of technology 4.0 application was carried out both in the plantation production subsystem (on-farm) and in the processing plant production subsystem (off-farm). The research was conducted from February to April 2021.

The data obtained in this study were primary data and secondary data. Primary data was obtained from the results of student interviews and direct observations during the implementation of field work practices in the oil palm plantation. Secondary data, apart from being obtained from reports from the management of the oil palm plantation office in the research area, also came from literature studies and the results of previous studies. Interviews were conducted with plantation officials, which the assistant head (deputy manager of the plantation) and the head of the plantation division.

Data analysis used quantitative and qualitative methods. Qualitative data was obtained by the interviews with palm oil production actors in the palm oil plantation area. Meanwhile, the quantitative data was the research results that processed into a result of understanding and analysis then added with the results of existing research. The data also used primary and secondary data which was obtained directly from respondents and literature that already existed before.

3 Results and Discussion

The results show that technology 4.0 has been applied in oil palm plantations, both in the plantation production subsystem (on-farm) and in the processing plant production subsystem (off-farm). The application of technology 4.0 in the on-farm palm oil production subsystem only includes in several palm oil cultivation activities, such as calculation/compilation of harvest density rate (HDR), recording of employee attendance, recording of harvest yields along with recording of field quality and fruit quality. The application of technology 4.0 in the palm oil production subsystem is even implemented starting from the initial station for processing Fresh Fruit Bunches (FFB) into CPO to the final station, the oil refining station. One of the considerations the palm oil plantations management in implementing technology 4.0, is to obtain the value of efficiency and effectivity. The results of this study indicate that the use of technology 4.0 in palm oil plantations, especially on-farm, makes the use of resources such as time and human resources (HR) more efficient and achieves the effectiveness of the objectives of the activities in the production subsystem.

3.1 Effectivity and efficiency

Recording of employee attendance register, harvest recording, measurement and calculation of harvest density rate including field quality and fruit quality in palm oil plantations that apply 4.0 technology requires smaller resources rather than doing it manually. Referring to the environment and the principle of paper less, the application of technology 4.0 in palm oil plantations makes minimal use of paper for recording and reporting. The manual system produces waste in the form of paper, while using the 4.0 system it does not produce waste so it is more efficient than the manual system. Further explanation of the efficiency and effectivity of the application of technology 4.0 compared to manual implementation is shown in Table 1 and Table 2.

Table 1 Implementation of technology 4.0 on harvest density rate calculation and recording of employee attendance register

Activity	Manual	Technology 4.0
Harvest density rate (HDR) calculation	**Time**	
	1,5 hours for 150 plants	1 hour for 150 plants
	HR	
	2 people per 300 plants	1 person per 300 plants
	Tools	
	Office stationery : Rp 100.000 - 150.000 per month	Gadget : Rp 2.000.000 Maintenance : Rp 100.000 - 150.000 per mont
Recording of employee attendance register	**Time**	
	Takes more time	Takes less time
	HR	
	> 2 people	1 person

Table 2 Implementation of technology 4.0 on harvest record, harvest field quality, and harvest fruit quality

Activity	Manual	Technology 4.0
Harvest record	**Time**	
	Input data in the office before it becomes harvest record information	faster data entry
	HR	
	3 people	2 people
	Tools	
	Office stationery : Rp. 150.000 per month	Gadgets : Rp. 1.950.000 Maintenance : Rp. 100.000 per 3 months
Harvest field quality	**Time**	
	Input data in the office before it becomes field quality information	field quality information is directly received by decision makers
	HR	
	2 people	1 person
	Tools	
	Office stationery : Rp. 150.000 per month	Gadgets : Rp. 1.950.000 Maintenance : Rp. 100.000 per 3 months
Harvest fruit quality	**Time**	
	Input data in the office before it becomes fruit quality information	fruit quality information is directly received by decision makers
	HR	
	3 people	2 people
	Tools	
	Office stationery : Rp. 150.000 per month	Gadgets : Rp. 1.950.000 Maintenance : Rp. 100.000 per 3 months

Based on the Tables 1 and Tables 2, it shows that the application of technology 4.0 makes the use of resources such as employees, time, and costs is more efficient 25 - 50% with the assumption that the application of technology 4.0 are already owned by palm oil plantations.

3.2 Problems and Solutions

The application of technology 4.0 will be efficient and effective if the management of palm oil plantations both on-farm and off-farm can provide solutions to the existing problem. Especially on the on-farm, the performance of the gadgets can be errors or glitch. Another problem that encountered in the application of technology 4.0 is unstable signal or lost signal for one reason or another. The results of Dahliani's research, show that signal problems that often occur on cellphones such as lost signal, changing signal display (signal drop), unstable signal or full signal but unable to make calls (pseudo signal) [6]. To find out signal problems that occur in cellphone hardware, it is necessary to check the Tx and Rx circuits, CPU circuits related to signal processing and voltage distribution from the IC Power Supply to those circuits. The inspection process includes checking the path between components, measuring voltage and frequency on components related to the process of transmitting and receiving signals. By analyzing the indications/characteristics of the problem and the results of the examination of signal problems with the theory and function of each cell phone component related to signal processing, it can be found out the cause of the problem and the appropriate treatment can be carried out according to the problem.

Telecommunications observers said that the signal condition was determined by the condition of the cellular operator's Base Transceiver Station (BTS). The BTS power source is electricity from PLN. BTS is a telecommunications infrastructure that is used as a transmitter and receiver of cellular networks in one area coverage. BTS converts the transmitted and received signals into digital signals. These signals are then sent to other terminals for the circulation of messages or data. PLN electricity is required to power the entire BTS system.

Errors in inputting data are also one of the problems encountered in the application of technology 4.0 in oil palm plantations. This error results in data inaccuracies, which one of the causes is poor or unstable signal conditions. Data is one of the 5 core technologies that are the main pillars in developing technology 4.0 to be a digital-ready industry, in addition to : Internet of Things (IoT), Artificial Intelligence (AI), Cloud Computing and Additive Manufacturing. The skills of employees or palm oil plantation officers in operating gadgets are also one of the problems that encounter in the application of technology 4.0. Employees must be reliable in the operation of gadgets, especially in inputting data from the field so that the right information will be produced to support management decisions.

Solutions for the performance of the gadgets used, related to unstable signal reception and even no signal at all, need to be done so that delays in uploading daily data information can be overcome. One of the solutions is developing internet network infrastructure for remote areas by collaborating with parties providers. Dahliani explained that signal problems that often occur in plantation areas including palm oil plantations are signal loss, signal display changes (drop signal), unstable signal and full signal but cannot make calls (pseudo signal) [6]. One solution is to install signal amplifiers in oil palm plantation areas.

The low performance of the gadget is sometimes caused by the specifications not match with the required capacity; therefore gadgets need to be upgraded with more complete specifications according to camera needs. The solutions of employees or workers who have not mastered gadget operations, they need to be given training in using gadgets and applications.

4 Conclusion

Based on the results of research and discussion in the previous section, it can be concluded that:

1. Technology 4.0 in oil palm plantations in the research locations, has been applied both in the plantation production subsystem (on-farm) and in the CPO processing plant production subsystem (off-farm).
2. Technology 4.0 in the plantation production subsystem (on-farm) at the research location was applied in the following activities: measurement and calculation of harvest density rate (HDR), recording of harvest yields along with recording of field quality and fruit quality, and recording of employee attendance register. The implementation of

technology 4.0 in off-farm is starting from the boiling station to the oil purification station.

3. The use of technology 4.0 is more efficient in the use of resources such as time, employees, tools, and is more effective in achieving goals. All of that efficiency and effectivity can be achieved if the management of palm oil plantations can provide solutions to the problems that encountered in the application of technology 4.0.

References

1. C.H. Lim, S. Lim, B.S. How, W.P.Q. Ng, S.L. Ngan, W.D. Leong, H.L. Lam, *A review of industry 4.0 revolution potential in a sustainable and renewable palm oil industry: HAZOP approach.* Renewable and Sustainable Energy Reviews, 135, p.110223 (2021)
2. Directorate General of Estate Crops, *Tree Crops Estate Statistics of Indonesia 2018 - 202,* 26 (2019)
3. F. Firman, A. Suyatno, D. Kurniati, *Analysis of Income Level And Welfare of Oil Palm Farmers in Merarai Village, Sungai Tebelian District, Sintang Regency,* Tropical Plantations and Land, *8*(2), pp.61-66 http://dx.doi.org/10.26418/plt.v8i2.29799 (2018)
4. I. Darwis, *Impact of the existence of oil palm companies on the social welfare of the community in Bulo Mario village Nort Mamuju Regency (*2015)
5. I. Siradjuddin, *Impact of palm oil plantation on the regional economy in the district of Rokan Hulu,* Journal of Agrotechnology, 5 (2) 7-14 (2015)
6. L. Dahliani, *Implementation of Study From Home Activities in IPB Vocational School Study Program Technology and Management of Plantation Production in the Covid and New Normal Era,* New Normal Multidisciplinary Study, Psycology Forum and AMCA (2020)
7. N. Haryanti, A. Marsono, M.A. Sona, *Implementation strategy for oil palm plantation development in the 4.0 industrial era*, Journal of sharia economic dynamics, 8 (1), 76-87 https://doi.org/10.53429/jdes.v8i1.146 (2021)
8. Z. Nasution, *Analysis of palm oil commodity and its economic impact on the wefare of the community in the district Labuhan Batu* . Journal of Ecobisma 2(1) (2015)

Milk Quality Improvement with AHP (Analytical Hierarchy Process) Based on SCOR (Supply Chain Operation References) Performance and Business Canvas Model in Giri Tani Milk Cooperative

Mela Nurdialy[1*], *Suhendi* Irawan[2)], *dan Sazli* Tutur Risyahadi[3)]

[1]College of Vocational Studies - IPB University, Accounting, Kampus IPB Cilibende Jln Kumbang No.14 Bogor 16151, Indonesia
[2]College of Vocational Studies – IPB University, Industrial Management, Kampus IPB Cilibende Jl Kumbang No 14 Bogor 16151, Indonesia
[3]Faculty of Animal Sciences – IPB University, Nutrition and Feed Technology, Kampus Dramaga Jl Agatis Bogor 16680, Indonesia

Abstract. KUD Giri Tani has faced varying quality of milk, resulting in several times rejection of milk by the industry. The programs that have been running are unable to provide the same quality milk. This research aims to prioritize programs that will be implemented by the top management of KUD Giri Tani. This research uses SCOR reference and Business Canvas Model to investigate proper improvement methods. The weighting level 1 shows that the top 3 significantly affect milk quality such as production, milk supply from farmers, and delivery. The weighting calculation for core activity shows that Plan Reliability (0.0439), Source Reliability (0.0825), Makes Effectivity (0.1513), Delivery Reliability (0.1160), and Return Reliability (0.0513) are the most significant weights of each criterion. The result shows that improving food quality is the highest score (0.393) followed by improvement hygiene (0.386) and milk grading process (0.221). The Farmer members can focus on those two things to improve the quality of milk produced. Those programs are appropriate to analyze the Business Canvas Model.

1 Introduction

Livestock and Animal Health Statistics data shows that the national dairy cow population in2019 was 561,061 heads with milk production of 996,442 tons, while the national milk demand in 2019 reached 4.3 million tons. To meet the demand for milk from the community, actors in the milk supply chain continue to improve their performance, especially the cooperatives. As the largest cooperative in Bogor Regency, KUD Giri Tani

* Corresponding author: melanu@apps.ipb.ac.id

has hundreds ofpassive farmer members and over a hundred active members. The cooperative receives milkfrom farmers, carries out a cooling process, and then distributes it to the Cimory milk as the processing company. In addition, to improve the quality and quantity of milk, the cooperative provides support services for member farmers, such as providing loans for operational, credit for animal feed, veterinary medicines, and other services to support dairy farming activities [1]. It is supported by the results of the SWOT analysis [2]. The strategy to improve dairy cooperatives can be achieved by increasing awareness of the important role of farmers and their function as cooperatives.

Only 6.8% of the total milk production is marketed directly to end customers. To gain greater profits and sustain, the farmer members must improve their product quality, provide cheaper products, and follow customer wishes. Measurement of supply chain performance for a company's needs aims to reduce costs, meet customer satisfaction, increasecompany profits, and determine the extent to which its supply chain performance [3, 4]. Performance measurement aims to support the design of goals, evaluate performance, anddetermine future steps at the strategic, tactical, and operational levels [5, 6], and an integrated measurement is obtained between suppliers, internal companies, and consumers [7].

Nasrudin and Rivana [8] revealed that 50 performance indicators could be assessed to increase milk production productivity with the SCOR approach. The AHP method is faster than manual calculations, it can be more efficient and accurate for the result analysis [9]. In contrast to Nasrudin and Rivana [8], this study is more focused on improving the quality of milk in a village cooperative, so that is the basis for the AHP method. For the weighting of performance indicators, the Analytical Hierarchy Process (AHP) method is used [10, 11].

2 Methodology

The research was conducted at the Giri Tani KUD for three months, from September to November 2020. It focused on improving milk quality by identifying supply chain performance indicators from farmers to the milk processing company. The study began by conducting interviews and direct observations regarding general data and describing the working methods of cooperatives, then continued with in-depth observations on the process of procuring milk from the farmer members, processing cooling units in cooperatives, and the delivery process to the milk processing company. It is to find out the problems on the supply chain in improving the milk quality process.

2.1 Data Collection

There were two types of data needed to conduct this research, namely primary data and secondary data. Primary data is data obtained from interviews, questionnaires, and direct observations at the Giri Tani KUD. Meanwhile, secondary data was data already exists or generaland historical data of KUD Giri Tani. Based on data collection, Bussiness Canvas Model wasconducted [12].

2.2 Data Processing and Analysis

Processing begins with supply chain identification, process decomposition based on the SCOR model [5, 6], KPI validation (Key Performance Indicator), weighting the KPI hierarchy using the AHP (Pairwise Comparison) [9] method until recommendations are given for improvement. Flowchart of data processing methodology in Figure 1.

2.3 The decomposition process of SCOR (Supply Chain Operation Reference) and Indicator Validation

Cooperatives used SCOR as a reference for detailed supply chain processes. SCOR defines and classifies each cooperative dairy process to develop measurement indicators needed to measure integrated supply chain performance between suppliers (breeders) and internal companies (Cooperatives) and Milk Processing companies.

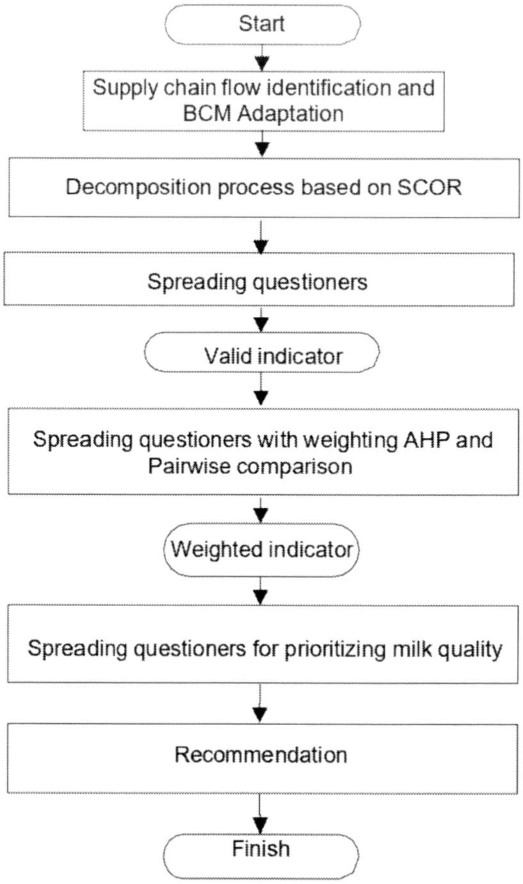

Fig. 1 Methodology

The supply chain processes in the cooperative are divided into five integrated processes such as planning (Plan), source of supply (Source), production (Make), delivery (Delivery), and return (Return). The scoring metrics in the SCOR model are expressed at two levels. The supply chain process model is a process hierarchy, and assessment metrics expressed in a ranking hierarchy. The number and levels of the metrics adjust to the type and number of processes in the cooperative. The researchers hold Focus Group Discussions (FGD) and ask questionnaires to identify and validate several Key Performance Indicators of the cooperative.

2.4 Weighing Indicator with AHP

The initial stage of this weighting was to make a questionnaire paired which the weighting filled in by the related respondents. The respondents were the chairman, the treasurer, the secretary, and two senior cooperative members. Then, the Analytical Hierarchy Process (AHP) calculates the data obtained from the questionnaire results. The weighting process consists of 2 levels. Level 1 was the weighting for each perspective such as plan, source, make, deliver and return. Level 2 was a weighting for each dimension of each supply chain perspective with the dimensions were reliability, responsiveness, cost, assets, flexibility. Each perspective adjusted to efforts to improve milk quality.

2.5 Prioritization program with AHP

The chairman, treasurer, secretary, and the senior farmer members of the cooperative attended the FGD to obtain three main programs to improve milk quality. There were three main programs, (1) Improving the quality of feed, (2) Improving the hygiene and sanitation of farmers (3) Payments to farmers according to the quality of milk.

This research compares the three alternatives in pairs by pairwise comparison, for each level 2 indicator was weighted previously. The process uses a questionnaire, filled out by five respondents and then combined respondents' opinions into one matrix. Then, multiplying the results of the assessment of each indicator for each alternative by the weight of each indicator carried out previously. The sum of the multiplication results becomes the final value of each option. The most significant value was the recommendation to become a priority cooperatives program.

3 Result and Discussion

The flow supply chain KUD Giri Tani identification was carried out directly and followed by interviews with the cooperative chairman and secretary. Figure 2 shows that the green line indicates the milk material flow from the farmers to downstream, the brown line indicates the flow of animal feed. In contrast, the red line indicates the flow of returns from the milk processor to the cooperative and the farmer. The supply chain flow cycle has several entities, including the Cimory Milk Processing Industry, Cooperatives, Breeders, feed suppliers, direct consumers, and milk beverage SMEs. This study focuses on the milk flow from the farmers to consumers and the return.

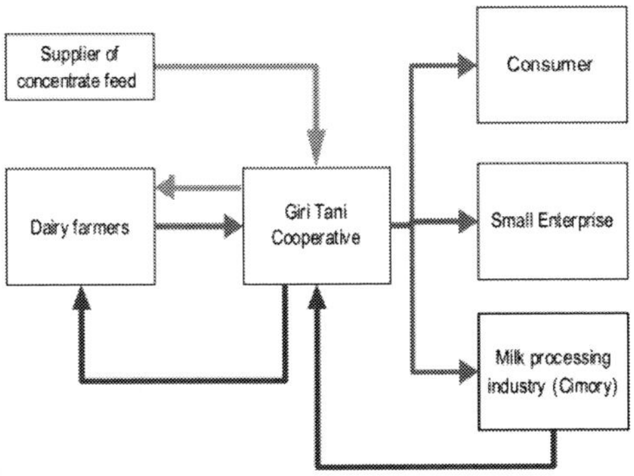

Fig. 2. Supply Chain Flow

The next stage describes the processes based on the SCOR model. The decomposition of supply chain flow processes from general to detail refers to the Supply Chain Operation Reference manual version 10 by the Supply Chain Council (SCC). It adjusted the vision and mission of the cooperative and the results of interviews with the cooperative's chairman.

The objective of using BCM was to identify key partners, activities, resources, value proposition, customer relationship, and channel and customer segment of Giri Tani in mapping existing conditions and develop an alternative strategy.

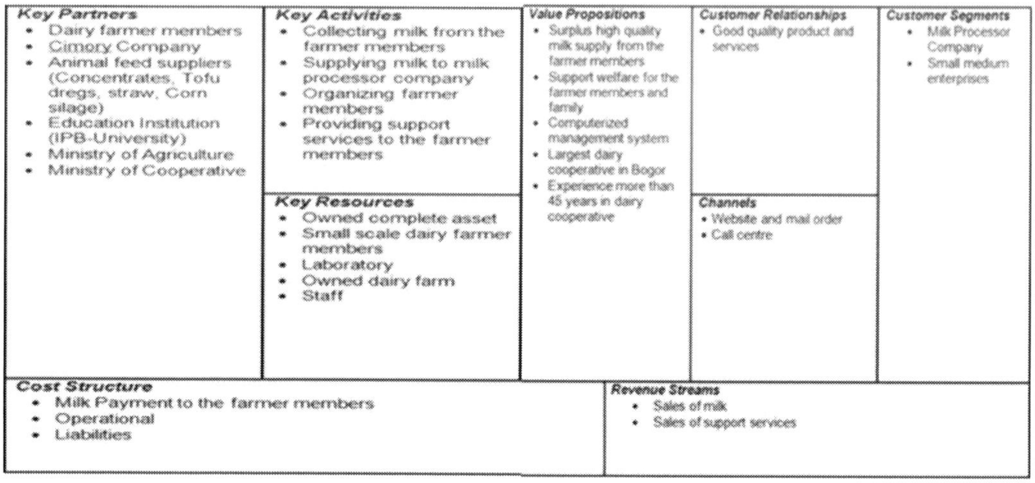

Fig. 3. Business Model Canvas

Therefore, the BCM identified that the key activities to improve milk quality are collecting milk from the farmers, supplying milk to milk processors, organizing farmer members, and providing support services Fig 3.

The SCOR model framework begins with decomposition into five core processes (plan, source, make, delivery, and return). Each of the five core processes contains five performance attributes (reliability, responsiveness, flexibility, costs, and assets). However, based on the FGD results, not all core processes have five performance attributes.

By carrying out the validation stages of the results from the decomposition of the SCOR process, these indicators can present the performance of the Giri Tani cooperative. The chairman of the cooperative validated the indicator metrics through discussions and questionnaires 19 indicators used, where previously there were 24 indicators in measuring supply chain performance to improve milk quality in the cooperative. In detail, validation produces two indicators for the planning process (plan), six indicators for the supply source process (source), three indicators for the production process (make), five indicators for the delivery process (delivery), and three indicators for the return process. Figure 4 shows the indicator validation result.

One of the crucial things in milk supply chain management is milk quality. The milk quality is the basis for paying the milk price received by the Giri Tani cooperative from Cimory. Meanwhile, Giri Tani paid the milk to all farmers at the same price based on the quantity dropped. Giri Tani cooperative set a minimum standard for the quality of milk from the farmers. However, due to one price for milk payment policy, it makes the farmer members deliver the milk to the cooperative by varying quality. They are not motivated to provide high-quality dairy. As a result, the price paid by Cimory to the cooperative is not very satisfactory. It causes some farmers to often sell their milk to other parties who are willing to pay higher than cooperative. Realizing this problem, the head of the Giri Tani cooperative intends to change the milk payment policy based on quality. For this reason, the measurement of supply chain performance emphasizes milk quality. The milk payment policies for the Giri Tani can use the result of supply chain performance measurement

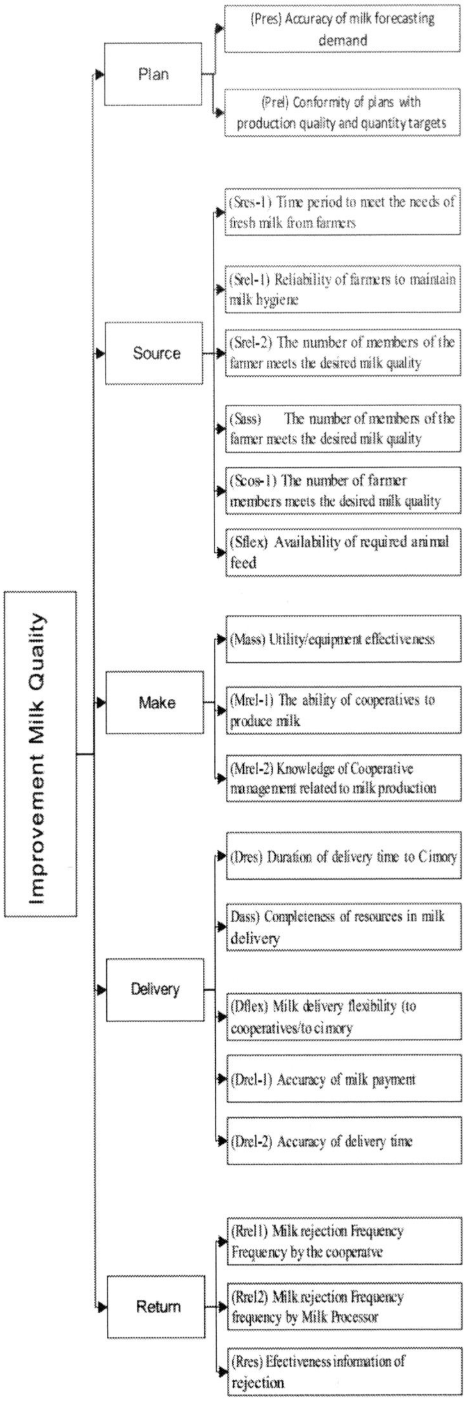

Fig. 2. The Indicator of SCOR Validation Giri Tani Cooperative

The weighting of indicators level 1 and 2 using the AHP method and supported by a pairwise comparison matrix [13]. It started from the AHP hierarchical structure to improve milk quality, and then continued by classifying levels 1 and 2, namely five core processes and 19 indicator metrics. This weighting to determine the weightiest value of the five core processes and 19 indicators.

The five people who influence the cooperative's policy (as experts) answered the questionnaire for the weighting. The next step was compiling the criteria in a paired matrix until the calculation was known to have a fixed eigenvalue. Before combining the questionnaire results, this study counted the five people separately to determine whether the level of importance was consistent. Then use the geometric mean formula to produce a combination matrix of the five experts in the Giri Tani cooperative environment. The results of the core process questionnaire are shown in the form of a paired matrix in Table 1.

Table 1. Comparison matrix

Combination	Plan	Source	Make	Delivery	Return
Plan	1	0.29	0.19	0.27	1.42
Source	3.21	1	1.12	1.32	4.84
Make	5.24	0.58	1	2	4.55
Delivery	3.55	0.76	0.50	1	4.93
Return	0.51	0.21	0.14	0.20	1

Then, the pairwise comparison matrix results for the combined five respondents were weighted on each core criterion. The process was to multiply each paired matrix and raise it according to the criteria, so the result follows.

Table 2. The result of the comparison matrix

Criteria		Weight
Plan	0.46349	0.07649
Source	1.87331	0.30916
Make	1.94700	0.32132
Delivery	1.46031	0.24100
Return	0.31521	0.05202
Total	6.0593	1.0000

The results of the level 1 weighting show that the top 3 that significantly affect the milk quality are production, milk supply from the farmers, and delivery. The high weight is because these three activities are the main activities in the Giri Tani Cooperative which sometimes have problems in its operational process. It is relevant to research, which states that several issues still occur to obtain milk with good quality and quantity are manual production methods, lack of availability of animal feed, and handling of milk distribution.

The return has the lowest weight because milk is seldom rejected, and until now, the cooperative has not implemented an evaluation system if the Cimory rejects the milk from Giri Tani Cooperative.

Every day, the farmer members supply the milk to Giri Tani in the morning and evening; the group's leader collects the milk, then from the leader group, it is handed over to the tank truck of the cooperative. Although the milk collection trucks were carried out every morning and evening by visiting each farmer member group, the milk delivered to milk processing was carried out every day only in the morning, and the cooperative's refrigeration machines usually used before going to Cimory.

The weighting of 19 quality improvement indicators and the weighting of the core process had the same calculation process. Beginning with questionnaire questions regarding pairwise comparisons through 19 questionnaires, then followed by combining the opinions of five respondents. Below are the results of the calculation of the weighting of 19 indicators.

Table 3. Weighting calculation results

Criteria	Subcriteria	Weight
Plan	Pres	0.0166
	Prel	**0.0439**
Source	Sress-1	0.0094
	Srel-1	**0.0255**
	Srel-2	0.0825
	Sass	0.0467
	Scoss-1	**0.0809**
	Sflex	0.0642
Make	Mass	0.1533
	Mrel-1	0.0511
	Mrel-2	0.0511
Delivery	Dress	0.0234
	Dass	0.0261
	Dflex	0.0765
	Drel-1	0.1160
	Drel-2	0.0466
Return	Rrel-1	0.0226
	Rrel-2	0.0513
	Rres	0.0122

In the core of planning activity, the most significant weight is the Prel indicator which explains the achieving of the plan, where the milk obtained by the cooperative is adjusted to supply customer demand. Priyanti and Soedjana [14] states that farmers still provide the dairy industry in Indonesia, and most women are members of the Village Cooperative.

On Sourcing milk from farmers, dairy cooperatives determine the weight of the most significant indicator, Srel-2, indicating that the number of farmers who meet milk standards is an important criterion that significantly affects the quality of milk collected by the cooperative. Based on Hafidh and Purwono [15], the farmer members strongly influence the cooperative's quality and quantity of milk.

In the form of cooling unit activity, the cooperative weighs the largest Mass. Mass is the utility effectiveness of the cooling unit asset. Unused refrigerator equipment maximally will quickly be damaged and directly impacts the quality of milk.

In shipping activities, out of 5 indicators, two indicators are quite large compared to the other 4, Dflex and Drel-2. Dflex describes the flexibility of milk delivery to Cimory, and cooperatives can deliver anytime to the milk processor as long as the quality requested by Cimory meets the requirements. The next most significant indicator is Drel-2 which represents the accuracy of milk payments; psychologically, farmers and cooperatives will be enthusiastic about maintaining milk quality if the milk processing company on time pays to cooperatives.

The return activity that describes the rejection of milk from Cimory to the cooperative, and the cooperative to the farmer members shows that Rrel-2 is an essential indicator of returns. Rrel- 2 is the frequency of refusal of milk by Cimory, because the primary source of income for both farmers and cooperatives is a payment of milk from Cimory, so if there is a refusal, of course, there will be no payment.

Table 4. Result of Prioritization

Subcriteria	Alternative weight for any subcriteria weight			
		Milk Grading	Improvement feed quality	Hygiene Improvement
Pres	0.020	0.094	0.642	0.264
Prel	0.057	0.094	0.642	0.264
Sress-1	0.023	0.084	0.750	0.166
Srel-1	0.055	0.084	0.750	0.166
Srel-2	0.005	0.084	0.750	0.166
Sass	0.051	0.084	0.750	0.166
Scoss-1	0.072	0.084	0.750	0.166
Sflex	0.103	0.084	0.750	0.166
Mass	0.103	0.308	0.222	0.470
Mrel-1	0.073	0.308	0.222	0.470
Mrel-2	0.145	0.308	0.222	0.470
Dress	0.020	0.330	0.129	0.541
Dass	0.042	0.330	0.129	0.541
Dflex	0.071	0.330	0.129	0.541
Drel-1	0.056	0.330	0.129	0.541
Drel-2	0.052	0.330	0.129	0.541
Prel-1	0.025	0.169	0.187	0.644
Prel-2	0.012	0.169	0.187	0.644
Rres	0.015	0.169	0.187	0.644
Total				

Determination of each final value for each alternative is calculated by multiplying the weight of each sub-criteria by each alternative weight, then adding up for the entire sub-criteria as many as 19 indicators from SCOR [16].

$$Alternative\ Score = \sum_{i=1}^{19} Weight\ of\ subcriteria\ \ i\ x\ Weight\ of\ alternative \quad (1)$$

Table 5. Final Score of Alternative

No	Alternative	Final score
1	Milk Grading	0.221
2	Feed Quality Improvement	0.393
3	Improve Hygiene of Milk Process	0.386

Based on the final value of each alternative, in addition to the SCOR criteria activities, itis known that improving feed quality and improving hygiene are priorities. Farmer membersand cooperatives can enhance milk quality. It can be part of the cooperative and member program.

4 Conclusion

Milk production in the farmer members must be the focus of cooperatives to improve the milk quality. Improving feed quality and improving hygiene are more priorities than milk grading. This research needs improvement by the cooperative's leadership and itsmembers to become the basis for payment policies and efforts to improve quality. Improvingmilk quality can be done through improving supply chain activities by focusing on specific activities. This research can continue with a broader scope of SCOR to distribute to consumers and implement a more updated AHP such as AHP Fuzzy. In addition, improving quality by increasing the income of farmers and cooperatives is an area that can be conducted after this research.

References

1. Z. Zhong, C. Zhang, F. Jia, eta J. Bijman, *Vertical coordination, and cooperative member benefits: Case studies of four dairy farmers' cooperatives in China*, J. Clean.Prod., libk. 172, zenb. (November 2018), or. 2266–2277, Available at : doi 10.1016/j.jclepro.2017.11.184 (2018)
2. U. I. L. Rahmah, *Strategies to Strengthen Dairy Cooperatives Performance through Cooperative Financial Performance Approach*, IOP Conf. Ser. Earth Environ. Sci.,libk. 466, zenb. 1, Available at: doi 10.1088/1755-1315/466/1/012029 (2020)
3. Stadtler H, *Supply chain management—an overview. Supply chain management and advanced planning*, 2008:9-36, (2008)
4. I. N. Pujawan eta Mahendrawathi, *Supply Chain Management Edisi 3*, Surabaya:Penerbit ANDI, (2017)
5. Asrol M, Marimin M, Machfud M, *Supply chain performance measurement and improvement for sugarcane agro-industry,* International Journal of Supply Chain Management., 2017 Sep 30;6(3):8-21 (2017)
6. Georgise FB, Thoben KD, Seifert M, *Implementing the SCOR model best practices for supply chain improvement in developing countries,* International Journal of u-and e-Service, Science and Technology. (2013 Aug 30); 6(4):13-25, 2013
7. Thomas DJ, Griffin PM, *Coordinated supply chain management. European journal of operational research*. 1996 Oct 11;94(1):1-5, (1996)
8. I. Nasrudin and R. Rivana, *Pengukuran Kinerja Supply Chain KPBS Pangalengan Dengan Pendekatan Supply Chain Operation Reference (SCOR) Untuk Meningkatkan Produktivitas, Rekayasa Ind. dan Mesin*, libk. 1, zenb. 1, or. 29–41, (2019)
9. Sutoni A, Subhan A, Setyawan W, Bhagyana FO, *Performance Analysis Using the Supply Chain Operations Reference (SCOR) and AHP Method*, In Journal of Physics: Conference Series 2021 Feb 1 (Vol. 1764, No. 1, p. 012155), IOP Publishing (2021)
10. International Finance Corporation, *Dairy Industry Development in Indonesia: Final Report* - May 2011, zenb. May, 2011, [Sarean]. Available at: http://www.ifc.org/wps/wcm/connect/93f48d00470e3bf883ffd7b2572104ea/Dairy+Industry+Development-2011.pdf?MOD=AJPERES (2011)
11. Saaty, TL, *Decision Making with Analytical Hierarchy Process*, Int. Journal Service Science Vol 1, No I (2008)
12. Braun AT, Oliver S, Bern R, *Adaptation of Business Model Canvas template to developa business model for the circular economy*, Procedia CIRP 99 (2021) 698-702, (2021)
13. Erdil A and Hikmet E, *Selection Strategy via Analytical Hierarchy Process: An Application for Small Enterprise in Milk Sector*, World Conference on Technology, Innovation and Entrepreneurship

14. A. Priyanti and T. D. Soedjana, *Indonesian Dairy Industry Perspective Within the ASEAN Economic Community*, Indonesia. Bull. Anim. *Vet. Sci.*, libk. 25, zenb. 4, or. 159–170, Available at : doi 10.14334/wartazoa.v25i4.1226, (2016)
15. M. Hafidh and J. Purwono, *Kinerja Koperasi dengan Pendekatan balanced Scorecard* (*Kasus: KUD Giri Tani Kabupaten Bogor*), J. Dep. Agribisnis Fakultas Ekonomi dan Manajemen Institut Pertanian Bogor, (2016)
16. Abdullaha RN, Yahyab S, Malimc MR. *Enhancing the decision-making process using AHP-SCOR Integrated Model* (*ASIM*), International Journal of Innovation, Creativity, and Change, (2019); 8(11):159-71, (2019)

Survival and growth performance the catfish *Clarias gariepinus* in high density nurseries using recirculating aquaculture system (RAS)

Cecilia Eny Indriastuti[1*], *Beata* Ratnawati[2], *and Ivone* Wulandari Budiharto[2]

[1]College of Vocational Studies of IPB University, Study Program of Production Technology and Management of Aquaculture, Cilibende Jalan Kumbang No 14, Bogor, Indonesia
[2]College of Vocational Studies of IPB University, Study Program of Environmental Engineering and Management, Cilibende Jalan Kumbang No 14, Bogor, Indonesia

Abstract. The recirculating aquaculture system (RAS) is a cultivation system that minimizes water changes and can optimize high stocking densities (SD). The purpose of this study was to determine the effect of nursery culture in high stocking density with RAS on survival rate (SR), specific growth rate (SGR), relative length (LR), weight gain (WG), length gain (LG)and feed utility efficiency (FUE) of juvenile catfish. The study used a completely randomized design with 4 treatments and 3 replications: R3000 (SD of 3,000 fish/m^2 with RAS); NR3000 (SD of 3,000 fish/m^2 without RAS); R4000 (SD of 4,000 fish/m^2 with RAS); NR4000 (SD of 4,000 fish/m^2 without RAS). Juvenile catfish measuring 3.3±0.2 cm were reared for 21 days and water changes were carried out every 2 weeks in the RAS and every 2 days in the non-RAS. The results showed that the SR was not significantly different for all treatments. The highest SGR, LR, WG, LG and FUE were shown in the R4000 treatment ($p<0.05$). Water quality parameters, especially total ammonia nitrogen in the RAS treatment had a lower value than without RAS. The abundance of total bacteria in the RAS treatment was more than without RAS. This study showed that the density of 4,000 fish/m^2 with the RAS was able to provide better catfish nursery production performance than without RAS.

1. INTRODUCTION

Catfish *Clarias gariepinus* is the leading freshwater fish commodity in Indonesia. This fish ranks second highest in production after tilapia. The demand for catfish from year to year always increases, as shown from the 2016 production data, which was 764,796 tons, increasing to 1,125,526 tons in 2017 [1]. The increase in catfish meat production will of course be accompanied by an increase in the need for catfish seeds as an input for enlargement. To meet catfish seeds ready to be raised, a series of maintenance is needed from larval rearing to rearing at the nursery level.

* Corresponding author: cecilia_eny@apps.ipb.ac.id

Nursery for catfish is still mostly done in the conventional way, namely raising catfish without changing water and using low stocking densities. So that the survival and productivity of catfish seeds is still low. In general, catfish nursery is done *outdoors* using soil ponds, fiber tanks or tarpaulin tanks, without changing water and using low stocking densities. This condition usually requires a large area of land and water conditions that are not ideal resulting in growth and survival that are not optimal, so technology with a more intensive maintenance system, saving land and water is needed, to get a better production value.

RAS is one of the applications of aquaculture which can control the discharge of waste into the environment [2]. The existence of filtering in the recirculation system will reduce metabolic results and reduce the accumulation of organic matter leftover feed, feces contained in the rearing container, thereby increasing fish appetite. The nitrification process is the main process in RAS to remove ammonia and nitrite which are toxic to fish. Ammonium is converted to nitrite and to nitrate which is low in toxicity so that water can be reused [3-4].

In order to offset production costs and make the recirculation system more efficient, the density of catfish fry per unit area should be as much as possible. According to SNI [5], the stocking density of juvenile catfish is 1,000-1,500 fish/m^2. The stocking density commonly used by conventional catfish nursery cultivators is in the range of 1,500-2,000 fish/m^2 .It is hoped that by using RAS, the stocking density can be increased.

With a recirculation system, besides being able to optimize narrow land, it can also save water use which affects production efficiency. Recirculation can also be an alternative for catfish farmers who live in cities, where the availability of water sources is limited. In a recirculation system, an optimal environment can be obtained for seed maintenance which has an impact on survival and optimum growth [6]. The purpose of this study was to determine the effect of RAS on catfish nursery using high stocking densities on the survival, growth performance, feed efficiency, abundance of bacteria and water quality of the media.

2. MATERIALS AND METHODS

2.1 Ethics statements

All experiments in this study associated with fish complied with animal welfare and were handled under Indonesia accreditation SNI 6484.2:2014.

2.2 Experimental design

The study was conducted for three weeks in Tania Akuakultur farm, Loji, Bogor. Water quality measurements were carried out at the Environmental Laboratory, Department of Aquaculture, Faculty of Fisheries and Marine Sciences (FPIK), IPB. Bacterial counts were carried out at the Microbiology Laboratory, Vocational School, IPB.

The study used a completely randomized design with four treatments with three repetition, were applied: R3000; R4000; NR3000; NR4000 , R=recirculation, NR= non recirculation, 3000= SD 3,000 fish/m^2, 4000= SD 4,000 fish/m^2.

2.3 Experimental procedure

The research procedure is divided into 4 stages. These stages are preparation of rearing containers and recirculation systems, stocking of juvenile catfish, maintenance of juvenile catfish and sampling for measurement of test parameters.

2.3.1 *Preparation of the rearing container and recirculation system*

Preparation of the container began by preparing 18 tanks made of tarpaulin 700 cm ×100 cm ×40 cm in size and volume of 320 L. For catfish rearing, 12 tanks were used which were divided into 6 tanks installed in a RAS and 6 tanks non-RAS. While the other 6 tanks are used for the filtration system. These 18 tanks are connected by plastic pipes for inlet or outlet water lines which are driven by electric water pumps. The water used is well water that is stored in a reservoir. System stabilization was carried out for one week after the container was filled with precipitated water. The system is ready for use when the water quality parameters are in ideal conditions and sizes for the life and growth of the test fish (temperature: 28-30 °C, DO: 5-7 mg/l, pH: 6.5-9.0, ammonia: < 1.0 mg/l, alkalinity: = 0.1mg CaCO3/l).

2.3.2 Preparation of the recirculation system and filter

The recirculation and filter system uses 6 tanks, which have different functions, are presented in Figure 1. Tank no.1 collects dirty water from the fish rearing tank and then the water is passed to tank no. 2 which contains a dacron to filter coarse particles, then flowed into tank no 3. Tank no 3 is divided into 3 chambers, each containing sand, charcoal and zeolite. Sand serves to add minerals, charcoal to absorb ammonia odors, and zeolite to neutralize the pH of the water and increase oxygen. Then, the water flows into tank no. 4 which contains bioballs which are a biological filter as a home for bacteria. The water is then flowed into tank no. 5 which contains a tassel of raffia rope that serves to house bacteria. Finally, the water is channeled into tank no. 6 and is ready to be distributed to catfish rearing tanks with the help of a water pump. Water flow rate was 0.7 $l.min^{-1}$ and the complete water cycle took 43 min.

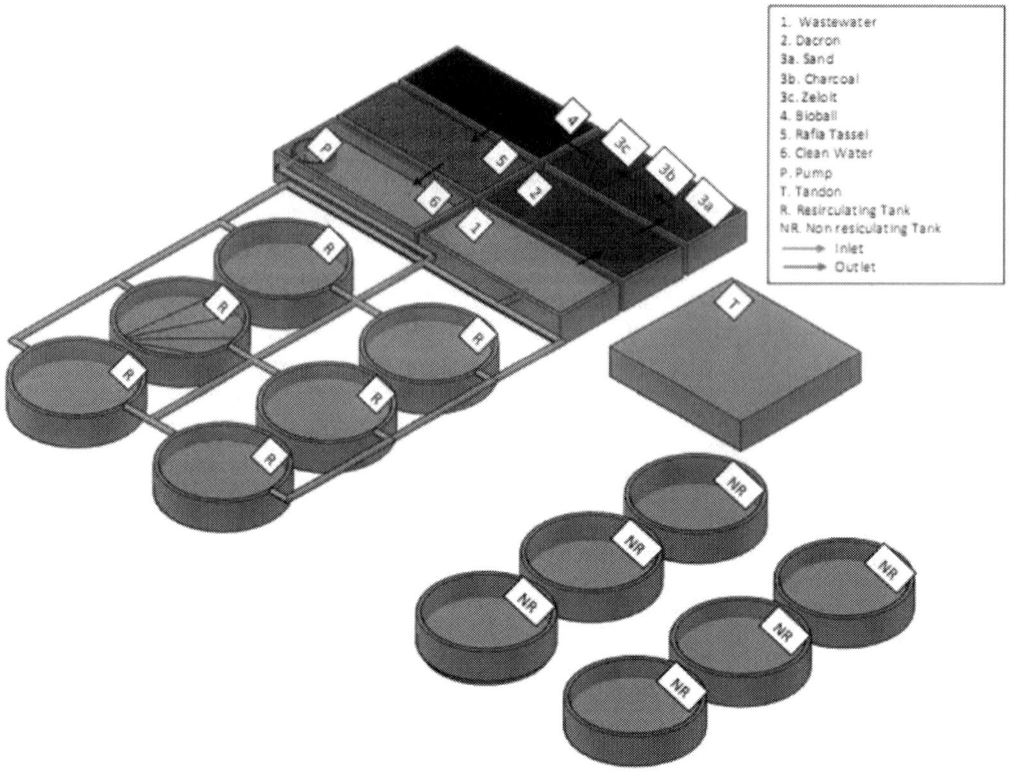

Fig. 1. Recirculating aquaculture system and rearing tanks of juvenile catfish

2.3.3 Juvenile catfish stock

The juvenile catfish used came from fish farmers around the IPB campus. The total juvenile catfish used amounted to 30,600 fish, three weeks old with a length of 3.2 ± 0.2 cm. Prior to stocking, the juveniles were acclimatized in an aquarium for 3 days, to adapt to new environmental conditions. Catfish stocked in each tank with different density levels, i.e., 3,000 fish / m^2 and 4,000 fish / m^2 were placed randomly. The R3000 and NR3000 treatments were filled with 2,250 fish/tank, while the R4000 and NR4000 treatments filled with 2,850 fish.

2.3.4 Catfish seed maintenance

Juveniles Catfish were reared for 21 days, given artificial feed with a protein content of 35-38%, as much as 5% by weight of biomass. Feeding was done twice in the morning and evening. Water changes were carried out every 2 weeks for the recirculation treatment and every 2 days for the non-recirculating treatment. During maintenance, several parameters were observed, namely weight, length and number of dead fish, bacterial density and water quality.

2.4 Variables evaluated

The variables measured were the survival rate (SR), specific growth rate (SGR), relative length (RL), weight gain (WG), length gain (LG), feed utilization efficiency (FUE), bacterial density and water quality parameters.

2.4.1 Survival rate (SR)

The survival rate (SR) was calculated at the end of the study (day 21), using the equation in Zonneveld [7].

$$\text{SR (\%)} = \frac{\text{final number of fish}}{\text{initial number of fish}}\ x\ 100$$

2.4.2 Weight gain (WG) and lLength gain (LG)

Weight gain (WG) and length gain (LG) are the differences between the final average fish weight or average fish length and the initial average fish weight or average fish length. The weight was measured using a digital scale with an accuracy of 0.0001 g, while the length was measured using a digital caliper with a precision of 0.01 mm. The weight and length of catfish were measured at the beginning of the study, the 7th day, 14th day and 21st day (the end of the study). The weight gain and length gain was calculated using this formula:

$$\text{Weight gain (g)} = final\ average\ fish\ weight - initial\ average\ fish\ weight$$

$$\text{Length gain (mm)} = final\ average\ fish\ length - initial\ average\ fish\ length$$

2.4.3 Specific growth rate (SGR)

Specific growth rate parameters of fish were measured once every seven days by weighing and measuring the fish length of 30 fish for each repetition. The specific growth rate (SGR) was calculated using the formula in [7]

$$\text{SGR (\% g day}^{-1}) = \frac{\text{ln final weight - ln initial weight}}{\text{days number of experiment}}\ \text{x}\ 100$$

2.4.4 Relative length (RL)

The relative length growth (RL) is the difference between the final average length and the initial average length divided by the initial average length, calculated using the following formula:

$$\text{LR (\%)} = \frac{\text{the average length of fish at the time t (mm)}-\text{the average length of fish at the beginning (mm)}}{\text{the average length of fish at the beginning of the study (mm)}}\ \text{x}\ 100$$

2.4.5 Feed utilization efficiency (FUE)

According to Tacon [8], the calculation of feed utilization efficiency (FUE) is as follows:

$$\text{FUE (\%)} = \frac{\text{fish biomass at the end of the study (gram)}-\text{fish biomass at the beginning (gram)}}{\text{amount of feed consumed during the study (gram)}}\ \text{x}\ 100$$

2.4.6 Total bacterial abundance and presence E. coli bacteria

Bacterial count was carried out at the end of the study. The bacterial parameters observed in this study included the parameters of *E. coli* bacteria and common carried out once a week. The method used is the *total plate count*, where a sample of treatment water is taken, then a gradual dilution is carried out from 10^{-1},$10^{-2,}10^{-3}$,10^{-4} and 10^{-5} from each sample to be tested as much as 1 ml. . 4, then 0.1 ml was taken from each sample using a micropipette at dilutions of 10^{-3},10^{-4} and 10^{-5} which were isolated on spread plates on PCA media (media for bacterial growth in general) and EMBA (special media for growth of bacteria). gram negative bacteria, namely *E. coli*). Next, incubation was carried out for 1x24 hours to determine whether or not there was bacterial growth as an indicator.

2.4.7 Water quality parameters

Water quality parameters observed during this study included temperature, pH and DO parameters which were measured twice a day using thermometer, DO-meter and pH-meter instruments. Ammonia, alkalinity, BOD, COD parameters were measured every seven days using titration, spectrophotometer and phenate methods.

2.5 Data analysis

Data were tabulated using Microsoft Excel 2010. Parameters of survival rate, specific growth rate, relative length, weight gain and length gain were analyzed for variance using one-way ANOVA at level P=0.05 with the help of SPSS 21 software. If the difference is significant, the data is further tested using Duncan's test. Inhomogeneous data, tested by nonparametric test using Kruskal Walis. Data on water quality and bacterial composition were analyzed descriptive in tabular form.

3. RESULT

3.1 Survival rate, specific growth rate, relative length, weight gain, length gain and feed utilization efficiency

Results demonstrated that there were no significant differences related to the survival rate of fish. Treatment R4000 resulted in the highest survival value (91.03±0.90%) and treatment NR3000 with the lowest value (89.93±0.40%), but all treatments were not significantly different (Table 1).

The highest SGR was demonstrated by the treatment R4000 at 7.71±0.19%, and the lowest was demonstrated by the treatment NR3000 at 5.92±0.31% (P< 0.05). The highest LR was demonstrated by the treatment R4000 at 0.61±0.02%, and the lowest was demonstrated by the treatment NR3000 at 0.44±0.03% (P< 0.05; Table 1).

The highest WG were demonstrated by the treatment R4000 at 1.67 ± 0.08g, and the lowest were demonstrated by the treatment NR3000 at 1.04 ± 0.09g (P< 0.05). The highest LG were demonstrated by the treatment R4000 at 20.86±0.57mm , and the lowest were demonstrated by the treatment NR3000 at 14.96±1.10mm (P< 0.05; Table 1).

The highest FUE were demonstrated by the treatment R4000 at 88.45±2.03%, and the lowest were demonstrated by the treatment NR3000 at 69.49 ± 4.39% (P<0.05; Table 1).

Table 1. Survival rate, specific growth rate, relative length, weight gain, length gain and feed utilization efficiency of juvenile catfish reared under condition of different stocking density and system recirculation for 21 days

Treatments	Survival Rate (%)	Specific growth rate (%)	Relative Length (%)	Weight Gain (g)	Length Gain (mm)	Feed Utilization Efficiency (%)
R3000	90.30 ± 0.20^a	6.68 ± 0.79^a	0.50 ± 0.06^a	1.30 ± 0.28^a	17.26 ± 1.88^a	78.43 ± 5.78^a
R4000	91.03 ± 0.90^a	7.71 ± 0.19^b	0.61 ± 0.02^b	1.67 ± 0.08^b	20.86 ± 0.57^b	88.45 ± 2.03^b
NR3000	89.93 ± 0.40^a	5.92 ± 0.31^a	0.44 ± 0.03^a	1.04 ± 0.09^a	14.96 ± 1.10^a	69.49 ± 4.39^a
NR4000	90.10 ± 0.53^a	6.10 ± 0.31^a	0.46 ± 0.06^a	1.10 ± 0.10^a	15.81 ± 2.21^a	71.08 ± 6.52^a

Different superscript letters within a single treatment column showed significant differences among treatments (Duncan's Test: $P < 0.05$).

3.2 Total bacterial abundance

The total bacterial density increased significantly with treatment R3000 and R4000 at each water sample. NR4000 treatment only few bacteria grow, whereas in the NR3000 treatment no bacteria grow. Meanwhile, for observations of bacteria *E. coli,* none grew in all treatment samples (Table 2).

Table 2. Abundance of total bacterial and presence of *E. coli* during 21 days rearing of juvenile catfish reared under condition of different stocking density and system recirculation

Treatments	Total bacteria			*E. coli* bacteria		
	H7	H14	H21	H7	H14	H21
R3000	+ (TMTC)	+ (TMTC)	+ (TMTC)	-	-	-
R4000	+ (TMTC)	+ (TMTC)	+ (TMTC)	-	-	-
NR3000	-	-	-	-	-	-
NR4000	+ (TFTC)	+ (TFTC)	+ (TFTC)	-	-	-

Noted :
(+) = bacteria detected
(-) = no bacteria detected
TFTC = Too few to count
TMTC = Too many to count

3.4 Water Quality

The results of water quality measurements showed that the quality of the media was still within the feasible range to support the survival and growth of catfish seeds (Table 3).

Table 3. Water quality range during 21 days of rearing juvenile catfish reared under condition of different stocking density and system recirculation

Treatments	Parameter					
	Temperature (°C)	pH	DO ($mg.L^{-1}$)	TAN ($mg.L^{-1}$)	Amonia ($mg.L^{-1}$)	Alkalinity (mg CaCO3/L)
R3000	25,58-27,12	7,26-7,36	3,60-4,10	0,30	0.033	104,59
R4000	25,58-27,12	7,25-7,26	4,10-4,70	0,25	0.027	122,96
NR3000	25,58-27,12	7,11-7,21	1,80-4,40	0,66	0.072	121,55
NR4000	25,58-27,12	7,09-7,22	1,60-4,20	0,73	0.080	135,68
Reference	28 – 31 [9]	6,5 – 8,5 [5]	Min 3,0 [5]	Maks 4 [10]	< 0,1 mg/L [11]	30-100 mg CaCO3/L [12]

4. DISCUSSION

Recirculating aquaculture system (RAS) is proven to increase the density of catfish seeds. In addition, RAS can also improve production performance and feed efficiency. Recirculating and non-recirculating treatments at all densities did not give a significant value to the survival rate. Increased density of 3,000 fish/m^2 to 4,000 fish/m^2 did not show any significant difference in the death of the fish, so the value of SR did not differ significantly. This is in accordance with a study conducted by Haylor [13] which found that catfish larvae reared with densities of 50, 100 and 150 L/larvae did not increase mortality. The same thing was shown by Siregar [14] and Rahmadiah [15] that the increase in stocking density did not cause mortality and cannibalism in juvenile catfish.

Specific growth rate of juvenile catfish reared in recirculation tanks with a high density of 4,000 fish/m^2 is better than the other interventions. The same results were also obtained for other growth parameters; weight gain, relative length and length gain. In general, SGR is influenced by stocking density (SD), this is related to competition for space, feed and maintenance water conditions [16].

Catfish have territorial aggressiveness, the tendency to attack their relatives to defend their territory which causes high levels of cannibalism. In this study, at a lower stocking density (3,000 fish/m^2), the rearing media still had space that allowed the catfish to attack their friends. At SD 4,000 fish/m^2 the room is getting narrower, catfish are more alert to other fish attacks, so that aggressive behavior decreases. Catfish with high stocking density reduces the territorial aggressiveness [17-18]. This also affects the diet so that at a stocking density of 4,000 fish/m^2 with recirculation, catfish are more efficient in utilizing their feed.

At SD 4,000 fish/m^2 metabolism results, feces will be more than SD 3,000 fish/m^2, but by using RAS it is able to have a significant effect on improving water quality than non-recirculating systems so that with SD 4,000 fish/m^2 can produce significantly better growth performance and feed efficiency.

This is possible because in the RAS, dirty water resulting from metabolism, fish feces and feed residues originating from the rearing media undergo physical, chemical and biological filtering so that they meet the quality standards of fish maintenance and can be used and channeled back to the rearing media. Poor water quality can cause stress for fish so that fish are susceptible to disease or reduce their growth rate [19].

The effect of RAS also has an effect on feed efficiency parameters. The recirculation treatment with a density of 4,000 fish/m^2 had the highest percentage of 88.45% which was significantly different from the recirculation treatment of 3,000 fish/m^2 and the non-recirculating treatment. This is in accordance with other studies which state that the use of a recirculation system in catfish has a significant effect on feed efficiency compared to water exchange systems and circulation systems [6]. This influence is thought to be due to the water quality of a good rearing medium that can affect fish in maximizing the utilization of well-digested feed. In addition, the rearing water in the recirculation system is possible for the presence of good bacteria and natural food in the water which comes from organic waste that is broken down by bacteria in the waters and flowed back into the rearing media. The water quality of the rearing media is an important factor related to feed efficiency, as stated by Effendi [16] that one of the factors that determine feed efficiency is water quality (especially oxygen, temperature, pH and ammonia) besides that it also depends on species (feeding habits, size/stadia), and type of feed (quality and quantity).

5. CONCLUSION

Juvenile catfish rearing using recirculating aquaculture system can increase stocking density so that it can optimize production with limited land. By using recirculating aquaculture system, there is a possibility that stocking density can be increased as evidenced by treatment with SD 4,000 with recirculating aquaculture system, giving significantly better results on growth and feed efficiency than other treatments.

REFERENCES

1. KKP, Kelautan dan perikanan dalam angka tahun 2018, Pusat data dan statistik, (2018)
2. N.M. Ramli, M.C.J. Verdegem, F.M. Yusoff, M.K. Zulkifely, J.A.J. Verreth. Removal of ammonium and nitrite in recirculating aquaculture systems by the epiphyte *Stigeoclonium nanum* immobilized in alginate beads. Aquacult Environ Interact. 9 :213-222 (2017)
3. J.V. Rijn. Waste Treatment in Recirculating Aquaculture Systems. Aquacultural Engineering. 53:49-56 (2013)
4. W. Pungrasmi, P. Phinitthanaphak, S. Powtongsook. Nitrogen Removal From a Recirculating Aquaculture System Using Apumice Bottom Substrate Nitrification Denitrification Tank. Ecological Engineering. 95:357-363 (2016)
5. SNI. Standar Nasional Indonesia. Ikan lele dumbo (*Clarias* sp.) Bagian 4: Produksi benih. SNI: 6484.4: 2014 (2014)
6. A.W. Primaningtyas, H.S. Subandiyono. Performa produksi ikan lele (*Clarias gariepinus*) yang dipelihara dalam sistem budidaya berbeda. Journal of Aquaculture Management and Technology. 4(4):51-60 (2015)
7. N. Zonneveld, EA. Huisman, JH. Boon, Prinsip-prinsip Budidaya Ikan. Terjemahan. Jakarta (ID): PT Gramedia Pustaka Utama. (1991)
8. A. Tacon, The Nutrition and Feeding of Farmed Fish and Shrimp, A Training Manual (2) Nutrient Source and Composition, FAO, Brasilia, (1987)
9. E. Baras, T. Raynaud, J. Slembrouck, D. Caruso, C. Cochet, M. Legendre, Interactions between temperature and size on the growth, size heterogeneity, mortality and cannibalism in cultured larvae and juveniles of the Asian catfish, *Pangasianodon hypophthalmus* (Sauvage), Aquaculture Research, 42, 260–276, (2010)
10. N.M. Stone, H.K. Thomforde, Understanding your fish pond water analysis report, Chicago (US): University of Arkansas Cooperative Extension, (2004)
11. G.A. Wedemeyer, Fish Hatchery Management, 2nd Edition, Bethesda, Maryland, American Fisheries Society, (2001)
12. C.E. Boyd, Water Quality in Warm Water Fish Ponds, Fourth Printing, Alabama (US): Auburn University Agriculture Experiment Station, hlm 359, (1988)
13. G.S. Haylor, Controlled hatchery production of *Clarias gariepinus* (Burchell 1822): growth and survival of fry at high stocking density, *Aquaculture Research*, 22(4), 405-422, (1991)
14. K.N. Siregar, Pengendalian kanibalisme benih ikan lele Afrika *Clarias* 24 *gariepinus* menggunakan hormon estradiol-17β dan pengaturan padat tebar, tesis, Bogor (ID): Institut Pertanian Bogor, (2017)
15. T. Rahmadiah MJr. Zairin, Alimuddin, I. Diatin, Aggressive and cannibalistic behaviour of african catfish larvae: effect of different doses of methyltestosteron injecting to female broodstock and larval stocking densities, *JAI,* 18(2), 182-192, (2019)
16. H. Effendi, Telaah Kualitas Air Bagi Pengelolaan Sumberdaya dan Lingkungan Perairan, Penerbit Kanisius, Yogyakarta. (2003)
17. T. Hecht, S. Appelbaum, Observations on intraspesific aggression and coeval sibling cannibalism by larva and juvenile *Clarias gariepinus* (Clariidae: Pisces), *Journal of Zoologi* 214, 21-44, (1988)
18. H. Kaiser, O. Weyl, T. Hecht, Observations on agonistic behavior of *Clarias gariepinus* larvae and juveniles under different densities and feeding frequencies in a controlled environment, *J of Appl Ichth*, 11(1-2), 25-36, (1995)
19. S. Diansyah, T. Budiardi, AO. Sudrajat, Kinerja pertumbuhan *Anguilla bicolor bicolor* bobot awal 3 g dengan kepadatan berbeda, Jurnal Akuakultur Indonesia, 13(1), 46-53, (2014)

The road ahead and the future of eliminating fisheries subsidies under the WTO

Song Soo Lim[1*], *Chang Min* Kim[1], and *Dae Eui* Kim[1], *Kyu Sung* Lee[2], *Eun Sang* Lee[1]

[1]Department of Food and Resource Economics, Korea University, 145 Anam-ro, Seungbuk-gu, Seoul, Korea

[2]Korea Fisheries Infrastructure Public Agency, 53 Gasan Digital 2-ro, Geumcheon-gu, Seoul, Korea

Abstract. The World Trade Organization has explored forging an agreement on eliminating trade-distorting fisheries subsides for twenty years. To hasten towards the compromise among WTO Members, the chair of fisheries subsides circulated the revised draft text eyeing on the conclusion of negotiations before its twelfth ministerial conference at the end of 2021. This study analyzes the draft agreement, focusing on prohibition on subsidies concerning illegal, unreported and unregulated fishing, overfished stocks and overcapacity and overfishing. In addition, cross-sectional models are specified to investigate the effects of subsidies. Results show that both direct and indirect payments contribute to catch while support aimed for resource management is largely decoupled from overcapacity. When countries are further divided into the developed and developing countries by the WTO and the HDI criteria, only indirect payments prove their effectiveness in boosting catch. These empirical results shed light on the importance and effectiveness of policy reform in global fisheries subsidies.

1 Introduction

The World Trade Organization (WTO) launched trade negotiations on fisheries subsidies in 2001 at the Doha Ministerial Conference. Its mandate is to clarify and improve disciplines on fisheries subsidies. The 2005 Hong Kong Ministerial Conference committed to prohibiting certain forms of fisheries subsidies that contribute to overcapacity and overfishing. In the midst of stalled talks, the 2017 Buenos Aires Ministerial Conference (MC11) renewed a sense of urgency and decided to conclude the agreement by the next MC12.

Based on this mandate, the WTO negotiators are expected to meet the United Nations Sustainable Development Goal Target 14.6 to "by 2020, prohibit certain forms of fisheries subsidies which contribute to overcapacity and overfishing, and eliminate subsidies that contribute to illegal, unreported and unregulated (IUU) fishing, and refrain from introducing new such subsidies." In preparation for the conclusion of the agreement at the MC12, originally scheduled for June 2020 in Kazakhstan, but had to be postponed to November 2021 in Geneva, members have been negotiating on the basis of a draft consolidated text first

* Corresponding author: songsoo@korea.ac.kr

introduced by the chair of the negotiations in June 2020, with revised versions introduced in November and December 2020, respectively. The chair introduced a new draft text in May 2021 and its revised draft in June 2021 [1]. Regarding subsidies reform, the latest draft text suggests prohibition on subsidies concerning IUU fishing (Article 3), overfished stocks (Article 4), and overcapacity and overfishing (Article 5).

While recognizing the adverse effects of harmful fisheries subsidies on the sustainable development of marine fisheries, negotiators have been debating how to ensure a fair and equitable discipline on these subsidies. The lack of transparency for domestic policy and differences in country-specific effects of fisheries subsidies, depending on economic and environmental situations make it difficult to clearly identify subsidies and their net effects [2]. Although there has been progressed in empirical studies on fisheries subsidies during the last decade, the number of studies is still limited and it is necessary to generate more insights into interaction between legal discipline and sustainable management of marine ecosystems [3].

Facing these limitations, this study aims to fill a gap in the existing literature by exploring the possibility and the interplay of negotiations on the basis of the latest chair draft text and analyzing the effects of fisheries subsidies on fisheries catch.

2 Evolution and assessment of the draft text

As seen in Figure 1, since the WTO embarked on the Doha Development Agenda (DDA) in 2001, Member states and other global institutions have discussed and explored ways to clarify and improve disciplines on fisheries subsidies under the Negotiation Group on Rules. Despite a broad agreement that countries should strengthen discipline on perverted fisheries subsidies, several unresolved issues are remained to be answered. One is an issue on enabling the transition to ensure sustainable objectives and commitments [4, 5].

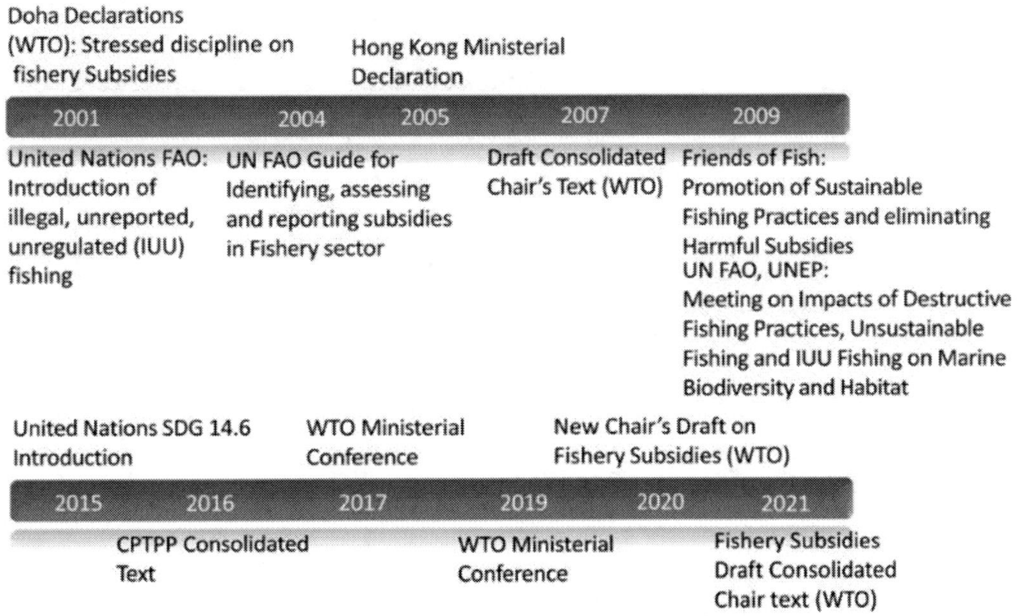

Source: Author's compilation

Fig. 1. Evolution of negotiations on fisheries subsidies

Table 1 shows a summary of negotiation proposals submitted by WTO Members. As for IUU fishing, a central issue is how it would be determined and then how the subsidy obligation would be triggered [6, 7]. The draft text explicitly links IUU fishing as activities set out in the FAO's 2001 International Plan of Action [8]. The draft also grants shared authority among a coastal Member, a flag state, or relevant Regional Management Organization (RFMO) in making an affirmative determination if a vessel or operator is engaged in IUU fishing. More specifically, governments should stop subsidizing a vessel or operator that is listed by its flag state or is identified by a coastal state as carrying out IUU activities within its maritime zones or by an RFMO within its geographical and species competence. The requirements of evidence-based IUU determination and notification by a coastal Member and the special and differential treatment (SDT) for developing countries are offered in the bracket.

Table 1. WTO Members' proposals on fisheries subsidies reform

Member states	WTO document #	Proposals
EU	TN/RL/GEN/ 181/Rev.1	Under the UN "SDG14", certain forms of subsidies contributing to overfishing and overcapacity as well as subsidies towards IUU fishing should be prohibited.
	TN/RL/GEN/ 188	Since '95, issued reduced percentage of WTO member level of subsidy notice (50→38%) and inadequate quality of notifications. Proposed incentives for simplifying monitoring and notification to ensure transparency.
LDC Group	TN/RL/GEN/ 184	The share of wild catch of LDCs are significantly small, thus, independent of overfishing. LDCs are directly affected by harmful subsidies towards large industrial fleets outside jurisdiction of EEZ. The negotiation proposes to discipline fisheries subsidies which contribute to overcapacity and overfishing as well as to eliminate subsidies for IUU fishing.
Cambodia on behalf LDC Group	TN/RL/GEN/ 193	Focused on prohibiting subsidies for overfishing and IUU fishing and suggested alternatives provisions for LDC.
Iceland, New Zealand and Pakistan	TN/RL/GEN/ 186	With SDG 14.6, certain forms of fishing subsidies that contributes to overcapacity and overfishing must be prohibited by '20 as well as effective and appropriate S&DT for developing and LDCs should be an essential part of the WTO subsidies negotiation.
ACCPPU (Argentina, Colombia, Costa Rica, Panama, Peru and Uruguay)	TN/RL/GEN/ 187/Rev.2	Certain forms of subsidies that contribute to overfishing and overcapacity should be prohibited and requires appropriate and effective measures for developing countries and LDCs . With the exception of IUU fishing, small fishermen should not be restricted in access to fisheries resources and markets in jurisdiction of member states.
Indonesia	TN/RL/GEN/ 189/Rev.1	Emphasized that subsidies contributing to IUU fishing should be strictly prohibited Along other major prohibited subsidies, the construction/improvement of ships and support of fishing management should be included. Argued the implementation of fisheries management plan should be accompanied by granting S&DT to developing countries to

		achieve desired results.
Norway	TN/RL/GEN/ 191/	Suggestion by RFMO in the negotiation seem to be progressing but including other list is a concern as it may add controversy. No subsidies of any kind shall be granted to IUU's ships or operators.
ACP (African, Caribbean and Pacific Countries)	TN/RL/GEN/ 192	Security Concerns in developing countries and promises to ban certain subsidies contributing to overcapacity, overfishing and elimination of IUU fishing and specify definitions once again until 2020 by SDG.
China	TN/RL/GEN/ 195	Agrees to ban of subsidies of IUU fishing but IUU fishing activities related to territorial, sovereignty and maritime jurisdiction disputes should be excluded from the scope of agreement.
	TN/RL/GEN/ 199	Subsidy cap system to be set at [X]% of the reference period and introduced adjustable subsidy cap as well as efforts to reduce. To encourage design of fisheries subsidy policies which conform to SDG, four types of categories (Green box) were presented that are not subjected to fisheries subsidy reduction arrangements.
Philippines	TN/RL/GEN/ 196/Suppl.1	No subsidies to fishing and related activities in disputed water Subsidies to overfishing and overcapacity and promises to prohibit IUU fishing should not be avoided due to concerns over disputed waters.
Argentina, Australia, the United States and Uruguay	TN/RL/GEN/ 197/Rev.2	Fexible approach is required. Divide countries into tiers 1, 2 and 3 by the proportion of fish caught in the world and prohibited subsides that exceeds limits. Prohibits IUU and fishing outside of state's jurisdiction despite subsidy cap system.
Argentina, Australia, Indonesia, Japan, New Zealand, the United States and Uruguay	TN/RL/GEN/ 201/Rev.1	Flag of Convenience (FoC) is ofen abused to avoid control and monitoring measures by RFMO, causing overfishing and IUU fishin. Prohibit subsidies on vessels that do not raise national flags of states.
Canada	TN/RL/GEN/ 198	Discipline for overfishing resource subsidies, including discipline for subsidies that leads to overcapacity should be stated and specified in Dispute Settlement Rules.

Source: Author's compilation

The draft agreement grants the coastal Member or a relevant RFMO to identify if fish stock is overfished under its jurisdiction or in areas and for specifies under its competence. Any subsidy that is aimed to promote the rebuilding of the fish stock to a biological sustainable level would be maintained without reduction or elimination.

Finally, the draft text lays out fisheries subsidies that contribute to overcapacity or overfishing, including price and income support, input and capital subsidies and cost recovery subsidies. A subsidy is not subject to the discipline if the subsidizing Member demonstrates its measure is implemented to maintain the fish stock at a biologically sustainable level. A critical point is how to "demonstrate" such a subsidy measure as beneficial or harmful to sustainability through a lens of overcapacity.

3 Empirical analysis on the effects of fisheries subsidies

As recognized by the draft agreement, the same types of subsidies can be used to maintain the fish stock or trigger overcapacity that may deteriorate the marine ecosystem. In fact, the draft illustrates "catch per unit effort" as one of the indicators for reference points. To earn these insights from actual data for most WTO Members and economic modeling, this study specifies an empirical equation and estimates the effect of different types of fisheries subsidies on fish catch.

3.1 Model specifications

As Equation (1) shows, the subsidy effect can be investigated by the relationship between a subsidy and fish catch. The total catch is shown as a function of fishing effort and fish stocks [9]:

$$Catch = F\,(fishing\ effort,\ fish\ stock) \tag{1}$$

where *Catch* is the total catch. The fishing effort can be proxied by various metric measures such as the number, gross tonnage, or the engine power of vessels. However, studies show that the number of vessels may not correspond to the total catch [11]. Thus, following FAO [12], this study adopts the engine power (kW) of vessels as an explanatory variable for catch. Since it is not feasible to obtain the amounts of stocks, they are not explicitly accounted for in this model.

Equation (2) is a modified version that incorporates the effects of different forms of fisheries subsidies on catch.

$$Catch = F\,(engine\ power,\ fisheries\ subsidies) \tag{2}$$

Equation (3) is the empirical model with a log-log specification.

$$\ln Catch_i = \beta_1 + \beta_2 \ln EngPow_i + \beta_3 \ln DirPay_i + \beta_4 \ln IndirPay_i + \beta_5 \ln Resmgt_i + \varepsilon_i \tag{3}$$

where *ln* indicates a natural log term, β_i refers to the coefficient to estimate, $Catch_i$ is the total catch for country i in metric tons, $EngPow_i$ is the engine power in kW, $DirPay_i$ and $IndirPay_i$ are aggregate amounts of direct and indirect subsidies in US dollars, respectively, $Resmgt_i$ is the support amount for fishery resource management in US dollars, and ε_i is an error term.

3.2 Data and subsidy classification

Based on the database compiled by Sumaila et al. [13], this study regrouped national fisheries subsides as direct, indirect and resource management payments in Table 2. Direct payments are supposed to have straightforward effects on catch while indirect payments are deemed to deliver their effects indirectly.

Table 2. Classification of national fisheries subsidies

No.	Classification	This study	Sumaila et al. (2019)
1	Vessel construction and renovation	Direct	Capacity-enhancing
2	Fisher assistance	Direct	Ambiguous
3	Fishing access	Direct	Capacity-enhancing
4	Fuel subsidies	Direct	Capacity-enhancing
5	Tax exemption	Direct	Capacity-enhancing
6	Fisheries development projects	Indirect	Capacity-enhancing
7	Fisheries management	Indirect	Beneficial
8	Fishery R&D	Indirect	Beneficial
9	Fishing port development	Indirect	Capacity-enhancing
10	Marking and storage infrastructure	Indirect	Capacity-enhancing
11	Rural fisher communities	Indirect	Ambiguous
12	Marine Protected Areas (MPAs)	Resource management	Beneficial
13	Vessel buyback	Resource management	Ambiguous

Table 3 provides descriptive statistics for each variable. It is the cross-sectional data on the basis of 2018. Data for *Catch* and *EngPow* are obtained from the FAO and Rousseau et al. [14].

Table 3. Descriptive statistics

	Catch (ton)	EngPow (KW)	Number of vessels	DirPay (USD)	IndirPay (USD)	Resmgt (USD)
Mean	616,076	966,034	17,381	1.08E+08	1.05E+08	26,498,920
Median	86,280	146,063	2,297	5,108,681	20,584,688	2,189,036
Max	13,148,442	14,559,387	453,328	4.84E+09	2.49E+09	8.92E+08
Min	254.74	781.42	3.0	0.0	147,008.7	0.0
Std.Dev	1,595,741	2,398,951	55,702.84	4.47E+08	3.26E+08	86,496,181
Obs	143	143	143	143	143	143

3.3 Estimated results and implications

Table 4 provides the estimated results of the regression model. The model is estimated by the Ordinary Least Squares (OLS) and the Two Stage Least Squares (2SLS) methods. The 2SLS method is relevant for addressing a potential problem of endogeneity [15, 16]. The number of vessels is used as an instrumental variable that highly correlates with the engine power but not with the dependent variable, fish catch.

Table 4. Estimated results

	Variables	(1) All countries	(2) Developed countries (WTO)	(3) Developing Countries (WTO)	(4) Developed Countries (HDI≧0.8)	(5) Developing Countries (HDI<0.8)
OLS	lnEngPow	0.291*** (0.099)	0.424 (0.406)	0.309*** (0.102)	-0.055 (0.216)	0.398*** (0.104)
	lnDirPay	0.038** (0.019)	0.243** (0.114)	0.029 (0.020)	0.125*** (0.020)	0.022 (0.018)
	lnIndirPay	0.762*** (0.095)	0.242 (0.291)	0.826*** (0.101)	0.890*** (0.161)	0.744*** (0.116)
	lnResmgt	0.005 (0.015)	0.127 (0.155)	-0.000396 (0.016)	0.090*** (0.031)	0.001 (0.017)
	Constant	-5.644*** (1.074)	-3.913** (1.725)	-6.750*** (1.176)	-6.436*** (1.469)	-6.242*** (1.410)
	R^2	0.743	0.785	0.735	0.800	0.748
	N	143	30	113	53	90
2SLS	lnEngPow	0.393*** (0.120)	-0.162 (0.380)	0.459*** (0.126)	-0.196 (0.216)	0.542*** (0.130)
	lnDirPay	0.033* (0.019)	0.210 (0.134)	0.021 (0.020)	0.134*** (0.022)	0.014 (0.019)
	lnIndirPay	0.701*** (0.108)	0.697** (0.302)	0.744*** (0.111)	0.990*** (0.162)	0.669*** (0.126)
	lnResmgt	0.001 (0.015)	0.211 (0.164)	-0.005 (0.016)	0.092*** (0.033)	-0.003 (0.017)
	Constant	-5.741*** (1.085)	-5.179* (2.525)	-7.026*** (1.218)	-6.535*** (1.491)	-6.569*** (1.451)
	R^2	0.740	0.752	0.729	0.797	0.742
	Hausman test	0.138	0.082*	0.024**	0.337	0.032**
	Weak instrument	223.558	19.824	212.123	64.782	185.793
	N	143	30	113	53	90

1) Korea and Singapore are regrouped as the developed countries under the WTO system, noting that they manifest to give up the developing country status.
2) The Human Development Index (HDI) over 0.8 is considered as a very high level.
3) *** p<0.01; ** p<0.05; and * p<0.1
4) ( ) is the standard errors

The 2SLS estimator for all 143 countries shows that direct and indirect payments have statistically significant effects on the total fish catch. Against prior expectations, the size of the estimated coefficient for indirect support with 0.701 turns out to be larger than that of direct support with 0.033. This means an 1% increase in direct support is likely to yield 0.7% increase in catch, exceeding 0.03% gain by direct payments. Even when countries are further divided into the developed and developing countries by the WTO and the HDI criteria, the models demonstrate statistically significant and substantial effects of indirect payments on catch. On the contrary, direct payments failed to prove their effectiveness in boosting catch except in model (4).

The estimated coefficients for resource management support are not statically significant in all models, which appear to approve that these beneficial or good fisheries subsidies are largely decoupled from fish production or catch. One can, however, imagine that their effects will be positive in the long run by enhancing stocks and contributing to a sustainable marine ecosystem.

4 Conclusions

The revised draft text tabled by the chair of negotiations for fisheries subsidies in the WTO has paved the way to strengthen enforceable discipline on harmful fisheries subsidies. While some of the draft text are remained to be resolved, including special and differential treatment provisions, the Members are edged closer to an improved and more balanced agreement for fisheries subsidies concerning IUU fishing, overfished stocks and overcapacity and overfishing.

The empirical results of this study shed light on the importance and effectiveness of policy reform in fisheries subsidies. The analysis shows both direct and indirect global subsidies contribute to an increase in catch or overcapacity. Although the catch response is inelastic, it could eventually deplete marine resources. More importantly, the fact that more than 80% of global fisheries subsidies went to the large-scaled industrial fishing sub-sector and the majority of the subsidies were in the form of capacity-enhancing subsidies is threatening livelihood security and sustainability for artisanal fishing in the developing countries [17].

References

1. WTO. Revised draft consolidated chair text. 30 June 2021, Geneva (2021)
2. D. Skerritt, U. Sumaila. Marine Pol. **128** (2021)
3. Y. Sakai, N. Yagi, U. Sumaila. Fisheries Sci., **85** (2019)
4. V. Le Brenne, L. Bisiaux, F. Le Manach. Marine Pol. **132** (2021)
5. A. Cisneros-Montemayor, Y. Ota, M. Bailey, C. Hicks, A. Khan, A. Rogers, et al. Marine Pol. **117**, (2020)
6. ICTSD. Fisheries subsidies rules at the WTO: A compilation of evidence and analysis. Geneva (2018)
7. I. Damme. Reflections on the WTO negotiations on prohibiting IUU fishing subsidies, IISD (2020)
8. M. Rosello. Croatian Int. Relations Rev., **22** (2016)
9. FAO. International Plan of Action to prevent, deter and eliminate illegal, unprotected and unregulated fishing, Rome (2001)
10. FAO. Report of the Technical Working Group on the Management of Fishing Capacity. FAO Fisheries Report No. 586 (1998)
11. D. Holland, E. Gudmundsson, J. Gates. Marine Pol. **23** (1999)
12. FAO. FAO Fisheries Report. No. 698 (2003)
13. U. Sumaila, N. Ebrahim, A. Schuhbauer, D. Skerritt, Y. Li, H. Kim, T. Mallory, V. Lam, D. Pauly. Marine Pol. **109** (2019)
14. Y. Rousseau, R. Watson, J. Blanchard, E. Fulton. PNAS **116** (2019)
15. D. Gordon. Environ Resource Econ, **61** (2015)
16. D. Gujarati. Basic Econometrics. McGraw-Hill (2009)
17. A. Schuhbauer, D. Skerritt, N. Ebrahim, F. Le Manach, U. Sumaila. Front. Mar. Sci. **29** (2020)

Development of Digital Learning Application to Support Smallholder Dairy Farmers in Indonesia

Yuni Resti[1*], *Sofiyanti* Indriasari[2], and *Bayu* Widodo[2]

[1]Livestock Management and Technology Study Program, College of Vocational Studies, IPB University, Indonesia
[2]Informatic Management Study Program, College of Vocational Studies, IPB University, Indonesia

Abstract. The Indonesian government has facilitated farmers through the role of agricultural extensionists (AE). One of the agricultural businesses that have a high interaction with AE is dairy farmers. AE have the responsibility to connect regularly with dairy farmers. However, since the Covid-19 pandemic, AE have limitations in coordinating with farmers directly. Therefore, a strategy is needed to overcome so farmers can continue to receive counselling and AE can still carry out their duties. In addition, several routine training activities carried out to improve the ability of farmers in practical management have also been delayed. The study aimed to develop an online application in collaboration with the triple helix concept through the interaction of multiple sectors fromgovernment, education, and business. The study used the collaboration between farmers and cooperatives as partners who cooperate with government agencies through agricultural extensionists and knowledge institutions to share information related to livestock business development. The application was built using the Extreeme Programming (XP) methodology. The research was held by conducting independent interviews with relevant stakeholders such as dairy farmers, cooperative boards, AE, the Ministry of Agriculture, and academicians. The study facilitated dairy farmers and AE to interact online with specific features and discussions about dairy farming. Moreover, educators such as livestock teachers, lecturers, and trainers can also take advantage of this digital application, especially the updated information related to livestock farming practices.

1 Introduction

1.1 Background

The Indonesian Government has made various improvements in strengthening the agricultural business, particularly for smallholder farmers. One of which is by facilitating farmers through the role of agricultural extensionists (AE) to increase farmers' productivity. AE play a significant role in the growth of livestock business [1] due to their responsibility to guide in solving problems to the community. Dairy farming is one of the agricultural

* Corresponding author: yuni.resti@apps.ipb.ac.id

businesses that have a high interaction with AE. The supervision given by AE is expected to help farmers in improving the performance of their farms. Generally, AE are divided into private and government extensionists. Private extensionists are mostly part of private companies as a supporter of the limited number of governments extensionists [2]. According to [3] some of the extension methods that have been implemented to the farmers are observing farmers groups; delivering information through resource persons/facilitators; demonstration by providing examples on a farm; field study by inviting farmers to visit other farms to exchange information with other farmers; field school extension (SL) by conducting short education to farmers such as training or short courses; and interview counselling by holding two-ways meetings between AE and farmer groups to discuss related issues.

Although many methods have been used in agricultural extension, there are still many obstacles that occurred during its implementation in the field. The study results of [4] concluded that distance is one of the main problems for AE in carrying out their duties. AE need to cover around 3 – 18 km to reach the farmer's house and 15 – 30 km to be at their farm location. The other thing is the facilities used during their fieldwork are sometimes not very attractive to farmers. Farming activities require lots of time which often overlap with the schedule of the visit. Therefore, AE must deliver interesting tools to support the farmers by applying various application-based technologies.

Many applications have been supported for extension activities through application-based technologies. According to [5], several extension applications have been developed by the Research and Development Agency of the Ministry of Agriculture for example Cyber Extension which is initiated by Agency for Agricultural Extension and Human Resources Development. It consists of extension materials and information that can be accessed by farmers in facilitating their agricultural business; Integrated Planting Calendar, an application that contains information about the right rice planting time, supported by the Indonesian Rice Consultation Service. It is a website-based application settled to assist farmers in rice cultivation designed by the International Rice Research Institute Indonesia and Indonesian Agency for Agricultural Research and Development; MyAgri is established by the Vegetable Research Institute and Wageningen University & Research through an android-based vegetable cultivation information system; Takesi is a technology founded by Indonesian Agency for Agricultural Research and Development, the Centre for Veterinary Research, to control cow health; and Coffee Experts introduced by Indonesia Research Institute for Industrial and Beverages Crops to provide information about coffe farming.

AE have the responsibility to interact regularly with farmers directly in their working areas which leads to high mobility. Since the implementation of working from home activities, several technical problems occurred. It is limiting the capacity of AE to visit farmers. A special strategy is needed to overcome this so that farmers can continue to receive counselling. Hence, the use of the internet can be one of the solutions so AE can carry out their duties which are in line with the current condition of Covid-19. One way that can be done is by promoting an online extension application specifically for dairy farmers. The development of this online extension application provides a prototype that facilitates farmers with a conversation menu where registered farmers can directly choose their AE. Farmers can discuss and improve their knowledge and practical skills through interactive learning. This application is expected to be a medium of learning and digital consultation for farmers during the Covid-19 pandemic.

According to [6], there is a positive relationship between the dynamics of the triple helix model in terms of the type and purpose of innovation, introducing new products and ecological innovations and their efforts to improve communication. The triple helix concept can be realized through interaction between farmers and cooperatives as partners who work with government - agricultural extensionists and educational institutions to share information related to the principle of "good farming practices" in assisting the development of farmers' businesses. Henceforth, the study objective was to develop an online application in collaboration with the triple helix concept that involves users from farmers, government, and educational institutions. The application was built to facilitate online consultation and education activities.

2. Method

The research was conducted by taking a case study of dairy farmers in Cisarua, Bogor District. Dairy farmers in Cisarua are organized as independent farmers and some are formed in farmers groups under the coordination of dairy cooperatives. The study focused on cooperative farmers. Farmers who are members of the cooperative were chosen with consideration because the average cooperative members are small farmers who need more support. The methodology used was extreeme programming (XP), one of the iterative and agile software development methodologies to deal with change [7]. The research was also held by conducting independent interviews with relevant stakeholders such as dairy farmers, cooperative boards, the Ministry of Agriculture, and academicians. The XP methodology aimed to produce software with higher quality and help in providing optimal solutions [8]. The methodology had some very good practices such as paired programming and unit tests [9].

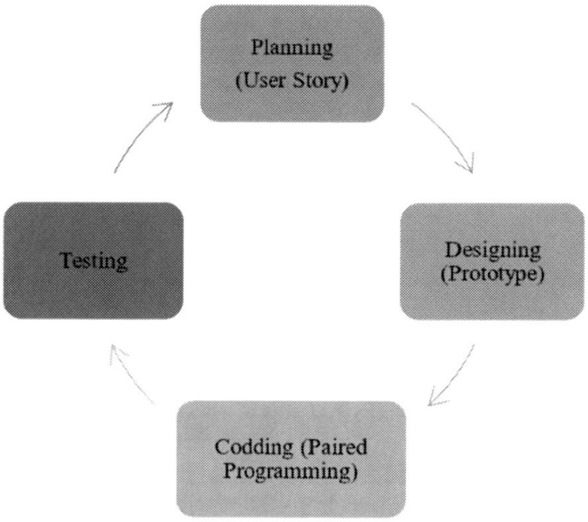

Fig. 1. Extreeme Programming Method

The stages passed during the study: Planning, Designing, Coding, and Testing. Figure 1 shows the stages of research carried out using the XP approach. The planning stage was done by collecting data to create a user story. User stories are sourced from users and are used to create time estimation for the planned release [9]. Techniques in collecting data for making user stories were interviews and brainstorming. Therefore, the application feature design was made as quickly as possible so it could be immediately used as a basis for implementation.

The design stage that had to be done was to make a prototype. A prototype is an image that resembles the original product to visually describe the flow of application functionality. It will make the implementation process easier. The next process was an implementation called coding. Coding is the implementation stage of creating program code based on the design results. The implementation reference used a prototype. Programming code was done in pairs (pair programming). Paired programming was done to speed up production and control product quality.

The final stage was testing. It was done to ensure quality. The test was done in two stages. The first test was a unit test conducted by the development team to ensure the program code that needed to be delivered had minimal bugs and errors. The second test was managed with the application together with the users. This second stage of testing was held based on a predetermined acceptance test scenario. The user acceptance test is the phase where the client validates the software to find out whether the solution provided is suitable or not for their use and meets their business needs. According to [10], the user acceptance test was a testing process by users which was intended to produce documents that were used as evidence that the application developed has been accepted by the users.

3. Results and Discussions

3.1 Planning

Planning is making procedures about application development activities based on the needs. Application requirements were collected from prospective application users which were then presented through user stories. Making user stories was done by collecting data during interviews and brainstorming techniques. Direct interviews with dairy farmers, ranging from experienced farmers who have started farming since the 1990s to young farmers. Along with this, data collection for user stories was done by brainstorming. The brainstorming aimed to dig deeper into the application needs of the experts. The brainstorming was attended by representatives of AE, educators from universities, and representatives from cooperative boards. User stories that had been formed were analysed to take values.

Visualization of the analysis results is needed to make it easier to understand the selected features, so they can be explained to users for approval. The visualization used to display the results of the requirement feature application analysis was a use case diagram. The use case describes the function provided by the system as units that exchange messages between units with actors expressed by using verbs [11]. The results of functional visualization of applications using use case diagrams can be seen in Figure 2. The main features of this application are related to education to farmers. The educational features included chats for farmers and AE; management of educational materials; displaying educational materials; managing livestock data; and accounts. In addition, prospective application users who register as users are verified by the admin through the activation and user management features. Table 1 shows the description of each feature.

There are 4 types of users in the application, namely, admins, extensionists, farmers, and educators. Admin is an administrator user who is only responsible for managing users' applications such as activating users who have just registered, refusing activation of unidentified accounts, deactivating users who are already in the application, and deleting registered users. AE are representatives of the government who carry out the task of providing consulting services to farmers. Farmers are users who are dairy farmers who must be verified by the cooperative. Educators are those who have the right to provide learning materials that can be posted as articles, documents, or videos. Educators consist of teachers from Animal Husbandry Vocational Schools, university lecturers, and researchers or animal husbandry experts who are experienced in the field of dairy farming. The four users have access to each feature in the application. Figure 2 shows the features that can be accessed by the user according to the specified access rights.

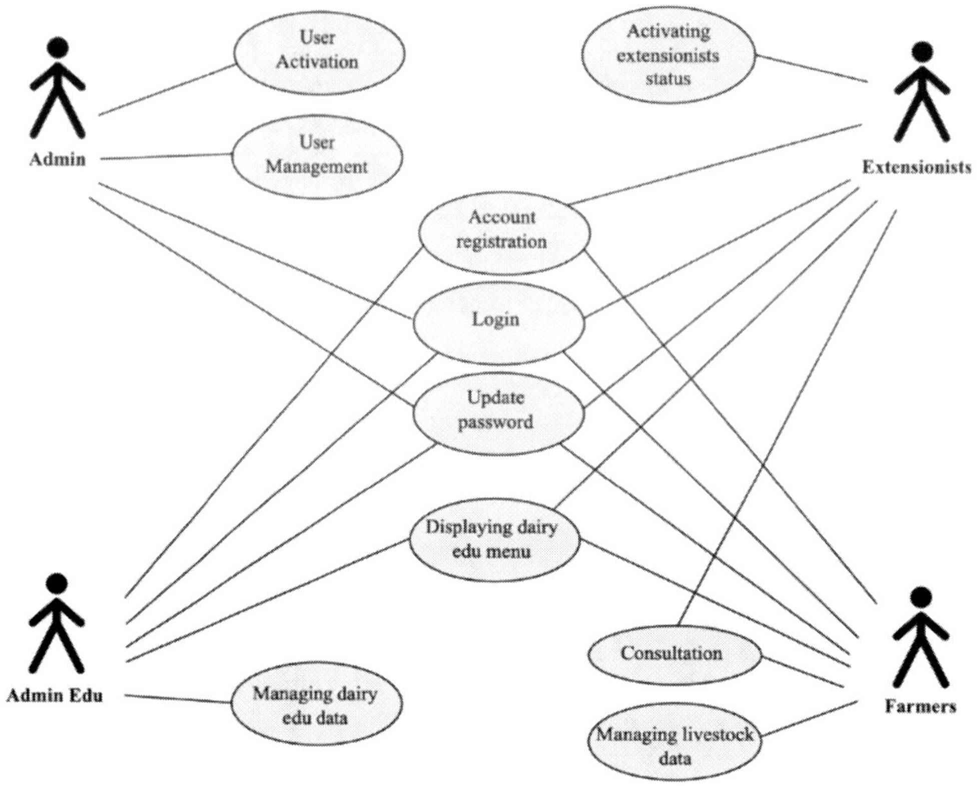

Fig. 2. Use case diagram

Table 1. Features description

No	Features name	Users	Descriptions
1	User activation	admin	The feature is designed to enable users who create new accounts to operate the app
2	User management	admin	It is established for user account management that has been registered. Admin can determine whether users will be activated or deactivated
3	Account registration	Users' candidate	Account registration is used for potential application users. Application users who have not registered can only view articles without being able to interact (as readers). If the user registers an account, they can choose which access rights are required. Whether as a farmer, PPL, or as an admin Edu. The account registration will be verified for activation by admin as the owner
4	Login	All registered and active users	Login is created for all users who have an active account and status. This feature is used to enter the application with access rights and interactivity according to the type of account.

No	Features name	Users	Descriptions
5	Update Password	All registered and active users	A feature is developed for users who have an active account and status to change the password to access the application.
6	Displaying dairy Edu menu	All users	A feature displays educational materials that have been uploaded to the application. Educational materials can be articles and attachments in which can be downloaded in pdf files, as well as learning video links
7	Managing dairy Edu data	Admin Edu	A feature shows all educational materials that can be uploaded by educators. In addition, educators can also improve existing content, and delete content that is no longer relevant
8	Activating extensionists status	Extensionists/ AE	The feature is used by AE for activation. Activation of the AE will affect their access to receive the consultation. If they choose to activate the status, the farmers can send a message. On the other hand, if AE choose to be inactive, the farmers cannot send any messages.
9	Consultation	Farmers	The feature can be a media for farmers to send messages (chat) to AE in consultation activities.
10	Managing livestock data	Farmers	The feature is planned to enter livestock data owned by each farmer. Data includes the number of livestock, age, and livestock categories. Farmers can also change the latest data if there is a change in the condition of their farm

3.2 Designing

Application design is an activity to visualize the flow of application functional requirements in more detail and more technically. The functional requirements of the application summarized in the use case diagram are detailed in its flow and visualized more technically. The flow of each application function is visualized by creating a prototype. The prototype is made by considering the function flow and then describing the layout of the application interface according to the needs and business processes described in the user story. The prototype provides an easier description for implementation reference. An example of a prototype showing the application page design can be seen in Figure 3.

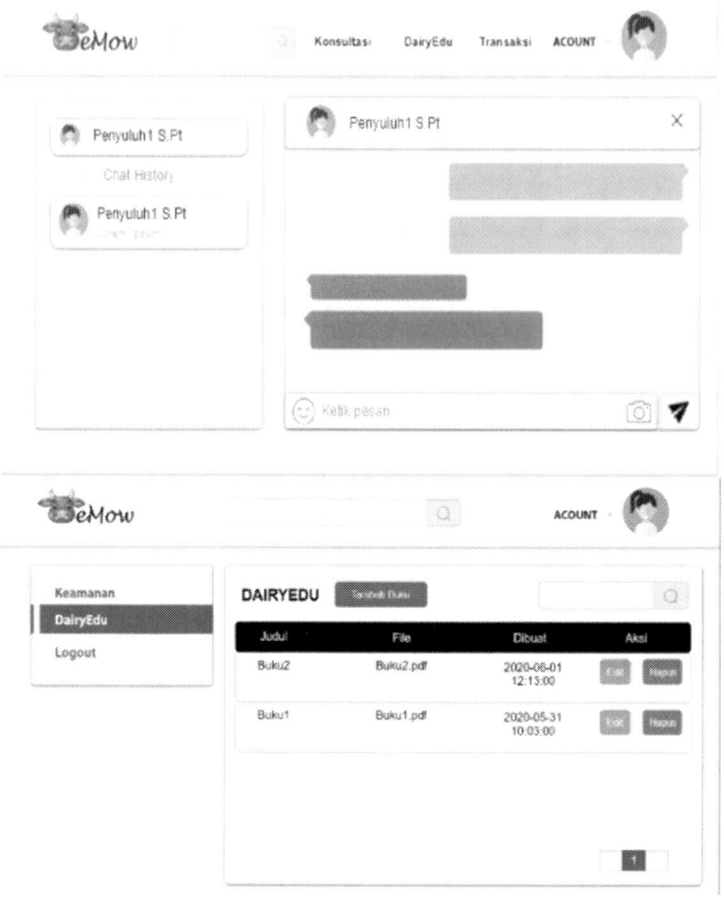

Fig. 3. Application prototype

3.3 Coding

The implementation stage is to create program code to form the application. This application development process uses a paired programming approach which is the hallmark of the Extreeme Programming methodology. Pair programming is one of the software developments practices in which two programmers work together in partnership using one computer, cooperating on the same problem. Paired programming shows a positive impact on the quality of the resulting product on the attitudes and feelings of the team [12]. Additionally, paired programming has a positive influence on the quality and duration of application completion [13]. The application was built in a web-based application. Implementation was referred to the design that has been made previously. The results of the application that have been established then were hosted on the IPB Vocational School server. The application could be accessed at the link http://klik.sv.ipb.ac.id/emoww. Screenshots of the application can be seen in Figure 4 and Figure 5.

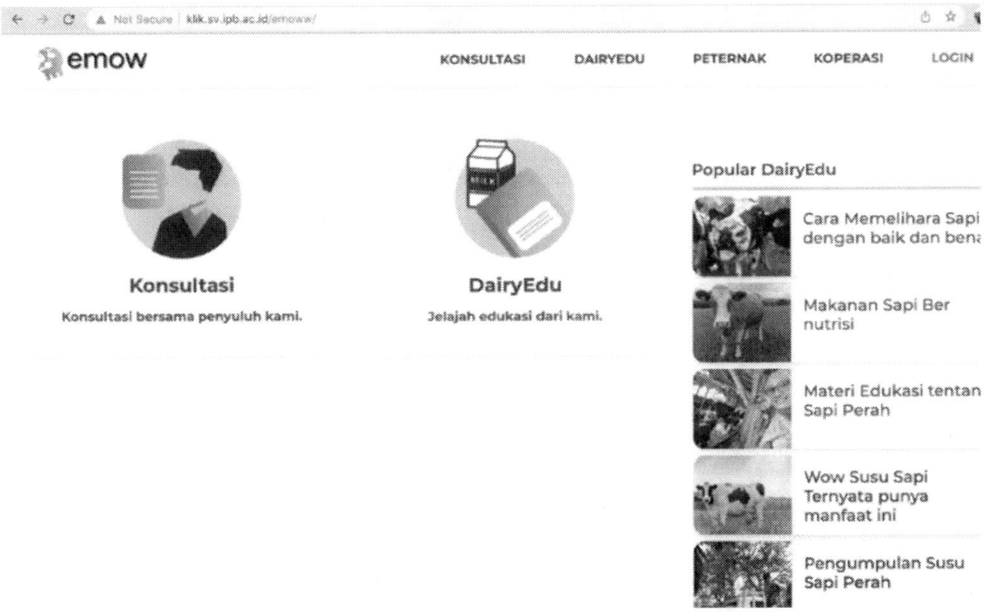

Fig. 4. Main page

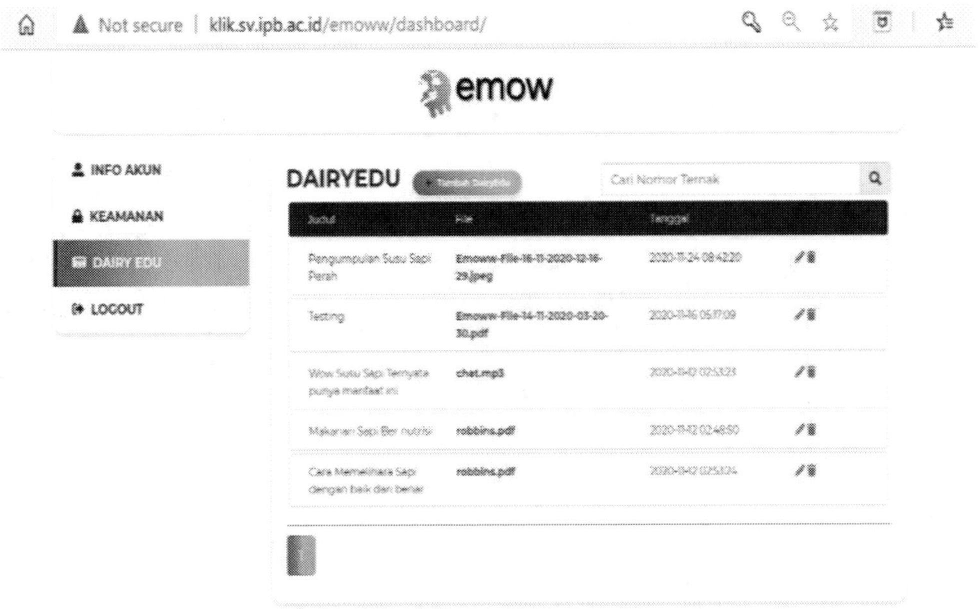

Fig. 5. Education data page

3.4 Testing

Testing was done to ensure the running process of application functions. The black box testing technique was used to test this application. Testing was conducted through the application system interface using a black box. The points were to check the functions in the application and evaluate the suitability of the application to user needs [14]. Examiners, in black box testing activities only need to know about the required inputs and outputs, or in other words, the examiners do not need to understand the internal system [15]. Test scenarios were designed based on experiments with various inputs. If the input was correct, then the application would display the correct response or information. If the input test case was wrong, then it would show how did the system respond to the wrong input. Table 2 presents the results of the application testing.

Table 2. Application features testing

No	Features Name	Testing Scenario	Testing results description
1	User activation	1. Admin displays a list of new users who have registered 2. Admin selects the accept button which means the new account is approved to be activated 3. Admin selects the reject button which means the new account is not approved to be activated and registration is rejected	1. The screen displays a list of registered users 2. A new user has an active account. A user can login and access features according to their access rights 3. The user gets a registration rejected message and cannot login to the application
2	User Management	1. Admin shows a list of users who already have an application account 2. Admin selects the activate button on users whose status is inactive 3. Admin selects the inactive button on users whose status is active	1. The screen displays a list of users who already have an account to access the application 2. Users whose accounts are inactive become active can enter the application and can access application features in accordance with the access rights granted 3. Users whose status is active become inactive and cannot access application features for interactivity
3	Account registration	1. The user registers an account by entering the required identity: full name, email, selecting access rights (farmer, extension worker, or admin Edu), username, and password in accordance with the	1. Account data is stored and will appear on the admin page for the new user list, to be accepted or rejected. 2. The new user account data is not saved, and an error message appears informing that the field is not complete or other

No	Features Name	Testing Scenario	Testing results description
		provisions. All parameters are required. 2. The user registers an account by filling out a wrong form or do not filling in the parameters	messages such as the wrong email format.
4	Login	1. Users who have an account log in by entering the correct username and password 2. Users who have an account log in by entering the wrong username and password or leaving it blank 3. Users who do not have an account try to login with any account 4. Users who successfully login then log out and try to return to the page link when logging in	1. The user has successfully entered the application and can access the application features in accordance with the access rights granted 2. A message appears on the screen that the username and or password fields and or is still empty 3. A message appears on the screen that the username and or password fields and or is still empty 4. The user cannot return to the account page after logging out even though pressing the back button on the browser. Session on the application works properly
5	Update Password	1. The user changes the password by entering the old password correctly and entering the new password and re-confirming the new password 2. The user changes the password by entering the wrong old password and or entering a new password but when re-confirming the new password entered is not the same	1. Password has been changed successfully and the user can login with her/his new password 2. Password failed to change; a message appears that the entry is not correct
6	Displaying dairy Edu menu	1. The educational material appears on the homepage of the application and when clicked, the details of the selected educational material appeared 2. Educational materials in files format can be downloaded by pressing the	1. The home page runs appropriately. It shows the educational materials according to the latest uploaded material 2. The feature for downloading materials works well

No	Features Name	Testing Scenario	Testing results description
		download button on the application 3. Video links on educational materials can be clicked and display videos of learning materials	3. All learning video links can operate properly and display videos that are relevant to the material
7	Managing dairy Edu data	1. Entering dairy Edu data by filling in the title, uploading images, materials, and uploading files 2. Updating dairy Edu data by changing one of the elements: the title, uploading images, materials, or uploading files 3. Deleting dairy Edu data by selecting the data to be deleted and clicking delete 4. Selecting dairy edu menu to display dairy Edu data	1. The new dairy Edu data has been successfully saved and displayed on the education page 2. Dairy Edu data has been successfully updated and displayed 3. The selected dairy Edu data has been deleted successfully 4. The application shows the dairy Edu page and its educational materials
8	Activating extensionists status	1. The AE select an active status on his account page 2. The AE select an inactive status on her/his account page	1. The farmers can send a message for consultation with the AE. 2. The farmer cannot send messages to the AE because they are inactive
9	Consultation	1. The farmer selects an active extension agent and sends a message 2. Farmers choose inactive extension workers 3. Extension workers reply to incoming messages from farmers	1. Message received by selected extensionists 2. The extension button is inactive and cannot send messages to inactive AE. 3. Farmers receive a message from the AE
10	Managing livestock data	1. Farmers enter new livestock data correctly 2. Livestock data is updated by replacing certain selected livestock data 3. Selecting the livestock data to be deleted and press the delete button 4. Selecting the livestock menu to display livestock data	1. New livestock data has been successfully saved and displayed 2. The livestock data has been successfully updated and displayed 3. The selected livestock data has been successfully deleted 4. The Livestock menu displays the livestock data

Based on the test results described in Table 2, it is shown that the application runs very well. Besides testing using the black box method, brainstorming sessions with users who have tried the application was also conducted. The brainstorming results indicated that there were features that were not yet needed, really needed and had just been revealed. After trying the application and having a discussion with farmers, it led to a suggestion to add forum features in the application. Users said that a forum can be a useful tool to share knowledge to solve problems faced by other farmers. This is related to a study conducted by [16] in the development of mobile learning applications for farmers in Tanzania which allows farmers to communicate and share the latest technology and innovations. Currently, many digital applications have been developed to facilitate farmers with extension services, agricultural information, and financial services. Several studies show a positive impact on the use of this technology that can improve livelihoods, especially in the global south. For example, m-Agri Service in Africa. Several educational and extension applications developed through m-Agri Service for farmers include iCow, Vet Africa, Agro-Hub, Modisar, uLima, and Agtag. The existence of this service was able to increase access to information for farmers. Particularly, sharing of information and knowledge can support the practical skills of farmers, which can increase their productivity [16].

4. Conclusion

AE have a responsibility to continue interaction regularly with farmers which cause high mobility. Since the outbreak of the Covid-19 and the enactment of working from home activities, it has limited the capacity of AE to provide counselling to farmers. The use of the internet is one solution to support AE in their duties. The application of this study can facilitate AE and farmers in interacting online with specific features and discussions. Moreover, this application can also maximize the role of educators by providing topics related to farming practices.

The application produced in this study has been tested by the technical side of the development team as well as tested by users to confirm the suitability of the features with their needs. The results showed that the application run properly. Besides, testing it, using the black box method, brainstorming sessions with users were also conducted. Direct consultation with AE was indeed very helpful. However, this application still needs further development. one of them is by adding a forum feature in counselling activities to share information and strengthen the relationship between farmers.

References

1. J.A. M. Lenzun, A. K. Rintjap, M. Najoan, H. O. Gijoh, *Hubungan Karakteristik Penyuluh Terhadap Keberhasilan Pengembangan Usaha Peternakan Sapi Potong Po Di Di Desa Tondegesan, Tondegesan Satu Dan Tondegesan Dua*, Zootec, Vol. 39, No. 1, pp. 1–12 (2019)
2. K. Sulandjari, G. Kurnia, T. J. Sugarda, H. Hapsari, *Paradigma Penyuluhan Pertanian Perusahaan Swasta di Kabupaten Bandung Barat,* Jurnal Penyuluhan, Vol. 16, No. 1, pp. 24–36 (2020)
3. A. Imran, Muhanniah, and B. R. W. Giono, *Metode Penyuluhan Pertanian dalam Meningkatkan Pengetahuan Dan Keterampilan Petani (Studi Kasus di Kecamatan Maros Baru Kabupaten Maros)*, AGRISEP, Vol. 18, No. 2, pp. 289–304 (2019)
4. H. F. Allen, M. M. Batubara, H. Iswarini, *Kendala Penyuluh dalam Melaksanakan Aktivitas Penyuluhan pada Usahatani Kopi di Kecamatan Dempo Utara Kota Pagar Alam,* SOCIETA, Vol. IV, No. 2, pp. 105–110 (2015)

5. Susilowati, *Transformasi Penyuluh Pertanian pada Era Industri 4.0., https://tabloidsinartani.com/detail/indeks/agri-penyuluhan/10143-Transformasi-Penyuluh-Pertanian-pada-Era-Industri-40* (2019)
6. M. M. Natário, J. P. A. Couto, C. F. R. Almeida, *The triple helix model and dynamics of innovation: a case study,* Journal of Knowledge-based Innovation in China, Vol. 4, No. 1, pp. 36–54 (2012)
7. D. Wells, *Extreme Programming: A gentle introduction*, www.extremeprogramming.org (2013)
8. K. S. Yadav, M. A. Yasvi, Shubhika, *Review On Extreme Programming-XP*, International Journal of Advances in Electronics and Computer Science, Vol 6 (2019)
9. A. Dalalah, *Extreme Programming: Strengths and Weaknesses,* Computer Technology and Application, Vol. 5, No. 1, pp 15-20 (2014)
10. M. A. Nurdin, I. Hermawan, *Analisis dan Pengembangan Aplikasi Inhouse Klinik Perusahaan Menggunakan Framework Codeigniter, Studi Kasus PT Reckitt Benckiser Indonesia, Jurnal Informatika Terpadu*, Vol. 3, No. 1, pp. 1–7 (2017)
11. Muhammad, F. N. W. Dame, *Perancangan Sistem Pengambilan Keputusan Karyawan Terbaik Menggunakan Metode Simple Additive Weighting Pada Pt. Riau Andalan Pulp and Paper Estate Logas,* Jurn Pengujian Black Box Testing pada Aplikasi Action & Strategy Berbasis Android dengan Teknologi Phonegap," *STRING* al Intra-Tech, Vol. 2, No. 2 (2018)
12. M. Akour, K. Al Radaideh, I. Alazzam, I. M. Alsmadi, *Effective pair programming practice: toward improving student learning in software engineering class,* International Journal of Teaching and Case Studies. Vol. 4, No. 4, p. 336 - 345 (2013)
13. J. E. Hannay, T. Dybå, E. Arisholm, and D. I. K. Sjøberg, *The effectiveness of pair programming: A meta-analysis,* Information and Software Technology, Vol. 51, No. 7, pp. 1110–1122 (2009)
14. W. N. Cholifah, Yulianingsih, S. M. Sagita, *Pengujian Black Box Testing pada Aplikasi Action & Strategy Berbasis Android dengan Teknologi Phonegap*, Jurnal String, Vol. 3, No. 2, (2018)
15. M. E. Khan, *Different approaches to white box testing technique for finding errors,* International Journal of Software and Its Applications. Vol. 5, No. 3, pp. 1–14 (2011)
16. C. Sanga, M. Mlozi, R. Haug, S. Tumbo. *Mobile learning bridging the gap in agricultural extension service delivery: Experiences from Sokoine University of Agriculture, Tanzania,* International Journal of Education and Development using Information and Communication Technology (IJEDICT), Vol. 12, Issue 3, pp. 108-127 (2016)
17. E. M. Emeana, L. Trenchard, K. Dehnen-Schmutz, *The Revolution of Mobile Phone-Enabled Services for Agricultural Development (m-Agri Services) in Africa: The Challenges for Sustainability*, MDPI, Sustainability (2020)

Application of The Multiple Intelligent Level Determination for Interest and Talent Development

Medhanita Dewi Renanti[1*], *Anggia* Chrisanti Darmawan[2]

[1]Lecturer, Informatic Management, College of Vocational Studies-IPB University, Bogor Indonesia
[2]Psychologist, Sekar Psychological Services, Bogor Indonesia

Abstract. This research focused on making applications to determine the dominant multiple intelligences. Everyone has eight different intelligences that reflect different ways of interacting with the world. The dominant multiple intelligence needs to be known so that the child's growth and development is in accordance with his or her talents and interests. Therefore, it is necessary to make an application with tools that have been tested by experts, which can determine the level of intelligence so that the potential of children can be developed. This application was created using the extreme Programming software development method. The reasons for the choice of this method were because the resources to create the application were readily available, the programming was conducted by a small team, and the application development time was short. This application used the PHP and MySQL by displaying a list of 24 questions that had to be filled out by the users. The application testing method used black box testing. The accuracy of the application was 100% from 150 respondents who used this application. The output of this application is a sequence of multiple intelligences owned by users ranging from linguistic, logical-mathematical, musical, spatial, kinaesthetic, intrapersonal, interpersonal abilities, and naturalist.

1 Introduction

The theory of multiple intelligences was delivered by Howard Gardner, professor and psychology of Harvard University for the first time, in his book which has title *frames of mine: the theory of multiple intelligences*. According to the book, Gardner said that every human has 8 different intelligences which is showed as the way they interact to the world that is linguistic intelligences, logical-mathematical, musical, spatial, kinaesthetic, intrapersonal, interpersonal abilities, and naturalist. This dominant multiple intelligences is needed to be recognize in order to make the child development suitable with child's talent interest.

Howard Gardner's theory of multiple intelligences has been hailed by educators for decades and applied in hundreds of schools worldwide. In Frames of Mind, Gardner

* Corresponding author: medhanita@apps.ipb.ac.id

challenges the widely held notion that intelligence is a single general capacity possessed by every individual to a greater or lesser extent. Amassing a wealth of evidence, Gardner posits the existence of eight different intelligences, each as important as the next, that comprise a unique cognitive profile for each person [1]. Gardner has discovered that human abilities are not perfectly correlated. Indeed, Gardner proposes that humans have talents in music, dance, and personal relations that are not measured by tests of vocabulary and block design. The psychological world is shaken by the revelation that the full range of human intelligence is not revealed by a single IQ score [2].

The research of the representation of multiple intelligences in the science textbook and the extent of awareness of science teachers at the intermediate stage of this theory had been done by Alsalhi in 2020. The criteria examined include eight of the intelligences from Howard Gardner's multiple intelligences theory. The study used the analytical descriptive method. The analysis found that the intelligences represented in the Science textbook were mostly verbal/linguistic, visual/Spatial, and logical/mathematical intelligence, with a combined percentage of 73.3 %. The other 26.7 % was distributed between the other intelligence types: interpersonal intelligence, intrapersonal intelligence, bodily/kinaesthetic intelligence, natural intelligence, and musical intelligence. The findings referred to moderate Awareness among science teachers about including the multiple intelligences in science textbooks, and there is no significant difference between their awareness according to gender variable. Vary depending on educational level (in favor of those teachers with Bachelor's level education), and depending on experience (in favor of teachers with more than 10 years' experience) [3].

Multiple Intelligences theory was one of the first formulations about intelligence to be based on neuroscience evidence. This research reviewed 318 neuroscience reports and extracted neural regions cited for the core cognitive components of each intelligence. This neural data was organized into four levels: primary regions, sub-regions, particular structures, and multi-regions per intelligence. Using descriptive statistics, the patterns of neural citations were compared and contrasted. This research conveys that in was determined there is robust evidence that each intelligence possesses neural coherence that is clear, distinct and aligned with accepted cognitive–neural correlates. Furthermore, these neural patterns are consistent with Gardner's 1983 hypothesis that general intelligence is most closely associated with the linguistic and logical-mathematical intelligences. Implications for using multiple intelligences theory as a bridge between cognitive neuroscience and instruction [4]. Multiple intelligences theory explores the idea that people learn and process information in different ways based on a combination of different areas of intelligence. In the areas of communication and education, this becomes important as a means to understand the listening process and creating class environments that offer instruction in more than just the typical linguistic or logical-mathematical areas of intelligence [5].

Actually, every intelligent person, human has intelligence dominance between the multiple intelligences. There is logical-mathematical dominant intelligence, and also there are linguistic or others. If we are able to approach our students based on the dominant intelligence potential it is going to make the talent interest accordance with the intelligence characteristic. This is in line with McCroskey and Chesebro statement that they delivered. "People can be intelligent in different ways". This is the premise of multiple intelligences theory as presented by Gardner. Because individuals have different intelligences, the natural extension of multiple intelligence theory is that people listen and learn in different ways, depending on each person's primary intelligence. As each person has his own unique blend of intelligences, each student then has his own unique way of learning [6].

"Design convergence talent" (DCT) refers to a person who is talented in creative-convergence design and possesses a degree in design or a related field. By analysing typical and distinctive characteristics of DCTs, this study explored the depth of intelligence-based creativity. Creativity was assessed using the something About Myself (SAM) test, and multiple intelligences were rated using the Latent Ability Detection (LAD) test, on the basis of the multiple intelligence theory of Gardner. The results of this study offer a quantifiable basis for a DCT model as a means of creative-talent activation. At the industry level, the model may be useful for scouting and developing DCTs in order to establish a method of sustainable creation and innovative values [7]. Implementation effectiveness of multiple intelligence is shown in Applying multiple-intelligence approach to education and analysing its impact on cognitive development of pre-school children. This study aimed to find out multiple intelligence approach for education's impact on cognitive maturity of pre-school children. Applying MI approach in the classroom as a educational method results increasing in all five domains of pre-school children's cognitive development with 99% significance [8].

The using of this application is to decide the dominant multiple intelligences which is needed in order to make everything effective and efficient. This thing is going to help the child's talent interest based on their intelligences When multiple intelligences theory is practiced with the benefits of technological tools, the implementation of multiple intelligences theory in educational communication across a wide variety of intelligences and learning levels emerges as the best way to communicate with and educate the most people in any educational environment [9]. This research is focused on application creation that has goal to decide dominant multiple intelligences quickly and easily so that the child's potential can develop.

2 Method

Method of software development that is used for making Application of the Multiple Intelligent Level Determination for Interest and Talent Development is extreme programming (XP). This method was chosen because the source for making the system is available, time to develop short application, and programming is done by small team. The stage of XP method is shown in Figure 1 [10].

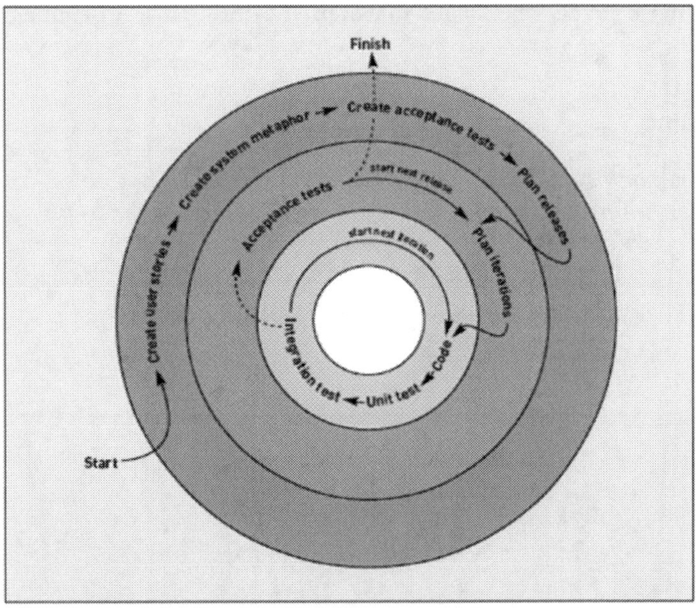

Fig. 1. XP development approach activities.

Activity level, system (outer circle) is system activity that happened once during the project development. The system is sent to the user in some stages which is known as release. Every release is functional system that is complete and it was from complete system requirements. A release is developed and tested in less than weeks or months. In this stage include user analysis and the making of use case in oriented analysis object to plan the release. Furthermore, the developer makes diagram class to present the object based on the need of the user. Next, developer and user make one set of acceptance test to every need of user analysis. If the acceptance test is considered complete, then the next step is making the development planning for release.

Activity level, release (middle part). Middle part activity is cycle multiple times once in every release. Release is divided to some iteration. Every iteration, the developer makes a code and do some functional specific test. Iteration is encoded and tested in days or weeks. First release is based on the first need of user and the next release is added with the need of other system specification. The first release activity was planning the series iteration. Every iteration focuses on small functional system which allow the developer to do code and test in few days which is integrated then. When all the release iteration has done, then the acceptance tests (release part level) begin. If the release fails, so the team back to the iteration level to fix it. If the release successes/ pass the acceptance test, so it is going to be sent to the final user and work starts on the nest release. If acceptance test from the final release is done, so the project development is done.

The third stage is iteration activity level (inner part). There are some iteration in every release, so that the inner iteration ring happens in some cycle. After iteration planning is done, the work starts on first iteration activity level. Code unit is divided among programming team and every team develop and test their own codes. XP recommend the first test approach to coding. The code test is done before the system is coded. When code module is done in unit test, then the next step is combining it into bigger unit to integration and test processes. When iteration pass the integration test, the work starts on the next release. When all the iterations are done, so acceptance test (release part level) is begun.

3 Results

Based on those levels, the stages that must be through are planning, design coding, and testing.

3.1 Planning

The planning test starts from contact the psychologist. After doing the discussion, they get the question information tools which has to be filled by the user to decide the artificial intelligence dominance for 24 questions that represent 8 intelligences: linguistic intelligences, logical-mathematical, musical, spatial, kinaesthetic, intrapersonal, interpersonal abilities, and naturalist. Question list is shown in Table 1.

Table 1. Question list to decide the dominance of multiple intelligences.

No	Question list
1.	I have sensibility in understanding someone's feeling and character.
2.	I am happy if I am involved in some helping job or proffesion for someone, such a teaching, counseling, and being a therapist.
3.	I am happy with harmony situation and tend to avoid some conflicts.
4.	I can influence some people to believe in what I believe in and what I am doing.
5.	I am confident in the middle of the crowd.
6.	I have an ideas, principal and being brave to express or deliver it to everyone.
7.	I have skill in using scissors, hammer, brush, pliers, needle, and others.
8.	I am expert in all physical activities such as sport, dance, or other games.
9.	I prefer study with direct practice rather than studying the theory only.
10.	I am confident to express myself and feeling with some writing or words.
11.	I am happy to read wherever and whenever it is.
12.	I am happy and I want to be fluent in many languages (foreign languages or local / cultural Indonesia language)
13.	My friends often ask my help to solve their counting problem and math.
14.	When facing some problems, I tend to use my logic and thinking analysis to solve it.
15.	I believe that all the things / incidents can be explained by scientifically and rationally.
16.	I like all music genre.
17.	My feeling is easy to get influenced when listen to some songs, music, or some holy verse (become happier / excited / sad)
18.	I am really easy to memorize song lyric and adjust the tone, rhythm, and song's intonation.
19.	I love to do some activities that related with livestock or pets or gardening or farming or fishery.
20.	I am interested in nature, plant and animal things.
21.	I can recognize and classify kinds of plant and animal
22.	I am able to explain what I imagine completely and in detail.
23.	I can read maps (google maps) well and correctly.
24.	I like drawing, painting, and sketch activity

3.2 Design

After through the planning stage, the next stage is designing. Designing here is include input design, process, output, and interface. The database design is shown in Figure 2.

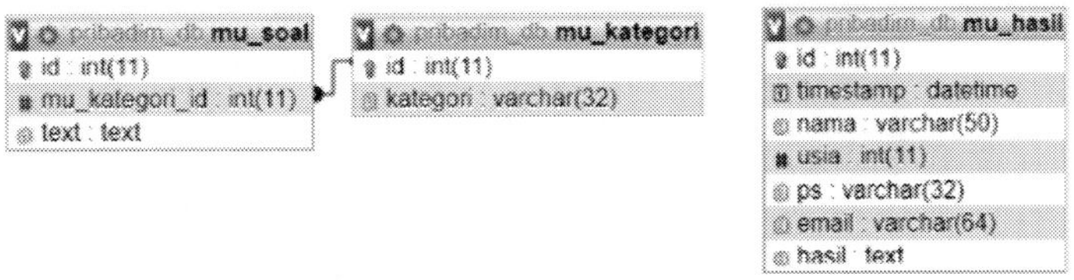

Fig. 2. Designing table in application database.

3.3 Coding

The next stage is coding. This application uses programming language PHP. PHP is programming language that is suitable to use for developing dynamic web. This is because PHP is open source, easy to install, and it can be used for many operational system, include Microsoft Windows and UNIX. [11]. The Program snippets that are applied in this application is shown as:

Result calculation logic

```
public function submit2()
  { $input = $this->input->post('jawaban');
    $data['nama']  = $this->input->post('nama');
    $data['usia']  = $this->input->post('usia');
    $data['ps']    = $this->input->post('ps');
    $data['mail']  = $this->input->post('email');

    $hasil = array(0,0,0,0,0,0,0,0);

    foreach ($input as $key => $value)
    { $hasil[$value['name']-1] += $value['value'];
    }

    foreach ($hasil as $key => $value) {
      $hasil[$key] = number_format((float)($value/12*100), 2, '.',
      '');
    }

    $keterangan = $this->M_post->saveHasil2($hasil, $data);
    usort($keterangan, array($this,'cmp'));
    if(!empty($data['mail'])) $this->sendMail2($keterangan,
    $data);

    echo json_encode($keterangan);
  }
function submitJawaban()
  { $('#part'+current).hide();
    $('#part'+current+'-line').addClass('bg-primary');
    $('#finish-bullet').addClass('bg-primary');

    $('#btn-prev').hide();
    $('#btn-fin').hide();
    var values = $("form").serializeArray();
    var nama  = "<?php echo $identitas['nama']; ?>";
    var usia  = "<?php echo $identitas['usia']; ?>";
    var ps    = "<?php echo
       (empty($identitas['ps'])?"":$identitas['ps']); ?>";
    var email = "<?php echo $identitas['email']; ?>";

    $.post("<?php echo base_url('survey/submit2') ?>", {
        jawaban:values,nama:nama,usia:usia,ps:ps,email:email },
        function( data ) {
        data = JSON.parse(data);

        var total = 0;
        $.each( data, function( key, value ) {
            if(key<8){
              $('#score-'+(key+1)).text(value.persen+'%'+'
```

```
        '+value.kategori);
       total += value;
     }
});
$('#total').text(total);

var k = data[0];
var m = data[1];
var s = data[2];
var p = data[3];

$('#kesimpulan').text(data[4]);

if(k>=13 && k<=20) $('#K1').removeClass().addClass('bg-
  primary text-white');
else if(k>=10) $('#K2').removeClass().addClass('bg-primary
  text-white');
else if(k>=7) $('#K3').removeClass().addClass('bg-primary
  text-white');
else if(k>=4) $('#K4').removeClass().addClass('bg-primary
  text-white');
else if(k>=2) $('#K5').removeClass().addClass('bg-primary
 text-white');
else if(k>=0) $('#K6').removeClass().addClass('bg-primary
 text-white');

if(m>=7 & m<=15) $('#M1').removeClass().addClass('bg-primary
 text-white');
else if(m>=6) $('#M2').removeClass().addClass('bg-primary
 text-white');
else if(m>=4) $('#M3').removeClass().addClass('bg-primary
 text-white');
else if(m>=3) $('#M4').removeClass().addClass('bg-primary
 text-white');
else if(m>=2) $('#M5').removeClass().addClass('bg-primary
 text-white');
else if(m>=0) $('#M6').removeClass().addClass('bg-primary
 text-white');

if(s>=7 & s<=17) $('#S1').removeClass().addClass('bg-primary
 text-white');
else if(s>=5) $('#S2').removeClass().addClass('bg-primary
 text-white');
else if(s>=4) $('#S3').removeClass().addClass('bg-primary
 text-white');
else if(s>=2) $('#S4').removeClass().addClass('bg-primary
 text-white');
else if(s>=1) $('#S5').removeClass().addClass('bg-primary
 text-white');
else if(s>=0) $('#S6').removeClass().addClass('bg-primary
 text-white');

if(p>=10 & p<=19) $('#P1').removeClass().addClass('bg-
 primary text-white');
else if(p>=7) $('#P2').removeClass().addClass('bg-primary
 text-white');
else if(p>=5) $('#P3').removeClass().addClass('bg-primary
 text-white');
```

```
        else if(p>=3) $('#P4').removeClass().addClass('bg-primary
         text-white');
        else if(p>=1) $('#P5').removeClass().addClass('bg-primary
         text-white');
        else if(p>=0) $('#P6').removeClass().addClass('bg-primary
         text-white');

    });

    $('#part-fin').show();
  }
```

Function send result

```
 public function sendMail2($hasil, $data)
   { $this->email->clear(TRUE);
     $this->email->from($this->email->smtp_user,"Pribadimu.id");
     $this->email->to($data['mail']);
     $this->email->subject('Multiple Intelligence Analysis');
     $data['identitas'] = $data;
     $data['hasil']     = $hasil;
     $html = $this->load->view('/mailer/majemuk', $data, true);
     $this->email->message($html);
     $this->email->send();
   }
```

The interface homepage Application of the Multiple Intelligent Level Determination for Interest and Talent Development is shown in Figure 3. A simple Assessment is shown in Figure 4 which shows 24 questions that are divided into 3 pages. This question list must be filled by the user. The output page for the dominance of multiple intelligences is shown in Figure 5.

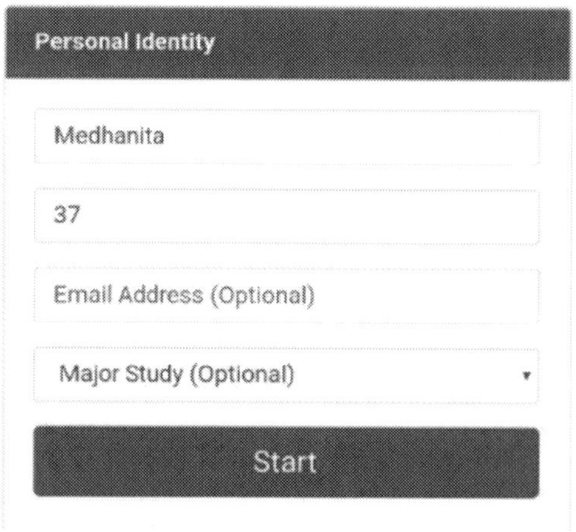

Fig. 3. The interface homepage application.

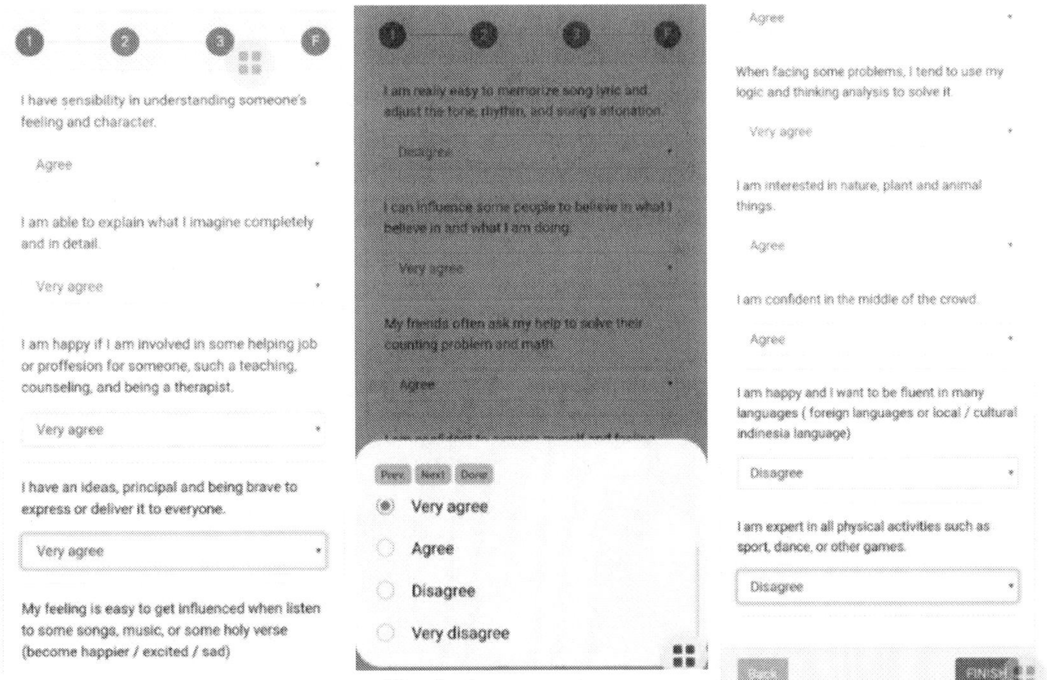

Fig. 4. Assessment page.

Each question contains four answer, namely very agree, agree, disagree, and very disagree. Output of this application is percentage score from each multiple intelligences. The display is sorted as descending like in Figure 5.

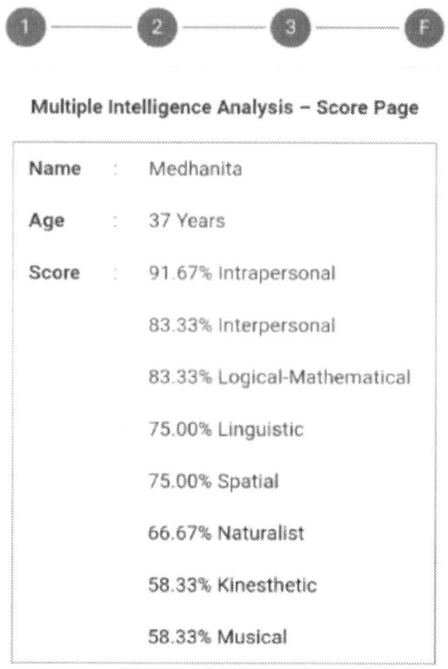

Fig. 5. The output page for the dominance of multiple intelligences.

3.4 Testing

The test stage is done by the application maker, psychologist, and user. The test is done by black box testing by making sure that the input and output which is resulted are suitable. Black box testing assesses a system solely from the outside, without the operator or tester knowing what is happening within the system to generate responses to test actions. A black box refers to a system whose behavior has to be observed entirely by inputs and outputs [12]. The test that is done include login feature, assessment filling process, and percentages multiple intelligences process, system accuracy is 100%, the result of classification uses paper filling suits with the application result for 150 respondents who are use this application.

4 Conclusion

The making of Application of the Multiple Intelligent Level Determination for Interest and Talent Development can be used by someone to decide his dominance of intelligence. The output of this application is a sequence of multiple intelligences owned by users ranging from linguistics, logical-mathematical, musical, spatial, kinaesthetic, intrapersonal; interpersonal abilities, and naturalist. This application is made by using programming language PHP and MySQL as Database Management System by showing 24 question list that must be filled by user. Accuracy of application is 100% from 150 respondents who have used this application. The existence of this application is expected to develop someone's potential based on his dominance of intelligences so he is more productive and creative suits to his interest and talent.

References

1. H. Gardner, *Frames of Mind: The Theory of Multiple Intelligences*. Basic Books, 2011.
2. S. Scarr, "An author's frame of mind: Review of Frames of Mind: The Theory of Multiple Intelligences by Howard Gardner, Basic Books (1984)," *New Ideas Psychol.*, vol. 3, no. 1, pp. 95–100, Jan. 1985, doi: 10.1016/0732-118X(85)90056-X.
3. N. R. I. Alsalhi, "The representation of multiple intelligences in the science textbook and the extent of awareness of science teachers at the intermediate stage of this theory," *Think. Ski. Creat.*, vol. 38, p. 100706, Dec. 2020, doi: 10.1016/J.TSC.2020.100706.
4. C. B. Shearer and J. M. Karanian, "The neuroscience of intelligence: Empirical support for the theory of multiple intelligences?," *Trends Neurosci. Educ.*, vol. 6, pp. 211–223, Mar. 2017, doi: 10.1016/J.TINE.2017.02.002.
5. H. F. Abenti, "How do I teach you? An examination of multiple intelligences and the impact on communication in the classroom," *Lang. Commun.*, vol. 73, pp. 29–33, Jul. 2020, doi: 10.1016/J.LANGCOM.2020.04.001.
6. J. C. McCroskey and J. L. Chesebro, *Communication for Teachers*. Allyn and Bacon, 2002.
7. J. H. Jung and D. R. Chang, "Types of creativity—Fostering multiple intelligences in design convergence talents," *Think. Ski. Creat.*, vol. 23, pp. 101–111, Mar. 2017, doi: 10.1016/J.TSC.2016.12.001.
8. Y. Delgoshaeia and N. Delavaria, "Applying multiple-intelligence approach to education and analyzing its impact on cognitive development of pre-school children," *Procedia - Soc. Behav. Sci.*, vol. 32, pp. 361–366, Jan. 2012, doi: 10.1016/J.SBSPRO.2012.01.054.
9. T. Kaewkiriya, N. Utakrit, and M. Tiantong, "The Design of a Rule Base for an e-Learning Recommendation System Base on Multiple Intelligences," *Int. J. Inf. Educ. Technol.*, vol. 6, no. 3, pp. 206–210, 2016, doi: 10.7763/ijiet.2016.v6.685.
10. J. Satzinger, R. Jackson, and S. Burd, *System and Analysis Design In A Changing World*. Boston USA: Course Technology-Cengage Learning, 2010.
11. S. Stobart and M. Vassileiou, *PHP and MySQL Manual*. London: Springer-Verlag, 2004.
12. A. Cabana, C. Charrier, and A. Louis, "Mono and multi-modal biometric systems assessment by a common black box testing framework," *Futur. Gener. Comput. Syst.*, vol. 101, pp. 293–303, Dec. 2019, doi: 10.1016/J.FUTURE.2019.04.053.

Analysis of The Best Method in Produced of Synbiotics Products for Shrimp Using Microencapsulation Techniques

Dian Eka Ramadhani[1]*, *Wida* Lesmanawati[1], *Erni* Sulistiawati[2], and *Widanarni* Widanarni[3]

[1]Production Technology and Management of Aquaculture Study Program, School of Vocational Studies IPB, Kumbang St. No. 14 Bogor, 16151, West Java.
[2]Veterinary Paramedics Study Program, School of Vocational Studies IPB, Kumbang St. No. 14 Bogor, 16151, West Java.
[3]Department of Aquaculture, Faculty of Fisheries and Marine Science, IPB University, Raya Dramaga St., Bogor, 16680, West Java.

Abstract. This research was conducted to find the best combination of coating material and microencapsulation method for synbiotics. This research consists of 3 chapters. Chapter 1, was conducted to obtain the best combination of 3 types of coating materials (whey protein, skim milk, and maltodextrin) with 3 doses of coating material (0%, 10%, and 20%). Chapter 2, was conducted to find the best microencapsulation method that can maintain viability, the density of probiotic bacteria, and physiological and biochemical properties of bacteria. The probiotic bacteria used was *Pseudoalteromonas piscisida* with the prebiotic *mannan-oligosaccharide* (MOS). A total of 9 treatments in Chapter 1 was microencapsulated using spray- and drum-drying methods. Chapter 3, was conducted to determine the economic value of the synbiotic microcapsule products. The best result from Chapters 1 and 2 is synbiotics coated with a combination of 20% skim milk and 20% maltodextrin (treatment 6) with the spray drying method compared to other treatments. Treatment 6 is produced the highest percentage of product, the physical characteristics of which were good white powder when stored in cold (4°C) and room (25-29°C) temperatures. The best result in Chapter 3 is treatment 6 that showed profitable, highest R/C ratio and B/C ratio than other treatments. However, even though the quantity of the product is profitable, the resulting product can not live. Our suggestion is to use another method namely, freeze-drying for microencapsulation of heat-resistant bacteria such as the bacteria used in this study.

1 Introduction

Various disease prevention in aquaculture commodities including shrimp have been carried out, one of which is the use of microorganisms known as probiotics. Probiotics are live microbes that, when administered in sufficient quantities, can have a beneficial effect on host immunity and can improve microbial balance in the digestive tract, feed efficiency, and environmental quality [1,2]. Recent studies reported that probiotic bacteria *Pseudoalteromonas piscicida* has the potential to be applied in aquaculture, this bacteria is isolated from Pacific white shrimp nauplii [3]. However, not many types of potential aquaculture probiotics from research results can be produced on an industrial scale because most probiotics are not resistant to high temperatures, resulting in cell viability that drops dramatically during the manufacturing process [4]. Therefore, the types of probiotics sold in the market are relatively limited. Commonly used probiotics in aquaculture include *Bacillus* sp., *Lactobacillus* sp., *Bifidobacteria* sp., and *Saccharomyces cerevisiae* [2,4].

Probiotics are often applied together with prebiotics to enhance the role of these probiotics. Prebiotics are food ingredients that cannot be digested and it have a beneficial effect on the host by stimulating the growth and activity of a number of bacteria in the digestive tract of the host [5]. Materials that are often used as synthetic prebiotics include mannan-oligosaccharides (MOS), fructo-oligosaccharides (FOS), galacto-oligosaccharides (GOS), and inulin while natural prebiotics can be obtained from honey, fruits, tubers, seeds, and so on. The combination of prebiotics with probiotics is called synbiotics, which have been shown to have a synergistic effect on the host [6]. Several studies reported that the administration of synbiotics could improve growth performance, survival, and immune response of Pacific white shrimp [7, 8, 9].

The potential of synbiotics is one of the way for disease prevention in aquaculture commodities, it is important to find methods or techniques that can be used to maintain cell viability in the minimum range of 10^6-

* Corresponding author: dianeka06@apps.ipb.ac.id

10^7 CFU/g of feed according to the recommended dose [2,10]. Microencapsulation is a new technology for packaging biomolecules or cells in encapsulated membranes to reduce cell loss during the manufacturing process and their release in the host digestive tract. Microencapsulation techniques that are widely used are spray-drying, freeze-drying, drum-drying, and so on. This method is most commonly used in the food, drug, and other feed additive industries [11,12]. In the microencapsulation process, the type of coating material that is often used as a coating material for probiotic bacteria is a combination of whey protein, skim milk, and maltodextrin. Several studies have reported that whey protein, skim milk, and maltodextrin can protect probiotic bacteria during the microencapsulation process [7, 13, 14, 15]. The right combination of coupling material is needed to obtain maximum viability of probiotic bacteria after passing through the microencapsulation process.

This research was conducted to find the best combination of coating material and the best microencapsulation method for probiotic bacteria *Pseudoalteromonas piscicida* mixed with MOS prebiotics so that synbiotic microcapsules with maximum viability can be produced to be applied in shrimp farming activities. The research is divided into 3 chapters. Chapters 1 was conducted to obtain a combination of coating materials that gave maximum results on the viability of probiotic bacteria in the synbiotic microencapsulation process. Chapter 2 was conducted to obtain the best method for the synbiotic microencapsulation process to produce synbiotic microcapsules. Chapters 3 research is to evaluate the economic value of the resulting synbiotic microcapsule product.

2 Method

2.1 Time and Place

This research was carried out from August-November 2020 at Laboratory of Microbiology, IPB Sukabumi Campus and SEAFAST (Southeast Asian Food & Agricultural Science & Technology) Center, Bogor Agricultural University.

2.2 Research Material

The materials needed in this research include cow's milk, rennet pills, skim milk powder, maltodextrin, $CaCl_2$, rifampicin, PBS (phosphate-buffered saline; 0,8 g NaCl, 0,02 KH_2PO_4, 0,15 g Na_2HPO_4, 0,02 g KCL, and aquadest) , bacto agar, and SWC media (seawater complete; 0,5 g bacto peptone, 0,1 g yeast extract, 0,3 mL glycerol, 75 mL seawater and 25 mL aquadest), probiotic bacteria *Pseudoalteromonas piscicida* 1 Ub from the Laboratory of Fish Health and Management, Department of Aquaculture and prebiotics mannan-oligosaccharide (MOS) (Alltech Inc., KY USA) with a minimum content of 30% crude protein, minimum 1,4% crude fat and maximum 13% crude fiber.

2.3 Research Design

This research is divided into 3 Chapters. Chapter 1, a combination of 3 types of coating material (whey protein, skim milk and maltodextrin) was used with 3 doses of coating material (0%, 10% and 20%), so there were 9 treatments in total (Table 1). Parameters observed were viability and density of probiotic bacteria before and after coating with various treatments. Chapter 2, the best result from Chapter 1 encapsulated with 2 microencapsulation methods, for a total of 18 treatments (Table 2). The encapsulation methods used are spray-drying and drum-drying. Parameters observed were viability and density of probiotic bacteria after passing through the microencapsulation process. Chapter 3, was conducted to determine the economic value of the three synbiotic microcapsules products so that the best and economical method could be determined to be applied in the field. Observation parameters in the form of profit R/C ratio and B/C ratio.

Table 1 Treatment in stage 1 research: the best combination and dose of coating material

No	Treatment	Description
1	PS 10	Synbiotic coating with whey protein + 10% skim milk
2	PM 10	Synbiotic coating with whey protein + 10% maltodextrin
3	SM 10	Synbiotic coating with 10% skim milk + 10% maltodextrin
4	PS 20	Synbiotic coating with whey protein + 20% skim milk
5	PM 20	Synbiotic coating with whey protein + 20% maltodextrin
6	SM 20	Synbiotic coating with 20% skim milk + 20% maltodextrin
7	PSM 10	Synbiotic coating with whey protein + 10% skim milk + 10% maltodextrin
8	PSM 20	Synbiotic coating with whey protein + 20% skim milk + 20% maltodextrin
9	K	Synbiotic coating with PBS only (Control)

Table 2 Treatment in Chapter 2: the best microencapsulation method

Combination coating material	Microencapsulation method	
	Spray-Drying (S)	**Drum-Drying (D)**
PS 10	PS 10-S	PS 10-D
PM 10	PM 10-S	PM 10-D
SM 10	SM 10-S	SM 10-D
PS 20	PS 20-S	PS 20-D
PM 20	PM 20-S	PM 20-D
SM 20	SM 20-S	SM 20-D
PSM 10	PSM 10-S	PSM 10-D
PSM 20	PSM 20-S	PSM 20-D
K	K-S	K-D

2.4 Procedures

2.4.1 Preparation of Synbiotic

Bacterial preparation was started by culturing the bacteria on SWC-agar media which already contained the antibiotic rifampicin 50 g/mL as a marker. Then the bacteria were cultured in 50 mL of SWC-liquid media and incubated in a water bath shaker at a temperature of 29-30°C for 18-24 hours at a speed of 140 rpm, then continued up scaling (1:10). After that, the bacteria were harvested and ready to be centrifuged 1 time and rinsed using PBS. The prebiotic that will be used is Bio-MOS. Synbiotic is made by combining probiotics and prebiotics in the same container.

2.4.2 Microencapsulation of Synbiotic

The first stage begins with the preparation of synbiotics and coating materials. According to the treatment, the coating material is prepared in a separate container. Whey protein coating is obtained by boiling 1L of cow's milk at a temperature of 50-60°C for 15 minutes and constantly stirring. Then wait until the temperature is around 30°C and give rennet pills to form lumps and then let stand at room temperature for 3-4 hours. Furthermore, it is filtered with sterile mori filter cloth and the resulting liquid (whey protein) is then stored in a refrigerator or can be used directly. After that, the coating material and synbiotic were mixed. The ratio of synbiotics, whey protein and maltodextrin is 1:1:0,1 (v/v/w) [16]. Furthermore, the synbiotic was dried using a sprayer with an inlet temperature of 50°C and an outlet temperature of 90°C. The results of the microencapsulation are then put in a closed container and stored in the refrigerator. For the drum-drying method, the sample is put into the drum dryer and the results are placed in a container and stored in the refrigerator.

2.4.3 Parameters

The product percentage parameter was calculated by comparing the synbiotic volume (fresh-culture) with the dry product produced after the encapsulation process (v/w). Parameters of viability and density of probiotic bacteria were calculated after coating, temperature incubation, before and after spray-drying and drum-drying as well as during storage. Measurement of bacterial viability was carried out using the spread plate method (TPC; total platting count) [17]. A sample of 1 g was homogenized in 0,9 mL of PBS then serially diluted. Samples were then taken as much as 50 µL and spread on SWC-agar + antibiotic rifampicin. Then incubated in an incubator at a temperature of 29-30° C for 18-24 hours. Then the bacterial colonies were counted one by one and added together. The formula for counting bacteria is below:

$$\sum \text{bacteria (CFU/g)} = \text{number of bacterial colonies (CFU)} \times \frac{1}{\text{dilution}-n} \times \frac{1}{spread\ volume\ (\mu L)}$$

The economic analysis parameters measured include profit, R/C ratio and B/C ratio, with the following calculation formula:

$$\text{Profit} = \text{total revenue} - \text{total cost}$$

$$\text{R/C ratio} = \frac{total\ revenue}{total\ cost}$$

$$\text{B/C ratio} = \frac{total\ profit}{total\ cost}$$

2.4.4. Data Analysis

The research data was tabulated in Microsoft Excel 2016 and data was performed using table and graph with descriptive analysis.

3 Results and Discussion

3.1 Product percentage

The results of calculating the percentage of probiotic bacteria products contained in synbiotics include: 1) the percentage of products after the spray and drum drying process; 2) the percentage of viability of probiotic bacteria after the spray and drum drying process. The following is the result of calculationg the percentage of the product (Figure 1).

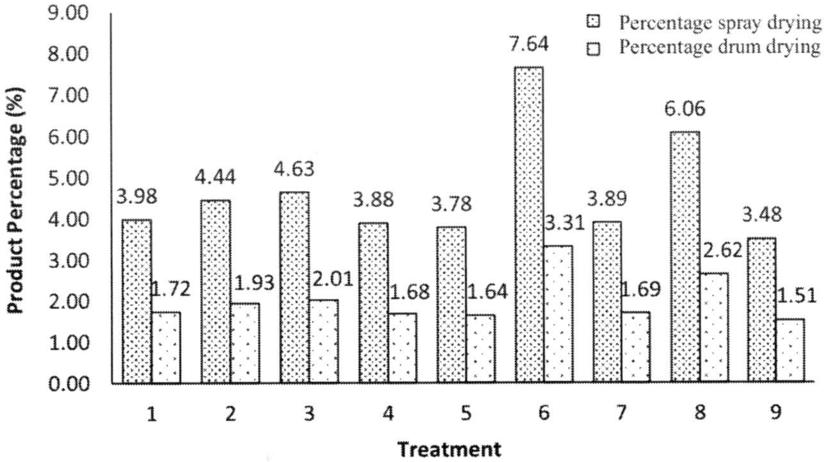

Fig. 1 Percentage of product after spray- and drum-drying

The results showed that the percentage of product in the spray drying method was found in treatment 6 (coating material skim milk 20% and maltodextrin 20%) which was 7,6% higher than other treatments. Meanwhile, the lowest product percentage was found in treatment 9 (Control), which was 3,4%. The results showed that the percentage of product in the drum-drying method was found in treatment 6 (coating material skim milk 20% and maltodextrin 20%) which was 3,31% higher than other treatments. While the lowest product percentage is in treatment 9 (control) which is 1,51%. The following is the viability of 1Ub probiotic bacteria after being coated with a coating material (Table 3). The results showed that the viability of probiotic bacteria in each treatment after being given the coating material still grew with a density range of 10^8-10^{10} CFU/mL, but after drying with the spray drying and drum drying methods, the viability of the bacteria became 0 meaning it did not grow at all.

Table 3. Density of 1Ub probiotic bacteria cells after being coated with various treatment materials

Treatment	Probiotic bacteria cell density (CFU/mL)
1	1,50 x 10^{10}
2	4,61 x 10^9
3	5,80 x 10^8
4	2,00 x 10^8
5	4,80 x 10^9
6	4,46 x 10^9
7	8,40 x 10^8
8	6,32 x 10^9
9	4,00 x 10^9

3.2 Synbiotic microcapsule physical quality

At the end of the in-vitro test, there was a change in the physical form of each treatment including a change in color and the presence of clumping. Table 4 shows the physical quality of dry synbiotic cultures from spray drying and drum drying during 1 month storage.

Table 4 Physical quality of dry synbiotic culture from spray drying and drum drying during 1 month storage.

Treatment	Storage	Physical form of spray drying		Physical form of drum drying	
		Day 1	Day 30	Day 1	Day 30
1	Room temperature (25-29°C)	Powder, white, slightly cream	Powder, white, lumpy	Creamy brown, slightly lumpy	Creamy brown, slightly lumpy
	Cold temperature (4°C)	Powder, white,	Powder, white,	Creamy brown, slightly lumpy	Creamy brown, slightly lumpy
2	Room temperature (25-29°C)	Powder, white,	Powder, white,	Creamy, slightly lumpy	Creamy, slightly lumpy
	Cold temperature (4°C)	Powder, white,	Powder, white,	Creamy, slightly lumpy	Creamy, slightly lumpy
3	Room temperature (25-29°C)	Powder, white,	Powder, white, slightly cream	Creamy, slightly lumpy	Creamy brown, slightly lumpy
	Cold temperature (4°C)	Powder, white,	Powder, white,	Creamy, slightly lumpy	Creamy brown, slightly lumpy
4	Room temperature (25-29°C)	Powder, white,	Powder, white, slightly cream	Creamy, slightly lumpy	Creamy brown, slightly lumpy
	Cold temperature (4°C)	Powder, white,	Powder, white, slightly cream	Creamy, slightly lumpy	Creamy brown, slightly lumpy
5	Room temperature (25-29°C)	Powder, white,	Powder, white, slightly lumpy	Creamy brown, slightly lumpy	Creamy brown, slightly lumpy
	Cold temperature (4°C)	Powder, white,	Powder, white, slightly lumpy	Creamy brown, slightly lumpy	Creamy brown, slightly lumpy
6	Room temperature (25-29°C)	Powder, white,	Powder, white, slightly cream	Creamy, slightly lumpy	Creamy brown, slightly lumpy
	Cold temperature (4°C)	Powder, white,	Powder, white,	Creamy, slightly lumpy	Creamy, slightly lumpy
7	Room temperature (25-29°C)	Powder, white,	Powder, white,	Creamy brown, slightly lumpy	Creamy brown, slightly lumpy
	Cold temperature (4°C)	Powder, white,	Powder, white,	Creamy brown, slightly lumpy	Creamy brown, slightly lumpy
8	Room temperature (25-29°C)	Powder, white,	Powder, white,	Creamy, slightly lumpy	Creamy brown, slightly lumpy
	Cold temperature (4°C)	Powder, white,	Powder, white,	Creamy, slightly lumpy	Creamy, slightly lumpy
9	Room temperature (25-29°C)	Powder, white, slightly cream	Powder, white, watery, yellowish white and lumpy	Brown, slightly lumpy	Brown, slightly lumpy, sticky
	Cold temperature (4°C)	Powder, white, slightly cream	Powder, white, watery, yellowish white and lumpy	Brown, slightly lumpy	Brown, slightly lumpy,

The results of microencapsulation (spray-drying) showed that the percentage of product in treatment 6 (addition of 20% skim milk coating material and maltodextrin) resulted in the highest percentage of product compared to other treatments. The higher composition of skim milk in the encapsulating material, the lower the water content of the microencapsulated produced. The water content of skim milk ranges from 3-7% [18]. According to [19] the range of good moisture content for microencapsulated products obtained from spray drying is 2-6%. Furthermore, carbohydrates such as starch, maltodextrin, are good coating materials because of their low viscosity at high solids and high solubility properties [20]. Microencapsulation using the drum-drying method showed the same results in treatment 6 (coating material skim milk 20% and maltodextrin 20%). This treatment produced the highest percentage of product compared to other treatments.

The concentration of addition of skim milk combined with maltodextrin can affect the viability of bacterial cells. This is in accordance with the research conducted by [21] which showed that the higher the coating concentration, the higher the encapsulation efficiency, the better and stronger the shell. This statement was again proven in the research of [22], encapsulation treatment with 10% skim milk and 20% maltodextrin resulted in bacterial cell viability of 97,76%. These results were significantly different from the other three treatments. The better the coating material used, it can protect the core material well and protect volatile substances when the drying process takes place, which results in increased retention of the core material.

The choice of coating material that will protect the core material is the most important factor to maintain the viability of bacterial cells in the microencapsulation process. According to [23] optimal efficiency can be obtained from the protein and carbohydrate matrix as a microencapsulated wall. The coating materials used in this study were whey protein, skim milk, and maltodextrin. Referring to the results of the study, the two encapsulation ingredients in the form of 20% skim milk and 20% maltodextrin (Treatment 6) were the best ingredients. The microencapsulated wall consisting of two encapsulated materials is able to provide good protection because the use of these two materials results in higher efficiency than one encapsulated material as a filler. This is due to the ability of the encapsulant to interact to form granules that can coat the encapsulated components better [22].

3.3 Economic Analysis

The economic analysis used includes profit, business feasibility analysis (R/C) ratio, and (B/C) ratio. The results of the economic analysis of the resulting product are shown in Table 5.

Table 5 Results of the economic analysis of synbiotic microcapsules using the spray drying and drum drying methods

Treatments	*Spray drying method*			*Drum drying method*		
	Benefit	R/C Ratio	B/C Ratio	Benefit	R/C Ratio	B/C Ratio
1	(335,444,444)	0.74	-0.256064461	(371,745,941)	0.72	-0.283775527
2	(221,111,111)	0.83	-0.168787108	(261,671,442)	0.80	-0.199749193
3	(174,833,333)	0.87	-0.13346056	(217,117,479)	0.83	-0.165738533
4	(359,944,444)	0.73	-0.274766751	(395,333,333)	0.70	-0.30178117
5	(384,444,444)	0.71	-0.293469042	(418,920,726)	0.68	-0.319786814
6	562,888,889	1.43	0.429686175	493,125,119	1.38	0.376431389
7	(356,541,667)	0.73	-0.272169211	(621,542,980)	0.53	-0.47446029
8	174,291,667	1.13	0.133047074	(238,247,851)	0.82	-0.181868589
9	(457,944,444)	0.65	-0.349575912	(489,682,904)	0.63	-0.373803743

Based on economic analysis, the most profitable treatment and has business feasibility is treatment 6. This can be seen in the results of the percentage of synbiotic products produced which have a greater value than the others. R/C ratio analysis is an analysis used to see the relative advantages that will be obtained in a business. Basically, a business will be said to be feasible to run if the R/C ratio value obtained is greater than 1. This can happen because the higher the R/C ratio value of a business, the higher the profit level to be obtained [24]. The results of the R/C ratio analysis showed that with the spray drying method, treatments 6 and 8 had values of 1,4 and 1.13, respectively. While in the drum drying method, the treatment with the highest R/C ratio value resulted from treatment 6. The next analysis is the B/C ratio, where the B/C ratio is the ratio between positive net benefits and negative net benefits in a business. In the application of the B/C ratio, a business/investment can be said to be feasible if the B/C value >1 is obtained, while a business is said to be unfeasible if the B/C value <1. In its application to a business, the analysis of the B/C ratio against a situation is an analysis that is needed to see to what extent the comparison between the value of benefits and the value of costs is seen in the present value condition. The results of the analysis of the B/C ratio on the spray drying and drum drying methods in all treatments showed a B/C value <1, but in treatment 6 the resulting value was still positive and if further improvements were made it could provide maximum business feasibility.

Conclusions and Suggestions

Synbiotic coated with a combination of 20% skim milk and 20% maltodextrin (Treatment 6) with spray drying and drum drying methods was the best result in this study compared to other treatments. Treatment 6 on the spray drying method resulted in the highest percentage of product, physical characteristics in the form of a white fine powder when stored at cold temperatures (4°C) or at room (25-29°C). Meanwhile, in the drum drying method, the characteristics are brownish beige and there are few lumps. The results of the economic analysis also show that treatment 6 is more profitable and from the analysis of the R/C ratio along with the B/C ratio, treatment 6 has business feasibility compared to other treatments. Research on drying methods that can maintain the viability of 1Ub bacteria needs to be done, one of which is the freeze drying method, but it needs a combination that matches so that the resulting product has high viability and the method used can be more efficient to use.

Acknowledgment

We express our gratitude to SV IPB which has funded this research and to the Study Program of Production Technology and Management of Aquaculture which has given permission to use the facilities for the implementation of this research activity.

References

1. [FAO] Food and Agriculture Organization. Evaluation of health and nutritional properties of powder milk and live lactic acid bacteria.Food and Agriculture Organization of the United Nations and World Health Organization Expert ConsultationReport(Availableat: http://www.who.int/foodsafety/publications/fs_management/en/probiotics.pdf. Last Accessed 19.03.134) (2001)
2. S. K. Nayak. Immunol, *Probiotics and imunity: a fish perspective*, **29**: 2-14. (2010)
3. Widanarni, I. Tepu, Sukenda, S. Mia, J Ris Akuak, *Seleksi bakteri probiotik untuk biokontrol vibriosis pada larva udang windu, Penaeus monodon menggunakan cara kultur bersama*, **4**: 95-105 (2009)
4. P. Chitprasert, P Sudsai, A Rodklongtan, Carbohydrate Polymers, *Aluminum carboxymethyl cellulose–rice bran microcapsules: Enhancing survival of Lactobacillus reuteri KUB-AC5*, **90**: 78–86 (2012)
5. R. Cerezuela, J. Meseguer, M. A. Esteban, J Aquac Res Develop. *Current knowledge in synbiotic use for fish aquaculture*, 2155 - 9546 (2011)
6. T. G. Huynh, Y. L. Shiu, T. P. Nguyen, Q. P. Truong, J. C. Chen, C. H. Liu, Fish Shellfish Immunol, *Current applications, selection, and possible mechanisms of actions of synbiotics in improving the growth and health status in aquaculture: a review*, **64**: 367-382 (2017)
7. D. E. Ramadhani, Widanarni, Sukenda. Jurnal Akuakultur Indonesia, *Microencapsulation of probiotics and its application with prebiotic in Pacific white shrimp larvae through Artemia* sp. **18**(2): 130-140 2019
8. Hamsah, Widanarni, Alimuddin, M. Yuhana, M. Z. Junior, Res J Microbiol, *Bacterial population, activity of enzymes and growth rate of pacific white shrimp larvae administered Pseudoalteromonas piscicida and Mannanoligosaccharide through bio-encapsulation of Artemia* sp., **12** (2): 128-136 (2017)
9. D. Febrianti, Yuhana M, Widanarni, J Fish Aquac Sci, *Dietary synbiotic microcapsule influence the immune responses, growth performance and microbial populations to white spot syndrome virus in pacific white shrimp (Litopenaeus vannamei)*, **11**(1): 28-42 (2016)
10. J. W. Collins, R. M. La Ragione, M. J. Woodward, L. E. J. Searle, Springer-Verlag, *Application of prebiotics and probiotics in livestock*, 57-88 (2009)
11. M. J. Martin, F. Lara-Villoslada, M. A. Ruiz, M. E. Morales, Innov Food Sci Emerg Technol, *Microencapsulation of bacteria: a review of different technologies and their impact on the probiotic effects*. **27**: 15-25 (2015)
12. K. Sultana, G. Godward, N. Reynolds, R. Arumugaswamy, P. Peiris, K. Kailasapathy, International Journal of Food Microbiology, *Encapsulation of probiotic bacteria with alginate–starch and evaluation of survival in simulated gastrointestinal conditions and in yoghurt*, **62**: 47–55 (2000)
13. D. Y. Ying, M. C. Phoon, L. Sanguansri, R. Weerakkody, I. Burgar, M. A. Augustin, Journal of Food Science, *Microencapsulated Lactobacillus rhamnosus GG powders: Relationship of powder physical properties to probiotic survival during storage*, **75**: 588–595 (2010)

14. C. B. Fritzen-Freire, E.S. Prudêncio, R. D. M. C. Amboni, S. S. Pinto, A. N. Negrâo-Murakami, F. S. Murakami, Food Research International, *Microencapsulation of Bifidobacteriaby spray drying in the presence of probiotics*, **45**: 306–312 (2012)
15. R. Paéz, L. Lavari, G. Vinderola, G. Audero, A. Cuatrin, N. Zaritzky, J. Reinheimer, Food Research International, *Effect of heat treatment and spray drying on lactobacilli viability and resistance to simulated gastrointestinal digestion*, **48**: 748–754 (2012)
16. W. Munaeni, M. Yuhana, Widanarni, *Microbiologi Indonesia*, *Effect of micro-encapsulated synbiotic at different frequencies for luminous vibriosis control in white shrimp* (*Litopenaeus vannamei*). **8**(2): 73-80 (2014)
17. J. Reynolds, M. Farinha, Richard College, *Counting bacteria*, (2005)
18. D. Setyaningsih, D Kurniawan, N. Muna, IOP. Conf. Series: Earth and Environmental Science, *Encapsulation of ginger oleoresin with a combination of maltodextrin and skim milk powder as wall material*, **472**: 012016 (2020)
19. M. I. Re, Drying Technology, *Microencapsulation by spray drying*, **16**(6): 1195-1236 (1998).
20. P. Balasubramani, P. T. Palaniswamy, R. Visvanathan, V. Thirupathi, A. Subbayaran, J. P. Maran, International Journal of Biological Macromolecules, *Microencapsulation of garlic oleoresin using maltodextrin as wall material by spray drying technology*, **72**: 210-217 (2014)
21. N. V. N. Joythi, P. M. Prasanna, S. N. Sakarkar, K. S. Prabha, P. S. Ramaiah, G. Y. Srawan, Journal of Microencapsulation, *Microencapsulation techniques, factors influencing encapsulation efficiency*, **27** (3): 187-197 (2010)
22. A. P. Cahya, M. Syaflan, Ngatirah, Indonesian Food and Nutrition Progress, *Probiotic (Lactobacillus casei) encapsulation the method of spray drying with combined prebiotic from iles-iles (Amorphopallus oncophyllys) and protectant agent (skim milk, gum arabic, maltodextrin)*, **15** (2): 0854-6177 (2018)
23. C. C. Lin, S. Y. Lin, L. S. Hwang, J. Of Food Sci, *Microencapsulation of Squid Oil with Hydrophilic Macromolecules for Oxidative and Thermal Stabilization*, **6** (1): 36-39 (1995)
24. J. Odeck, A, Kjerkreit, Transportation Research Part A, *The accuracy of benefit-cost analyses (BCAs) in transportation: An ex-post evaluation of road projects*, **120**: 277-294 (2019)

Production of Local Microorganism by Utilizing Organic Matter in PT Ultra Peternakan Bandung Selatan

Janatin Alifiah Gunawan[1], *Pria* Sembada[1*], *Suryo* Firmanto[2], and *Bagus* Ibnu Soewondo[2]

[1]College of Vocational Studies, IPB University (Bogor Agricultural University), Jl. Kumbang No. 14 Kampus IPB Cilibende, Bogor, Jawa Barat, Indonesia 16128.
[2]PT Ultra Peternakan Bandung Selatan, Jl. Raya Pangalengan No.340, Margaluyu, Kec. Pangalengan, Bandung, Jawa Barat 40378.

Abstract. Unmanaged waste in livestock can cause pollution, but it can benefit the environment if managed properly. PT Ultra Peternakan Bandung Selatan reported that effluent produced every day from this farm is about 150.000 liter. The application of local microorganisms (MOL) could be practical because used as a starter for decomposition. However, the most efficient way to produce MOL to manage waste effectively in this company needs to be carried out. This study aimed at improving effluent management by using MOL most efficiently from some organic matters. The analysis performed three treatments for making 10 liters MOL: M1 (Main MOL + molasses + water), M2 (Main MOL + mastitis milk), M3 (Main MOL + milk MOL + water). Each treatment was given a dose of 10% and 15% of 10 liters for main MOL, molasses, and milk MOL. The fermentation experiment was carried out by mixing all the ingredients, that is, main MOL, molasses, and water for M1. For the second ingredient, M2 is mixing mastitis milk and main MOL. Then M3 combines the main MOL, milk MOL, and water. After seven days fermented, observations in local microorganism characteristics were pH, density, and brightness value. The result suggested that treatment M2 (Main MOL + mastitis milk) 10% was the most efficient local microorganism reproduce based on the pH is 4,36, density has a value of 531 gram, and brightness value is 897 lux. The study underlined the importance of mastitis milk to produce MOL and seems effective and efficient to improve the effluent before being evacuated to the land.

1 Introduction

Waste is an organic or inorganic material that is no longer used, so it can cause severe problems for the environment if not handled properly [1]. If this waste is managed correctly, it can provide added value. Waste generated by cattle consists of solid waste in livestock manure and leftover feed, liquid waste in barn washing wastewater, livestock sanitation wastewater, cow urine, and air waste in the form of an unpleasant odor [2].

* Corresponding author: priasembada@apps.ipb.ac.id

PT Ultra Peternakan Bandung Selatan reported that it produced 150.000 liters of irrigation every day, including liquid waste, urine, water for washing tools, water for cleaning barns, and other activities. PT Ultra Peternakan Bandung Selatan wants to improve the quality of its liquid waste or effluent by mixing local microorganisms to get higher nutrients before irrigating it to the land. In addition, the production of local microorganisms helps utilize cow's milk affected by mastitis so that it is not immediately thrown away, and integrated farming can be applied. Local microorganisms are materials available in the environment and can decompose organic matter containing degrading microorganisms [3]. The addition of MOL solution has been shown to increase the number of microorganisms so that the decomposition process of organic matter is faster [4].
Therefore, if the concentration of MOL for irrigation on land is required 5% in a day according to the determination of the concentration in the previous experiment, the local microorganism used as a mixture is 7.500 liters per day, so efficient local microorganism production is needed.

2 Materials and Methods

This experiment was conducted at PT Ultra Peternakan Bandung Selatan, Jl. Raya Pangalengan No 340, Margaluyu, Pangalengan, Bandung from February 8 to February 13, 2021. The materials used in this experiment were water, mastitis milk, main MOL, milk MOL, and molasses. The tools used to produce the MOL were a jerry can, hoses, pH meters, lux meters, scales, and measuring cups. The formula for the production of Local Microorganisms (MOL) is:

1. M1 = Main MOL + molasses + water.
2. M2 = Main MOL + mastitis milk.
3. M3 = Main MOL + milk MOL + water.

Each formula consisted of 10% and 15% doses, so six treatments. These doses were used for the main MOL, molasses, and milk MOL from 10 liters of water and mastitis milk. So that obtained 1 liter and 1,5 liter respectively in the main material MOL, molasses, and milk MOL. The fermentation experiment was conducted with an anaerobic concept and mixed all the main MOL, molasses, and water for M1. For the second ingredient, M2 is mixing mastitis milk and main MOL. Then M3 combines the main MOL, milk MOL, and water. Fermentation aims to break down complex compounds into simpler compounds.

The selection of these materials to multiply the main MOL is based on organic materials that exist in the surrounding environment and have large quantities. Organic matter can be used as a culture medium (inoculant) for certain local microorganisms (MOL), which can degrade organic waste [5]. In making MOL, fermentation was carried out for seven days and the method used is the anaerobic concept. Anaerobic is a microbiological process without air. In multiplication with this concept, the bacteria that develop may be anaerobic. Observations on MOL production to compare formulas are pH, density, and brightness values.

Making local microorganisms is relatively easy to do by the wider community and farmers. In addition, the role of MOL as an essential component of fertilizer, microorganisms are not only beneficial for plants; they are also helpful as decomposers of organic matter, agricultural waste, household, and industrial waste [6]. The use of local microorganism doses as the decomposition of liquid waste is done with 5% of the liquid waste. According to [7], adding 10% MOL in livestock urine can produce the highest nutrient for plants.

2.1 pH

The MOL sample was measured using a digital pH meter in the pH test. pH or degree of acidity is used to express the level of acidity or alkalinity possessed by a substance, solution, or object [8]. pH also plays a significant role in controlling the microbial decomposition rates of organic matter [9]. The pH observation using a pH meter is shown in Figure 1.

Fig. 1. The pH observation of MOL by taking samples on a pH meter

2.2 Density

Density is defined as mass per unit volume [10]. There are various ways to measure the density of liquids, one of which is using a digital balance and measuring cup (see Figure 2). The purpose of observing density is to see the activity of microorganisms that turn MOL into lumps to increase their density. Density observations were carried out by taking a 500 ml MOL sample in a measuring cup and then weighing it using a digital scale.

Fig. 2. The density observation of MOL by taking sample 500 ml on a digital balance

2.3 Brightness value

Observation of the brightness value is carried out to determine the electrolyte content in MOL (see Figure 3)). A liquid with high acidity, large or smaller pH value can conduct large currents of electricity [11]. Measurements were carried out using a lux meter to measure light intensity in a dipped using a soil fertility tester on a 500 ml MOL sample. Measurements are taken at a distance of 20 cm and parallel to the light from the bulb.

 https://doi.org/10.1051/e3sconf/202234800018

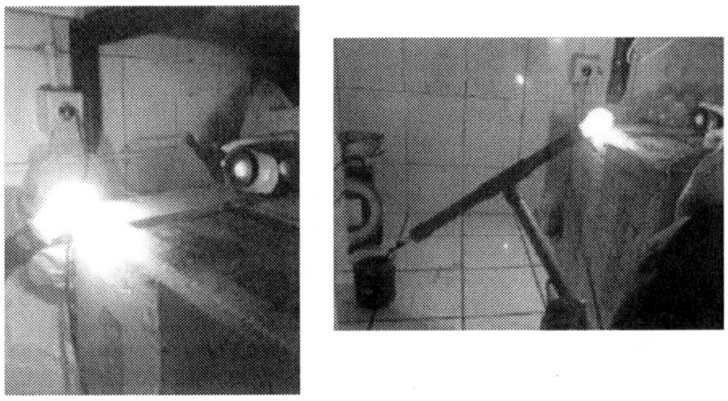

Fig. 3. The brightness value observation by dipped using soil fertility tester to conduct electricity

3 Results and Discussion

3.1 pH

This high to low pH process flow indicates the decomposition of organic matter. Following [12] research, the pH value tends to be acidic or down, revealing the decomposition process is still running. The pH of the main MOL obtained a value of 7,55; this shows that the pH has increased and is alkaline. According to [13], an increase in pH indicates an autolysis process. This process can increase the pH value because it produces compounds that are autolysis in main MOL is probably due to time storage and air contamination. The lowest pH value is obtained in the MOL M3 formula: main MOL + milk MOL + water; both doses are 3,47 (Fig. 4).

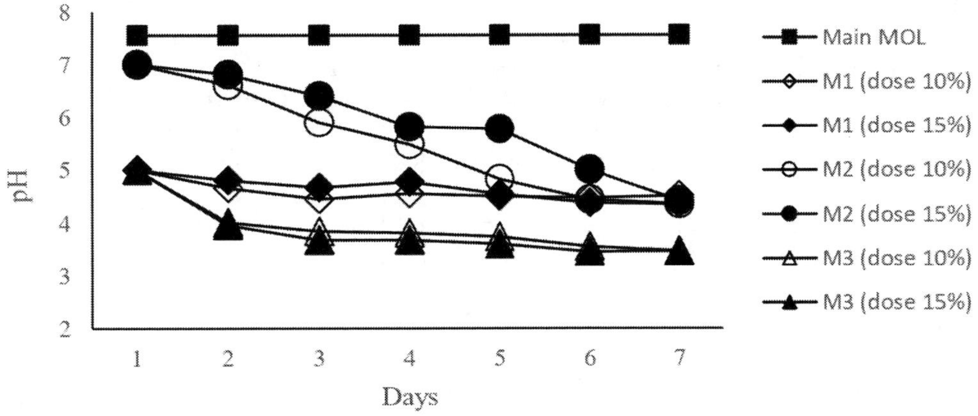

Fig. 4. The graphic of pH showing pH values of days 1 to 7

This might be the mixture of ingredients in the MOL of milk has a pH of 3,74, allowing the decomposition process in M3 to be faster. According to [14], the degree of acidity is closely related to the production of organic acids by microbes, especially lactic acid, which can lower the pH to 5,0 or more minor. Observing the pH value for all formulas M1, M2, and M3 on the 7th day had values below 5,0. This means that the organic acids produced by microbes are increasing. In other words, the whole method in the production of the MOL tends to be successful because of the microbial growth in the process. According to [15],

Acidity changes could be evaluated as an indirect characteristic of the growth of lactic acid bacteria.

3.2 Density

Figure 5 shows that the density of MOL after days 5 to 7 increased. However, on days 2 and 4, the density of the MOL decreased compared to day 1. This was due to an increase in temperature. According to [16], when the temperature increases, density will decrease. The rise in temperature in this MOL was because of the activities of microorganisms that decompose organic materials, and fermentation produces organic acids, CO2, and heat.

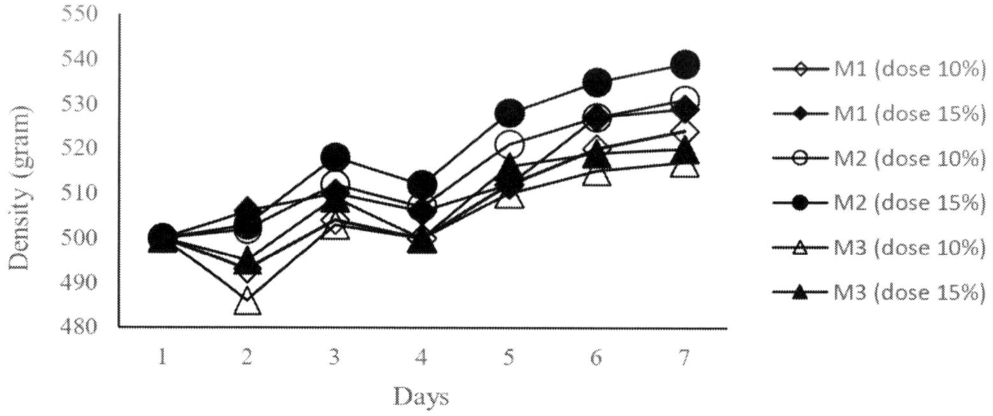

Fig. 5. The graphic of density days 1 to 7

One of the factors that cause this is the presence of microbes that decompose organic matter. It decreases pH and changes the character of organic matter such as milk that turns into lumps, causing the density to increase. According to [17], these clots are caused by lactic acid bacteria. The highest increase in specific gravity in this MOL was M2 with a dose of 15%. This is because the ingredients of the M2 formula are milk that turns into lumps. The lowest density is formula M3 with an amount of 10% due to the decomposition process of decomposition of organic matter, namely milk MOL, so the density decreases.
The density analysis results show that the overall formula for MOL tends to be successful because the MOL density increases every day, indicating microbial development and changes in the characteristics of organic matter in the process.

3.3 Brightness value

The brightness value is measured using a lux meter. [18] explains that illuminance is many light currents that come in one unit plane, measured by Lux or Lumen/m^3. At the same time, the process is called illumination (illumination), namely the arrival of light to an object. Observation of brightness value is influenced by distance and angle during measurement. So distances and angles must be equalized for every measure. The brightness value was taken using a lux meter and a soil tester by taking 500 ml of the MOL sample, then the soil tester was immersed in the MOL sample, and the distance to see the Lux meter brightness value was carried out at 20 cm from the light source.

Based on figure 6, the most considerable lux value on the 10th day was found in M1 at a dose of 15%. This is related to the material from M1; there are molasses. Observations brightness values were also performed on molasses, and the results obtained a value of more than 900. It proves that molasses have an acid content, and [19] mentioned that acidity could generate electrical energy because it is an electrolyte. This observation has a pH value of the brightness and inversely comparison, seen in figures 1 and 3 for the 10th day pH decreased while the brightness value is increased. The expected result of observing the brightness value is to find out the electrolyte produced by the solution associated with acid production.

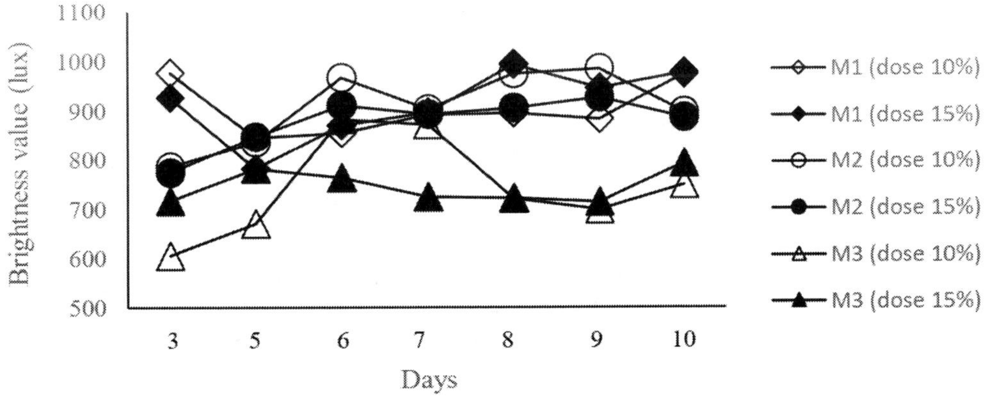

Fig. 6. The graphic of brightness value days 3 to 10.

4 Conclusion

The study underlined the importance of using the efficient method to produce MOL. The analysis found that the efficient way of producing MOL was the M2 (main MOL + mastitis milk with a dose of 10%). With this method, the pH value has entered below 5; the density increases and the brightness level shows a reasonably good value. In addition, the raw material can be made from mastitis milk so that the MOL propagation target of 7.500 liters/day will be easy and cheap to produce.

References

1. A. Adityawarman, Salundik, C. Lucia, *Pengolahan Limbah Ternak Sapi Secara Sederhana di Desa Pattalassang Kabupaten Sinjai Sulawesi Selatan*, **3**, 171-177 (2015)
2. H. Hutahaean, *Optimasi recovery ammonium dari limbah cair peternakan sapi menggunakan metode presipitasi struviet*, (2019)
3. F. Firdaus, B. Purwanto, Salundik, *Dosis Penggunaan Mikroorganisme Lokal (MOL) Ragi Tempe dan Isi Rumen Untuk Pengomposan*, **2** 257-261 (2014)
4. N.L. Widyasari, I.W.B. Suyasa, I.G.B.S. Dharma, *Upaya Pengolahan Limbah Kotoran Babi Menjadi Kompos Menggunakan Komposter Rumah Tangga*, **4** 104-116 (2018)
5. Muliyadi, D.W. Purwiningsih, *NPK Level in Anaerobic and Aerobic Composting Using Spoiled Rice MOL*, **10** 817-825 (2021)

6. R.R. Manullang, Rusmini, Daryono, *Kombinasi Mikroorganismem Lokal Sebagai Biokativator Kompos*, **4** 180-190 (2017)
7. K. Mamuda, Salundik, P.D.M.H. Karti, *Penggunaan Mikroorganisme Lokal dari Berbagai Formula terhadap Kualitas Biourine Kambing Terfortifikasi*, **8** 1-7 (2020)
8. D. Yusuf, Aswardi, M. Amin, *Alat Pendeteksi Kadar Keasaman Sari Buah, Soft Drink, dan Susu Cair Menggunakan Sensor pH Berbasis Mikrokontroler Arduino UNO*, **2** 1-11 (2018)
9. S.A. Wakelin, L.M. Macdonald, S.L. Rogers, A.L. Gregg, T.P. Bolger, and J.A Baldock, *Habitat selective factors influencing the structural composition and functional capacity of microbial communities in agricultural soils. Soil Biol. Biochem*, **40** 803-813 (2008)
10. M. Abdullah, *Fisika Dasar 1*, (2016)
11. H. Kholida, Pujayanto, *Hubungan Kuat Arus Listrik dengan Keasaman Buah Jeruk dan Mangga,* **6** 42-46 (2015)
12. R. Wijayanti, B. Prasetya, *Pengaruh Pemberian Urea Terhadap Laju Dekomposisi Serasah Tebu di Pusat Penelitian Gula Jengkol, Kabupaten Kediri*, **5** 793-799 (2018)
13. M. Santoso, E. Liviawaty, E. Afrianto, *Efektivitas Ekstrak Daun Mangga Sebagai Pengawet Alami Terhadap Masa Simpan Filet Nila Pada Suhu Rendah*, **8** 57-67 (2017)
14. N. Marsiningsih, A. Suwastika, N. Sutari, *Analisis Kualitas Larutan Mol (Mikroorganisme Lokal) Berbasis Ampas Tahu*, **4** 180-190 (2015)
15. S. Urbiene, D. Leskauskaite, *Formation of Some Organics Acids During Fermentation of Milk*, **15** 277-281 (2006)
16. Y. Damayanti, A.D. Lesmana, T. Prihandono, *Kajian Pengaruh Suhu Terhadap Viskositas Minyak Goreng Sebagai Rancangan Bahan Ajar Petunjuk Praktikum Fisika*, **7** 307-314 (2018)
17. Suprihana, *Pengaruh Lama Penundaan dan Suhu Inkubasi Terhadap Sifat Fisik dan Kimia Yoghurt Dari Susu Sapi Kadaluwarsa*, **6** 94-102 (2012) fir
18. P. Satwiko, *Fisika Bangunan sics 2*, (2004)
19. Atina, *Tegangan dan Kuat Arus Listrik Dari Sifat Asam Buah*, **12** 28-42 (2015)

 https://doi.org/10.1051/e3sconf/202234800019

Determination of LC50 and Clinical Symptoms of *Aeromonas hydrophila* Infection on the Fingerlings of Semah (*Tor soro*), the Indonesian Native Freshwater Fish

Firdausi Amalia Putri[1], Rahman[2], and *Mulya* Muhammad Arif[1]

[1]Production Technology and Management of Aquaculture Study Program, College of Vocational Studies, IPB University, Bogor
[2] Department of Aquaculture, Faculty of Fisheries and Marine Sciences, IPB University, Bogor

Abstract. *Aeromonas hydrophila* is a bacterial pathogen in freshwater aquaculture causing the motile aeromonad septicemia (MAS) disease and inflict significant economic losses. This study aimed to determine the lethal concentration (LC50-48h) of *A. hydrophila* on *Tor soro's* fingerlings and investigated the clinical signs that appeared following the challenge test. Briefly, 150 fingerlings (3,4 ± 0,18 cm) were challenged by immersion means with 0 (control), 10^4, 10^5, 10^6, and 10^7 CFU/mL of live cells of *A. hydrophila*. Reed and Muench's method's determination of LC50-48h was calculated and all clinical signs that appeared during the test were recorded. The result showed that the LC50-48h by immersion was 1.0 x $10^{6,66}$ CFU/mL with clinical signs observed as hemorrhages, depigmentation, lesions, anorexia, loss of balance, and enlargement of the abdominal cavity. Furthermore, based on the LC50-48h and the clinical symptoms, this study reveals that even though all symptoms are consistent among the fish, the severity of the disease tends to be prevalent and intense as the concentration of pathogen is higher.

1 Introduction

Semah (*Tor soro*) is a native freshwater fish in Indonesia. This species is known as a premium freshwater commodity because of its high demand but its existence is scarce. Domestication for *Tor soro* is made to explore its potential for both aquaculture purposes and conservation efforts. The major obstacle faced in *Tor soro* aquaculture is the presence of opportunistic pathogens such as *Aeromonas hydrophila* causing the motile aeromonad septicemia (MAS) disease. MAS disease is common in freshwater-aquaculture facilities and may inflict

* Corresponding author: rahmanaqua@gmail.com

significant economic losses due to high mortality and morbidity [4].

The pathogen is a rod-shaped, motile, gram-negative bacteria with a diameter of 0.3–1 μm and a length of 1–3.5 μm, without a spore phase, usually lacking a capsule, growing optimum at 28 °C but can also grow at extreme temperatures (4 °C and 37 °C). Its cosmopolite nature in the aquatic environment allows contact with fish and amphibians, and even enters the animal. Such contact can cause infection depending on its species and its virulent level [9].

Pathogenesis is the study of the course of diseases ranging from the influx of the pathogen to the onset of clinical symptoms. This study is needed to scrutinize the basic information in preventive control as well as treatment of disease. In this study, we aimed to evaluate the pathogenicity of *A. hydrophila* on *Tor soro* and observed any clinical symptoms resulting from the artificial infection [10].

2 Material and Method

2.1 Fish and acclimatization

One hundred fifty fingerlings of *Tor soro* (3,4 ± 0,18 cm) were purchased from a state-owned aquaculture facility (BPPBAT Bogor, Cijeruk Installation) and were acclimated to the Laboratory of Aquaculture, IPB Vocational School, Sukabumi (Indonesia) in tanks of 100 L of clean freshwater. Stocking density refers to the result of Priadi's *et al.* research (2021) [1]. During the rearing, fish were fed with commercial feed containing 38% crude protein for two weeks.

2.2 Pathogen preparation and Virulence Enhancement

Bacterial isolates came from the collections of the laboratory of aquatic organism health, Department of Aquaculture, Bogor Agricultural University, and was maintained in slant agar before the propagation. The liquid culture medium of trypticase soy broth (TSB, Sigma Aldrich™, USA) was prepared according to the instructions of the manufacturer by dissolving 37 g in 1 L of distilled water. The sterilization was processed by autoclaving at 121°C for 15 min (Gea, LS50LJ). Bacterial suspensions were prepared by transferring a loop containing three to five colonies of *A. hydrophila* isolated from Petri dishes of trypticase soy agar (TSA) medium after 24 h of cultivation at 30°C to test tubes containing 10 mL of TSB and reincubated for another 24 h in a rotary incubator at speed of 140 rpm. Virulence enhancement was done by injecting a suspension culture of *A. hydrophila* bacteria that have been washed with phosphate buffer saline (PBS) as much as 0.1 mL into intramuscular catfish (*Clarias garipienus*). After the clinical symptoms such as ulcers and hemorrhagic appeared, the bacteria was reisolated to TSA medium from the kidney of infected fish and incubated for 24 hours at a temperature of 29 °C to 30 °C. The result from reisolated bacteria was purified into four quadrants in a plate and recharacterized. Characterization of bacteria was conducted using software the API 20E KIT. The results of the characterization using the API 20E KIT showed that the bacteria were *Aeromonas hydrophila* with a similarity level of 94%.

2.3 Determination of LC50-48h

The pure cultured of *A. hydrophila* bacteria were isolated by picking a single colony from the stock culture to TSB medium as much as 50 mL and incubated in a rotary incubator at a speed of 140 rpm for 24 hours in 29 °C to 30 °C. After incubation, the logarithmic growth was measured by total plate counting. This data was used to obtain different concentrations of the pathogen by a ten-fold serial dilution for the LC50-48h bioassay test. Bioassay was done by immersing 10 fingerlings into each different series of final concentrations 10^5, 10^6, 10^7 CFU mL^{-1} for 48 hours long. All mortalities and any clinical symptoms were recorded, while the LC50 value was generated referred to modified of Reed and Muench's (1938) method [8].

2.4 Data analysis

Data of LC50-48h were analyzed with microsoft excel software descriptively. The LC50-48h were calculated. Meanwhile, appearing clinical symptoms were analyzed descriptively.

3 Results

Mortality of *Tor soro* fingerlings did not appear after 48 hours of artificial infection with a suspension of *A. hydrophila* on 1.0 x 10^4 CFU/mL. Only when the concentration raised to 10^5, 10^6, and 10^7 CFU/ml the mortality was beginning (Table 1.) and followed the onset of the clinical symptoms. Based on the Reed and Muench modified method (1938), it was concluded that the LC50-48h was 1.0 x $10^{6,66}$ CFU/ml.

Table 1. Mortality of *Tor soro* fingerlings after 48 hours challenged by *Aeromonas hydrophila*

Pathogen concentration (CFU/ml)	Response		Accumulation		Ratio	% Mortality
	Dead	live	Dead	Live		
1.0 x 10^7	6	4	11	4	11/15	73,4
1.0 x 10^6	3	7	5	11	5/16	31,25
1.0 x 10^5	2	8	2	19	2/21	9,52
1.0 x 10^4	0	10	0	29	0/29	0

The virulence of *A. hydrophila* was able to regain after enhancement by passaging the pathogen into the body of African catfish. The clinical symptoms appearing after 24 hours on the body of catfish including inflammation, hemorrhagic, body discoloration, anorexia, and lethargy. Meanwhile, infection on *Tor soro* fingerlings exhibits moderate clinical symptoms after 48 hours where body discoloration, hyperemia, and swelling in the body cavity were prevalent (Fig 1). The fingerlings had a reduction of respiratory, loss of balance, erratic swim-movement anorexia, erosion of the caudal fin, and dropsy containing yellowish ascites fluid.

 https://doi.org/10.1051/e3sconf/202234800019

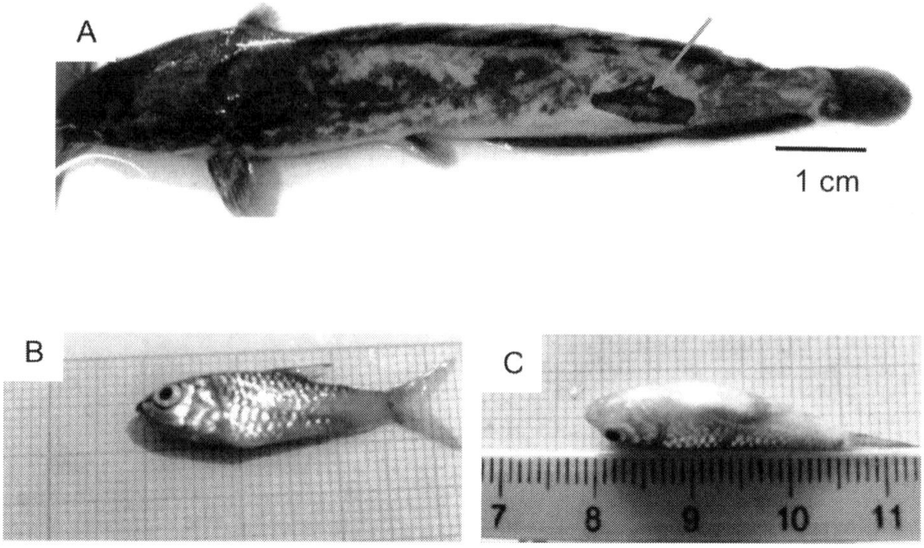

Fig. 1. The clinical symptoms of African catfish during enhancement of *A. hydropila*. (A) an open hemorrhagic (orange arrow) and body discoloration of African catfish showed after 24 hours. (B) Scales loss and hyperemia showed in the body of Tor soro fingerlings. (C) Dropsy also showed from moribund fingerlings.

4 Discussion

Motile aeromonads septicemia (MAS) is one of the major diseases that attack various species of fish [6]. The disease is caused by the bacteria *A. hydrophila* which is Gram-negative and lives with normal flora in freshwater [11]. The impact of the pathogen is hemorrhagic and often ends in death [3]. The fingerlings are the prone stage because do not have a sophisticated immune response compared to the adults that a massive disease outbreak will lead to higher mortality [7, 10]

Aeromonas hydrophila can attack the fins, tegument, and abdominal cavity, and may be able to break the blood vessels, resulting in the manifestation of hemorrhagic, hyperemia, wound, ulcer, severe scale, and blackened skin [2, 5, 12, 13]. In our bioassay with *A. hydrophila* on *Tor soro* fingerlings, the parameter of water quality was set constant and considered as suitable for most of the freshwater fish. However, the severity of the disease was increased along with the concentration of the pathogen thus influencing the mortality rate. We found that even particular clinical symptoms were consistent in all assays, the higher of pathogen concentration, the more prevalent and intense the clinical symptoms among the fish. All findings in this study provide important information not only on the pathogenesis of *A. hydrophila* but also for the development of treatment and crisis strategies for this bacteriosis. Furthermore, additional studies should be conducted in the search for improvements in the immune response against the pathogenicity of *A. hydrophila* using maternal immunization, vaccination, or stimulation by environmental enrichment.

References

1. B Priadi, D Irawan. Effect of Different Stocking Densities on the Growht of Tor Soro Fish. Aquaculture Engineering Bulletin, **19,** 85-88 (2021)

2. EH Hardi, CA Pebrianto, T Hidayanti, RT Handayani. Phatogenicity of *Aeromonas hydrophila* via some port entryin cultured Nile tilapia *Oreochromis niloticus* from Loa Kulu Kutai Kartanegara Kalimantan Timur. Journal of Veterinary Medicine **8**, 130-133 (2014)
3. GH Fu, F Liu, JH Xia, GH Yue. The *LBP* gene and its association with resistance to *Aeromonas hydrophila.* International Journal of Molecular Sciences **15**, 22028-22041 (2014)
4. J Hayes. *Aeromonas hydrophila* in Oregon State University, 24 Nov 2020, Corvallis (2000)
5. JA Plumb, A Larry, LA Hanson. Health Maintenance and Principal Microbial Diseases of Cultured Fishes in Blackwell Publishing **1**, 492 (2011)
6. JB Fernandez, AV Yambot, O Almeria. Vaccination of Nile tilapia *Oreochromis niloticus* using LPS prepared from *Aeromonas hydrophila*. International Journal of Fauna dan Biological Studies **1**, 1-3 (2014)
7. K Molnar, C Szekely, M Lang. Field Guide to Warmwater Fish Disease in Central and Eastern Europe, The Caucasus and Central Asia in Food and Agriculture Organization of The United Nations **1**, 128 (2019)
8. Reed LJ, Muench H. A simple method of estimating fifty per cent endpoints. Journal of Epidemiology **1**, 493-497 (1938)
9. RF Floyd. Aeromonas Infection in University of Florida, 27 Nov 2020, Florida (2002)
10. S Sukenda, O Carman, R Rahman, D Hidayatullah, SN Yumaidawati. Vaccination in Nile tilapia broodstock with whole-cell vaccine and disease resistance in its fry against *Aeromonas hydrophila.* Journal of Indonesia Aquaculture **2,** 268-276 (2017)
11. S. Dehghani, M Akhaghi, M Dehghani. Efficacy of formalin-killed, heat-killed and lipopolysaccharide vaccines against motile aeromonas infection in rainbow trout *Oncorhynchus mykiss,* Global Veterinaria **9**, 409-415 (2012)
12. Song, X., Zhao, J., Bo, J., Liu, Z., Wu, K., Gong, C. Aeromonas hydrophila induces intestinal inflammation in grass carp (*Ctenopharyngodon idella*): an experimental model. Aquaculture 434, 171–178 (2014)
13. T. Aoki. Motile aeromonas (*Aeromonas hydrophila*) **3**, 427-453 (1999)

Estimated Yield Potential of Robusta Coffee (*Coffea canephora* Pierre ex A. Froehner) at Bogor District

Ade Astri Muliasari[1*], *Helianthi* Dewi[2]

[1]Technology and Management of Plantation, College of Vocational Studies. Bogor Agricultural University, 16128 Indonesia
[2]Ecotourism, College of Vocational Studies. Bogor Agricultural University, 16128 Indonesia

Abstract. Bogor District is one of the Robusta coffee (*Coffea canephora* Pierre ex A. Froehner) producing areas in West Java. To estimate crop yield of Robusta coffee, planters usually count the amount of a given crop harvested in a sample area. Then the harvested crop is weighed, and the crop yield of the entire field is extrapolated from the sample. Recently by utilizing the geographic information system (GIS) used to determine the level of regional suitability combining with land productivity. It will help in analysing, planing and taking various policy for development Robusta coffee. This study aimed to estimate of yield potential of Robusta coffee. This research was conducted from November to December 2020 at Bogor District. The land suitability class for Robusta coffee in Bogor which was in the moderately suitable (S2) class of around 2% (5,227.78 ha), marginal (S3) 33% (99,189.20 ha) and not suitable (N) 65% (194,808.40 ha). The yield potential of Robusta coffee is obtained 105,030,473.1 kg or 105,030.47 tons. The highest yield potential of Robusta coffee is in Cigudeg and Sukamakmur subdistricts.

1 Introduction

Coffee has been viewed as a tropical commodity that links producing countries of coffee along the equator that's affects substantial policies and investments to sustainably manage economic and environmental resources for the coffee sector, increase export earnings, and ensure stable production have had a tremendous role in country's integration into and success in the global market. Two main species of coffee are of economic importance worldwide: *Coffea Canephora* Pierre ex A. Froehner and *Coffea arabica* L. Viet Nam, Brazil, Indonesia, Uganda and India are the five largest producers of Robusta coffee. Robusta coffee is considered to be easier and less costly to produce. It produces fruit more quickly and yields more per tree than Arabicas. Robusta has a stronger, harsher, and bitter taste, and contains twice as much caffeine as Arabica beans [1].

About 60% of the total coffee production has been exported to various countries. Currently, the national coffee production is still dominated by robusta coffee which reaches 90% [1]. Total production of robusta coffee in Indonesia reached 534,360 tons from 896,680 hectares (ha) with an average productivity value of 0.562 tons/ha/year in the last five years

[2]. In many tropical and subtropical regions, climate change is fast becoming an environmental disaster for farmers, with decreased water availability, new or altered insect and pest pressures, and increased risks of extreme events threatening crop yields and farmer livelihoods [3].

Bogor district is one of the Robusta coffee producing areas in West Java Province, Indonesia. The area of coffee plantations continued to increase to 6,407.70 ha in 2019. Currently, the area of Robusta coffee reaches 5,833.70 ha with produce 3,698,381.19 kg coffee beans. Areas of Robusta coffee are located in Sukamakmur, Tanjungsari, Cariu, Tenjolaya, Babakan Madang, Cigudeg, Jasinga, Pamijahan [4].

Estimated of crop yield is important for planing and taking various policy decisions. To estimate crop yield of Robusta coffee, planters usually count the amount of a given crop harvested in a sample area. Then the harvested crop is weighed, and the crop yield of the entire field is extrapolated from the sample. Generally, use the conventional technique of data collection for crop monitoring and yield estimated based on ground-based visits and report. Recently by utilizing the geographic information system (GIS) used to determine the level of regional suitability combining with land productivity. It will help in analysing, planing and taking various policy for development Robusta coffee. This study aimed to estimated yield potential of Robusta coffee based on potential land and productivity of Robusta coffee.

2 Materials and Method

The research was held from July to December 2020 at Bogor, West Java. This research was conducted based on Digital Elevation Model (DEM), agroclimate, soil physical/chemical properties, land use and socioeconomics parameter such as protected area, lake, road, river by utilizing GIS. The input data is processed using an interpolation and classification process based on the land suitability criteria of each parameter (Table 1) [5]. Overlays are maps of various themes that can be overlapped to produce new mapping units with new information [6].

After obtaining land suitability, then the appropriate land potential, the next step is to combine it with land productivity data so as to produce an estimate of the potential yield of Robusta coffee. Figure 1 is the flowchart illustrating and procedure for the data analysis land suitability of Robusta coffee. Figure 2. Flowchart and procedure for estimated yield potential of Robusta coffee.

Table 1. Land suitability class criteria for Robusta coffee

Land characteristics	Land Suitability Class			
	Very suitable (S1)	Suitable (S2)	Marginal suitable (S3)	Not Suitable (N)
Agroclimate				
Rainfall (mm)	1,500-2,000	2,000-2,500	2,000-3,000	<1,000 >3,000
Dry month (<60 mm/month)	2-3	3-4	4-5 1-2	<1>5
Altitude (m above sea level)	300-500	500-600	600-700	> 700
Slope (%)	0-8	8-25	25-45	>45
Soil Physical properties				
Effective depth (cm)	>150	100-150	60-100	<60

soil texture	sandy loam	loamy sandy	clay	sand
Drainage	well drained	moderately well drained	somewhat poorly drained, somewhat excessively drained	very poorly drained, excessively drained
Soil Chemical properties				
pH	5.5-6.0	6.1-7.0	7.1-8.0	>8
CEC (me/100 g)	>15	10-15	5-10	<5
Base Saturation (%)	>35	20-35	<20	

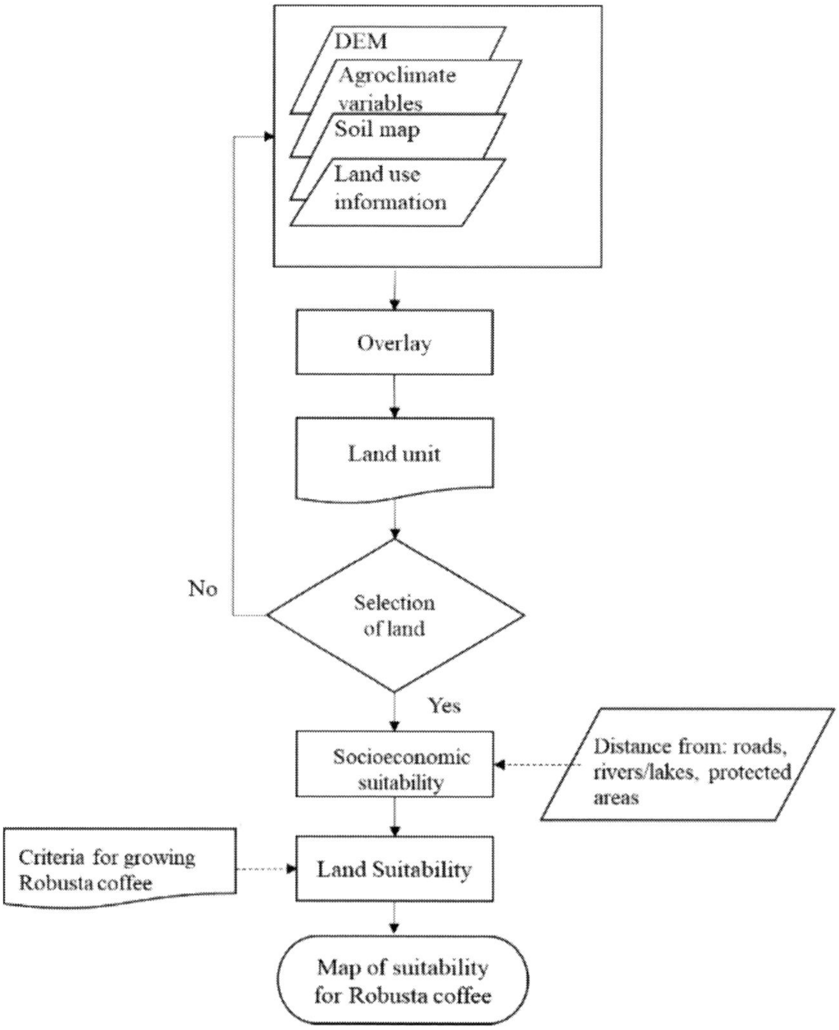

Fig. 1. Flowchart for the study method and data analysis

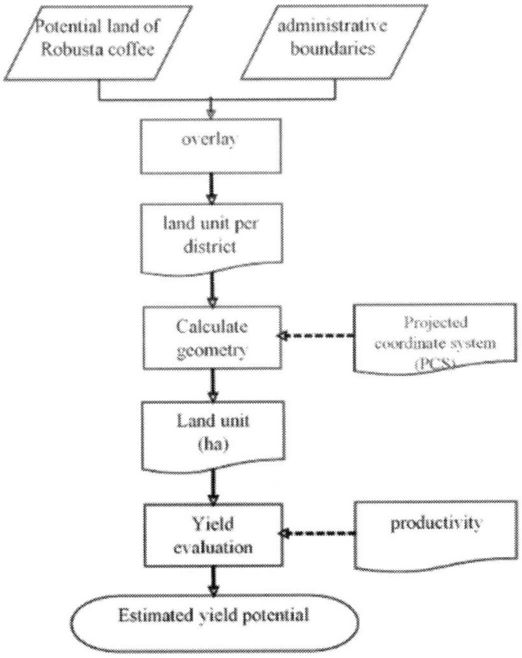

Fig. 2. Flowchart of estimated yield potential of Robusta coffee

3 Results and discussions

The area of land suitability class for Robusta coffee which was based on each parameter (agroclimate, soil physical and chemical properties) is presented in Table 2. The suitability class for Robusta coffee in Bogor which was in the moderately suitable (S2) class of around 2% (5 227.78 ha), marginal (S3) 33% (99 189.20 ha) and not suitable (N) 65% (194 808.40 ha) is shown in Figure 3.

Table 2. The suitability of land in area (Ha) based on the category of very suitable (S1), quite suitable (S2), marginal (S3), and not suitable (N) for each parameter of Robusta coffee in Bogor

Parameter	Land Area Suitability Class (Ha)			
	S1	S2	S3	N
Rainfall (mm/year)	8,466.18	94,017.16	99,465.85	97,276.22
Dry Month (mm/month)	97,760.18	111,592.64	77,703.10	12,169.49
Altitude (meters above sea level)	48,608.30	111,296.86	80,314.49	59,005.77
Slope (%)	105,120.00	106,266.57	53,763.79	34,075.05
Effective depth (cm)	114,129.59	137,250.63	46,900.03	945.17
Soil texture	51,212.78	173,578.10	64,347.20	10,087.33
Drainage	264,649.90	4,984.59	19,505.08	10,085.84
pH	125,087.21	18,484.69	154,708.34	945.17
Base Saturation (%)	201,220.51	97,059.73	0.00	945.17
CEC (me/100 g)	187,513.12	110,767.13	0.00	945.17

Climate is a key determinant of successful coffee production. Rainfall and dry months data in the study area were based on agroclimatic variables required criteria for Robusta coffee. The average rainfall in Bogor district ranging from 232-517 mm/month, so it was considered the rainy season, with the highest peak of rain was in April. Meanwhile, the lowest average rainfall happened in June-September, which reached 32.6-99.5 mm/month, with the lowest rainfall occurring in July. Various coffee bean characteristic subcomponents (e.g. insect damage and mouldy beans) and different bean sizes were also examined and were affected by a range of rainfall and temperature predictors across the flowering, growing and harvest seasons [7].

Indonesia especially bogor district is subject to high variability in rainfall, like Vietnam with droughts and floods common and anticipated to intensify under climate change [8]. Climatic conditions that reduce coffee bean profitability (e.g. because of increased defects) has widespread and high cumulative impacts. The average dry month ranging 1-6 month, its can be affects the process of primordial flower formation, initiation of flowering, pollination, and fertilization [9]. During the flowering process, coffee requires a dry month for the success of the pollination process so that it can produce fruit. If rain drops during the pollination period, usually the pollen will clump, and the flowers will be damaged so that pollination will not occur or fail to become fruit.

Robusta coffee is grown between 300-700 m above sea level. Altitude, a known surrogate variable for temperature, is the main driver of the epidemics of coffee leaf rust (CLR) diseases. Incidence and severity were highest in the lowland fields than highland, where poorly managed plantations of local varieties which are grown under the open sun were also more dominant. CLR intensity decreased with the increase in altitude at the highlands, where well-managed and improved varieties are grown under the shade [10]. The elevation a coffee is grown at directly affects the physical characteristics of the beans [11].

The parameters of the observed physical properties of the soil included effective depth, soil texture, and drainage speed. Soil texture was determined by the size of the soil particles represented by the percentage of sand, silt, and clay in the soil [12]. The texture was the relative ratio between the fractions of sand, silt, and clay, i.e., soil particles which effective diameter was 2 mm. This parameter was one of the most essential soil characteristics that affected soil moisture, drainage, infiltration, and retention of nutrient and water capacity [13].

The land suitability class was the S2 class which the highest limiting factor was rainfall and drainage speed parameter. The area for this suitability is 5,227.78 ha for the S2 class, 99,189.20 ha for the S3 class and 194,808.40 ha for not suitable (Fig. 3). The highlands areas are not recommended for Robusta coffee growing.

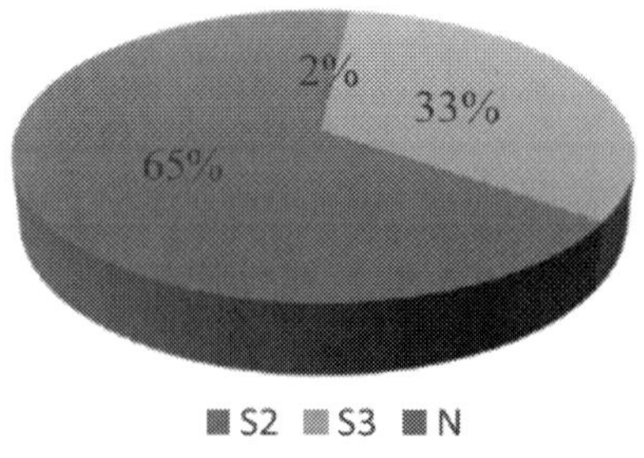

Fig. 3. Proportion of land suitability class for Robusta coffee

Based on the results of the analysis of potential land suitability maps and the availability of land for the development of Robusta coffee plants in Bogor Regency, it can be seen that the land that can be used as coffee commodity development is included in the S2 and S3 class categories because there is no S1 land suitability class. Referring to the data in Table 2 regarding the average productivity of Bogor coffee where several sub-districts that are centers of Robusta coffee production have their respective productivity, while for other districts using the average productivity converted to the potential area of suitability, it will produce potential results such as in Table 3.

Table 3. Yield potential of Robusta coffee in Bogor district

Sub District	Productivity (kg ha^{-1})	Land area (ha)	Yield Potential (kg)
Babakanmadang	800.0	1,809.356	1,447,485.144
Bojonggede	874.5	238.225	208,339.819
Caringin	874.5	3,136.408	2,742,945.235
Cariu	950.0	5,308.086	5,042,681.661
Ciawi	874.5	851.006	744,247.921
Cibinong	874.5	22.312	19,513.767
Cigombong	874.5	2,043.351	1,787,012.493
Cigudeg	912.1	9,549.057	8,709,694.909
Cijeruk	874.5	601.676	526,195.417
Cileungsi	874.5	2,261.127	1,977,468.615
Cisarua	874.5	17.072	14,930.873
Ciseeng	874.5	2,313.424	2,023,204.632
Citeureup	874.5	2,309.388	2,019,675.476
Gunungputri	874.5	2,021.208	1,767,647.686
Gunungsindur	874.5	2,738.552	2,395,000.563
Jasinga	800.0	10,726.810	8,581,446.342
Jonggol	874.5	9,036.056	7,902,482.796
Kemang	874.5	262.340	229,430.122
Klapanunggal	874.5	6,363.919	5,565,565.036
Leuwisadeng	874.5	222.925	194,959.045
Megamendung	874.5	428.815	375,020.502
Nanggung	874.5	1,785.176	1,561,226.029
Parung	874.5	1,159.888	1,014,380.247
Parungpanjang	874.5	5,205.681	4,552,628.234
Rancabungur	874.5	5.486	4,798.381
Rumpin	874.5	5,468.600	4,782,563.976
Sukajaya	874.5	4,534.958	3,966,047.780
Sukamakmur	891.6	9,630.534	8,587,066.079
Sukaraja	874.5	255.154	223,145.117
Tajurhalang	874.5	904.382	790,927.577

Tanjungsari	911.2	8,486.797	7,733,254.037
Tenjo	874.5	4,719.208	4,127,182.979
Total			105,030,473.100
Yield potential			105,030.473 ton

Based on Table 3 Yield potential lands for Robusta coffee development were located in Cigudeg, Sukamakmur, Jasinga, Tanjungsari, Jonggol, Klapanunggal, Cariu, Rumpin, Parung panjang, Tenjo, Sukajaya, Caringin, Gunung Sindur, Ciseeng, Citeureup.

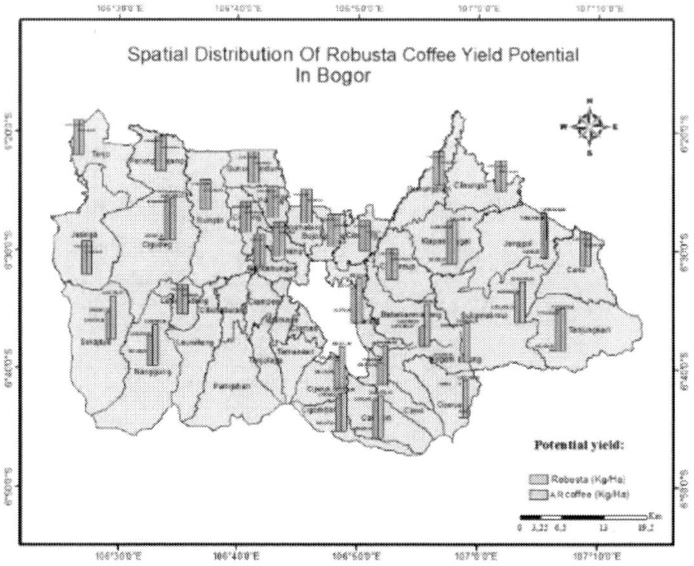

Fig. 4. Spatial distribution of Robusta Coffee Yield potential in Bogor

4. Conclusion

Based on data analysis, land suitability class for Robusta coffee in Bogor which was in the moderately suitable (S2) class of around 2% (5,227.78 ha), marginal (S3) 33% (99,189.20 ha) and not suitable (N) 65% (194,808.40 ha). The potential yield of Robusta coffee in Bogor district reached 105,030,473.1 kg or 105,030.47 ton. The highest potential for Robusta coffee is found in Cigudeg and Sukamakmur subdistricts.

Acknowledgments

We would like to thank Dr. Ir. Arief Darjanto, M.Ec as the Dean of the vocational school, Bogor Agricultural University as the sponsor of this research through a competitive grant program for vocational school applied researches.

References

[1] International Coffee Organization Coffee Development Report the Value of Coffee: Sustainability, Inclusiveness, and Resilience (2020)
[2] Badan Pusat Statistika Statistika Kopi Indonesia Direktorat Statistik Tanaman Perkebunan (2018)
[3] M.L. Parry, O.F. Canziani, J.P. Palutikof, P.J. van der Linden and C.E. Hanson. Climate Change : Impacts, Adaptation and Vulnerability. Contribution of Working Group II to the Fourth Assessment Report of the Intergovernmental Panel on Climate Change. Cambridge University Press, Cambridge, UK, 976 (2007)
[4] Dinas Tanaman Pangan Hortikultura dan Perkebunan. Rekapitulasi Luas Areal dan Produksi Perkebunan Rakyat Kabupaten Bogor Tanaman Tahunan Tahun 2018. Distanhorbun Kabupaten Bogor (2018)
[5] Ministry of Agriculture Directorate General of Plantation 2014 *Good Agriculture Practices /Gap on Coffee)*
[6] Klee P 2009 *Principles of Geographic Information Systems* Vol **4,** ed O Huisman and R A de By (Netherlands: ITC)
[7] J. Kath, V.M. Byrareddy, S.M. Mushtaq, A. Craparo, M. Porcel. Climate Risk Management. 32 100281 (2021)
[8] Schmidt-Thome, P., Nguyen, T.H., Pham, T.L., Jarva, J., Nuottim ̈aki, K. Preface. In: Climate Change Adaptation Measures in Vietnam. Springer Briefs in Earth Sciences. Springer, Cham. (2015)
[9] Kandari A M, Safuan L A O D E and Amsil L M 2013 *J. AGROTEKNOS* **3** 8–13
[10] Belachew K, Senbeta G A, Garedew W, Barreto R W and Del Ponte E M 2020 *Trop. Plant Pathol.* **45** 511–21
[11] A.A. Tassew, G.B. Yadessa, A.D. Bote, T.K. Obso. Heliyon. 7 e07790. (2021)
[12] Karimi F, Sultana S, Shirzadi Babakan A and Royall D 2018 *Appl. Geogr.* **4** 326–42
[13] Eponon E C, Snoeck D, Konaté Z, Kassin K E, Camara M, Legnaté H, Konan A, Chérif M and Koné D 2019. *J. Agric. For. Sci.* **7** 15–24

Consumer Behavior on Beef Purchasing Decision in West Java

Intani Dewi[1], *Khoirul* Aziz Husyairi[1], *Doni* Sahat Tua Manalu[1]

[1] Agribussiness Management Program, College of Vocational Studies (SV) IPB University, Indonesia

Abstract. The aim of this study was to determine consumer behavior in making beef purchasing decisions. The study was carried out for 3 months from April to June 2021. Research data were obtained from 300 respondents who used to buy beef from several areas in West Java. Sampling was carried out by convenience sampling method and data collection techniques using questionnaires. The analytical method used was descriptive analysis. The results of the analysis concluded that some respondents bought meat to meet family needs (as side dishes), with an average consumption requirement of less than 1 kg per week. Purchases of beef have been planned and most of them buy meat from traders in traditional markets who have become customers. Information related to meat prices was obtained by respondents from relatives or close neighbors. Even so, the respondent has no influence from other people in making their meat purchasing decisions. The portion of beef that often purchased was tenderloin, and if the price of beef rises the respondent will buy chicken as a substitute. The beef usually purchased in the form of fresh meat with a custom cut based on the needs. Mornings and weekends are the most preferred time for consumers to buy beef. Most purchases were made in cash. Only a small part is done in non-cash (debit). Most of the respondents also already trust the existing regular meat traders and willing to recommend them to their relatives/others. The reason was that the meat sold fit to the price and quality.

1 Introduction

Beef is one of very popular food in Indonesia, this can be seen from the increasing demand for beef in Indonesia. The Ministry of Agriculture noted that beef production centers in Indonesia with an average beef production in 2016-2020 are in 3 (three) provinces, namely East Java with a contribution of 20.00% (100.91 thousand tons), West Java contributing 15 .45% (77.97 thousand tons), and Central Java contributing 12.43% (62.73 thousand tons). During the last five years (2016-2020) household beef consumption increased by 3.98% per year [1].

Household consumption of fresh beef in 2019 was 0.485 kg/capita/year, an increase of 4.48% from 2018 of 0.464 kg/capita/year. There are various reasons why Indonesians consume beef, one of that is beef tastes better than other meats. Seen from the aspect of the nutritional content of beef and the change in Indonesian people's awareness of animal protein for health, beef become the second position after poultry meat in meeting the animal protein needs of the Indonesian people [2]. Then, another thing that happens to the behavior of beef consumers is the culture inherent in the people of Indonesia where the demand for beef increases, especially during religious holidays such as Eid Fitri, this makes the presence of beef not replaced with other meats so that it affects the price increase [3].

In 2020 there was a Covid-19 pandemic in Indonesia which had an impact on changes in consumer behavior, especially in purchasing beef. Changes in consumer behavior will certainly affect the demand and supply conditions for beef, especially in beef production centers. Consumer behavior that occurs in Indonesia, especially West Java as the province with the second largest beef production and consumption in Indonesia, needs to be analyzed further, the dynamics of consumer behavior will show how consumers are in the process of making beef purchasing decisions. Based on this description, the aim of this study is to determine consumer behavior in making beef purchasing decisions.

2 Methods

The research was carried out for 3 months, from April to June 2021. The research data was obtained from 300 respondents who used to buy beef from several areas in West Java. The sampling technique was carried out by the convenience sampling method where the sampling carried out by the researcher was free to choose individuals as respondents based on the author's criteria so that they could be used as samples [4]. Data were collected using a questionnaire. The analytical method used was descriptive analysis.

The analytical method used was the stages of consumer decision making. There were 5 stages that consumers went through when decided to buy a product: problem recognition, information search, evaluation of alternatives, purchase decisions and post-purchase evaluation [5].

3 Results and Discussions

3.1. Motivation and Introduction

Respondents in deciding the purchase of beef were mostly already planned. But some were done based on the situation. The purchase of beef was generally used to welcome certain moments such as the fasting month and Eid Fitri. While the purchase depends on the situation made when there were events in the family.

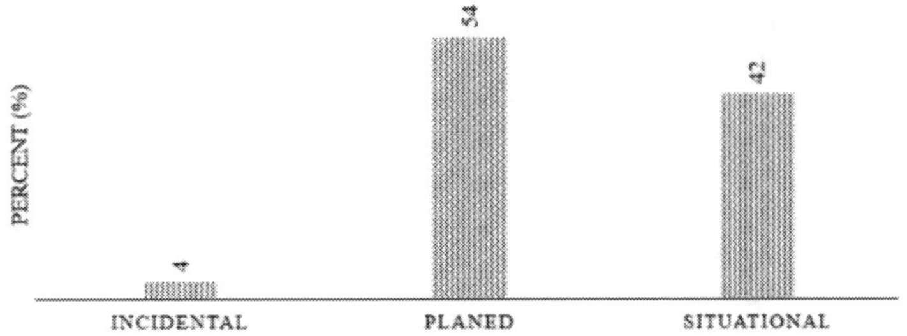

Fig. 1. Percentage of respondents' answers related to how to decide to buy beef

Respondents in buying beef were intended for family consumption. This is also because most of the respondents are housewives. Only a few respondents answered for business (restaurants). The family is also the biggest external factor in purchasing beef, where the desire to consume beef becomes the respondent reference to buy the product [6].

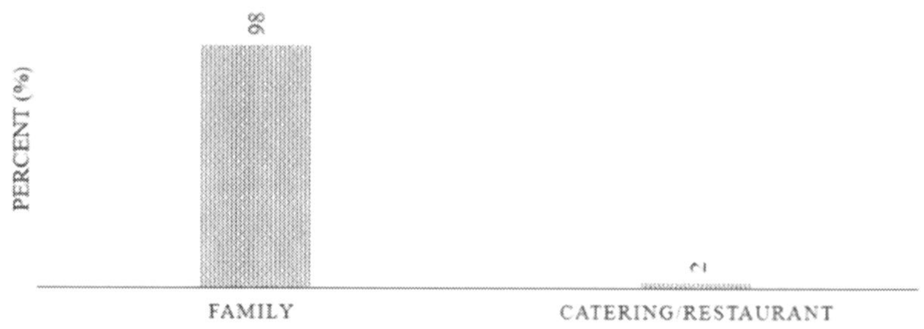

Fig. 2. Percentage of respondents' answers regarding the need to buy beef

The purchase of beef was usually done at certain moments and for family consumption, so the reason for consuming beef was as a food menu or family side dish [7]. Several respondents answered that beef has a distinctive and delicious taste, and as a source of animal protein. Taste was also the biggest driver for sustaining meat consumption, even in situations of uncertainty [8].

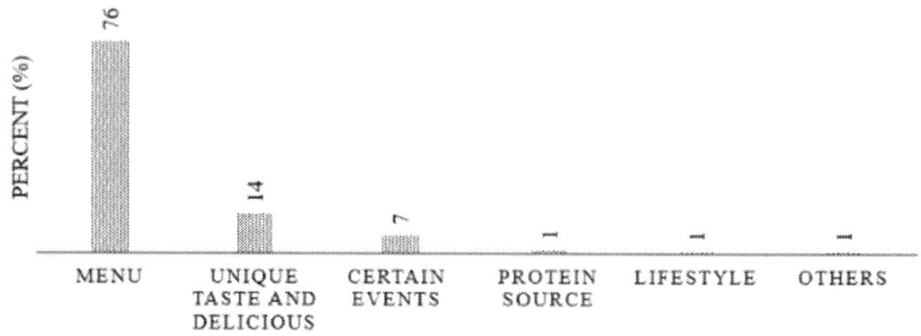

Fig. 3. Percentage of respondents' answers about the reasons for consuming beef

The frequency of respondents in consuming beef was categorized as rare. Most of the respondents consume 1-2 times a week. Even so, there are 12 percent of respondents who stated that they often consume beef. Respondents who often consume beef often do not only processing it themselves, but also from buying side dishes at food stalls.

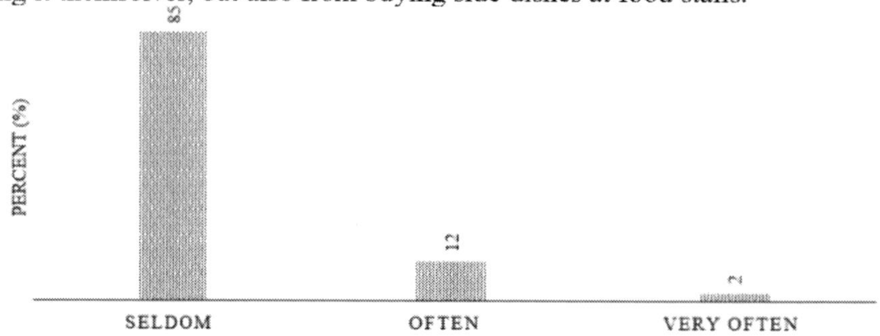

Fig. 4. Percentage of respondents' answers regarding the frequency of consuming beef

The average consumption of beef for most respondents was less than 1 kg per week. Consumers who consume between 1-2 kg a week are usually consumers who process beef not only for their own consumption, but also for sale. Likewise with respondents who consume 2 kg of beef per week.

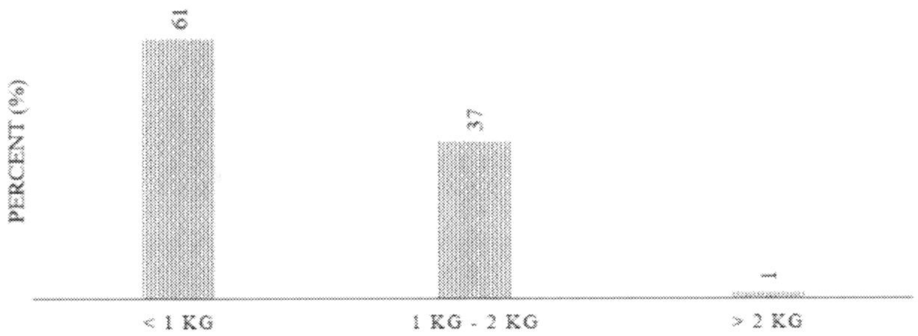

Fig. 5. Percentage of respondents' answers related to the amount of beef consumption per week

3.2. Information Search

The process that was carried out after the identification of needs is information seeking. One source of respondent's information was promotion/advertising. Respondents who have seen beef promotion/advertising were 71 percent.

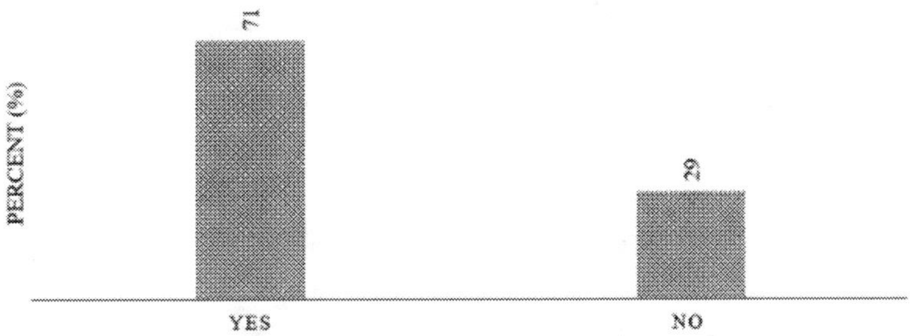

Fig. 6. Percentage of respondents' answers related to beef promotion/advertising

Although 71 percent stated that they had seen advertisements/promotions related to beef, most of the sources of beef information came from other people. The second source was electronic media, especially advertisements on social media. Only a small proportion of respondents get their source of information from the print media.

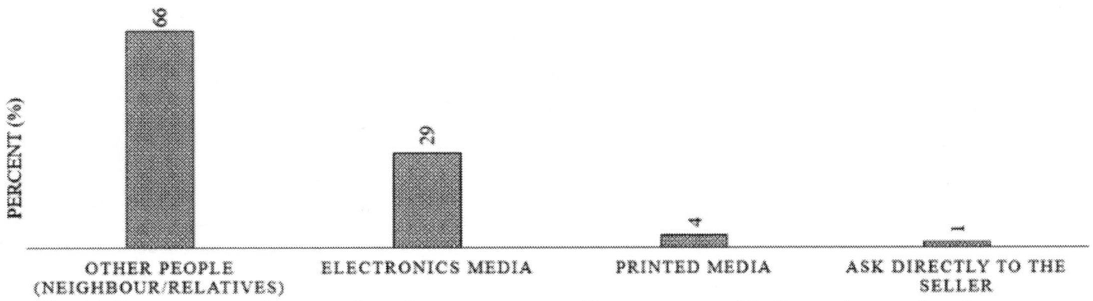

Fig. 7. Percentage of respondents' answers regarding sources of information on beef products

In making the decision to buy beef, respondents decided on personal influence. This means that the sources of information obtained by respondents both from advertisements and other people were used as consideration, but the decision remains in the hands of the respondent.

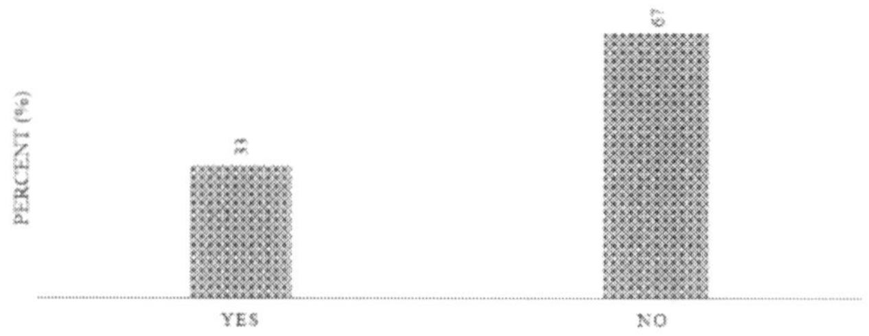

Fig. 8. Percentage of respondents' answers related to the influence of others in buying meat

3.3. Evaluation of Alternatives

The next stage after the information search was the evaluation of alternatives. In conducting the evaluation, the thing that becomes the evaluation material was related to the part of the meat to be purchased. The meat that many respondents were looking for is the tenderloin portion, where tenderloin was also one of the attributes that were consumer preferences in buying beef [9].

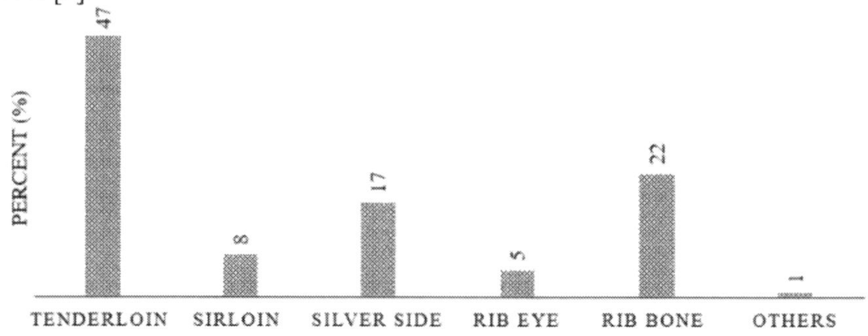

Fig. 9. Percentage of respondents' answers related to the most sought-after part of beef

The search results were based on previous buying habits, that the portion of beef that the respondents often buy was shown in Figure 10.

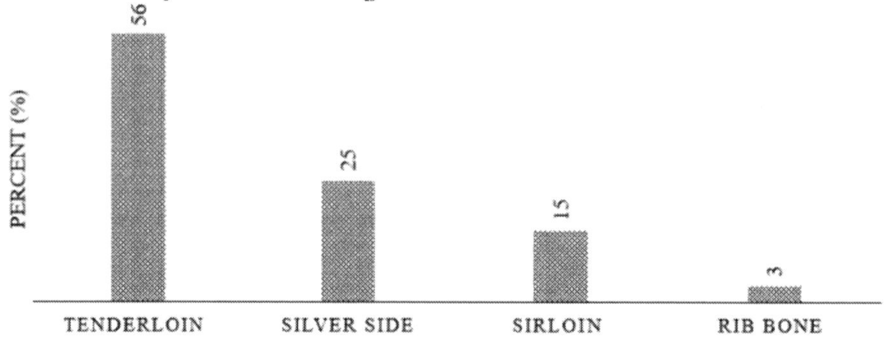

Fig. 10. Percentage of respondents' answers related to the part of beef that is often purchased

Tenderloin is a part of beef that is tender, dense but tends to be a little fat, so it is suitable for processing to various kinds of dishes. There were also some respondents who buy beef bones (ribs). It was very suitable for making soup or gravy for dishes such as meatballs.

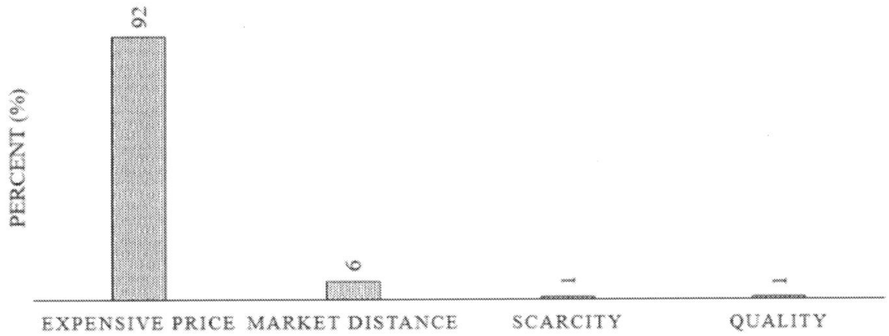

Fig. 11. Percentage of respondents' answers related to obstacles in buying beef

The price of meat was the main factor that is considered by respondents to buy meat [10] as shown in Figure 11. Especially during the pandemic where people's purchasing power were low [11]. However, this does not apply at certain moments. This means that even though the price of beef is expensive, consumers will still buy it as before fasting or Eid Fitri even though the quantity of purchases was reduced. This condition occurs due to the high demand to meet consumption needs at the beginning of the fasting month and before Eid Fitri [12][13].

3.4. Purchase

Traditional markets were the places most respondents visit to buy beef. The reason was because the beef sold in the market is fresh meat. Apart from traditional markets, some respondents also buy from modern markets, online and mobile traders.

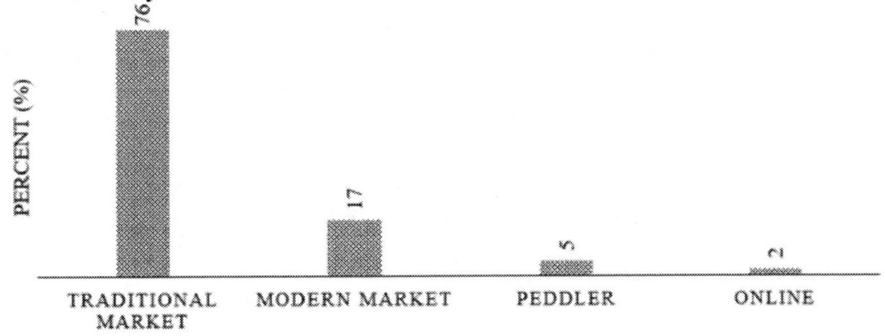

Fig. 12. Percentage of respondents' answers regarding where to buy beef

The reason consumers bought meat in these places was that the quality of the product is guaranteed. This showed that although there are traditional markets, consumers already believe in the quality of the products sold. The second reason was the location where the purchase is near their home. The third most common reason was the lower price. More information related to this information can be seen in Figure 13.

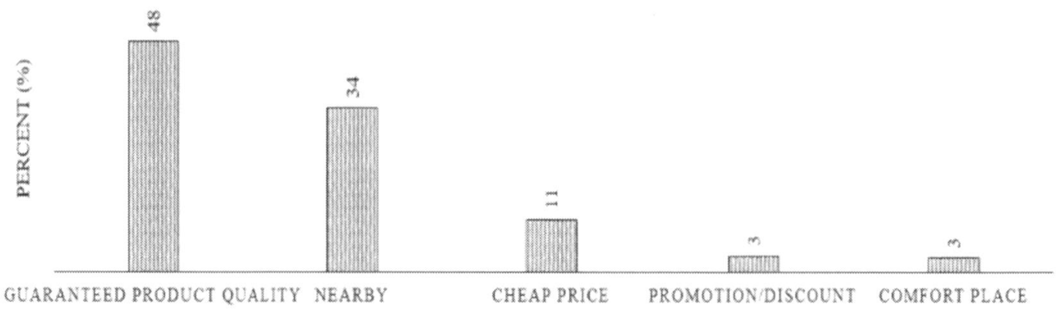

Fig. 13. Percentage of respondents' answers regarding the reasons for choosing where to buy beef

The beef purchased is usually in the form of fresh meat, while the size of the pieces is adjusted to the needs. Some consumers also buy meat in the form of cuts. In this regard, consumers usually buy meat in traditional markets where the cuts can be adjusted. In contrast to the modern market where the meat is sold already in the form of pieces.

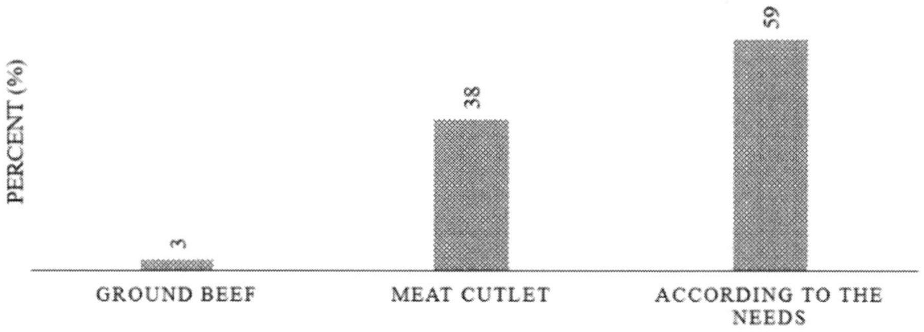

Fig. 14. Percentage of respondents' answers related to the form of meat they usually buy

The amount of meat purchased by consumers in one purchase is 1-2 Kg. Even so, the purchased meat didn't seem to be consumed all at once.

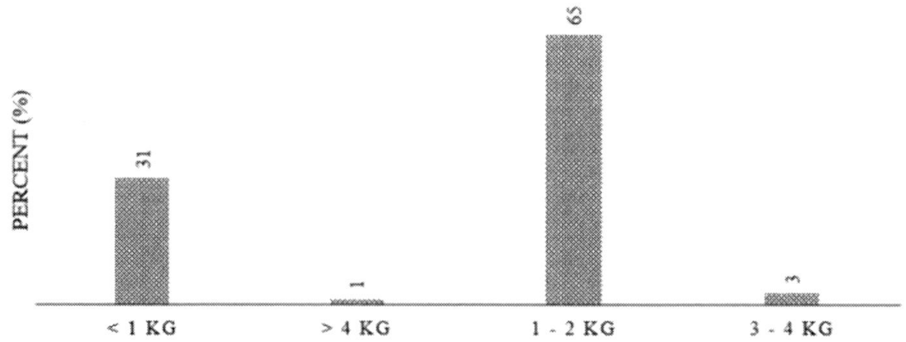

Fig. 15. Percentage of respondents' answers related to the amount of beef purchased in one purchase

Morning was the time that consumers choose to buy beef. Apart from the fact that most traditional markets operate in the morning, this purchase was also used to buy other necessities. Respondents usually buy beef at the end of the week to make dishes for family holidays.

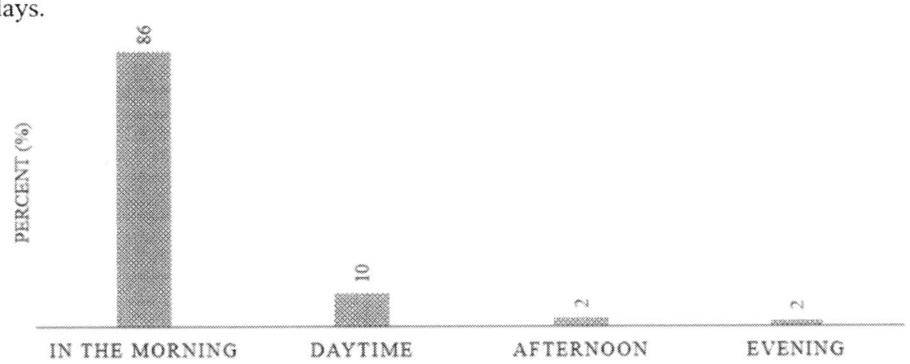

Fig. 16. Percentage of respondents' answers related to the time to buy beef

Most of the purchases were made in cash, considering that most of the consumers buy meat in traditional markets and at mobile traders. Only a small part was done in non-cash (debit). This type of consumer buys beef in modern markets and online.

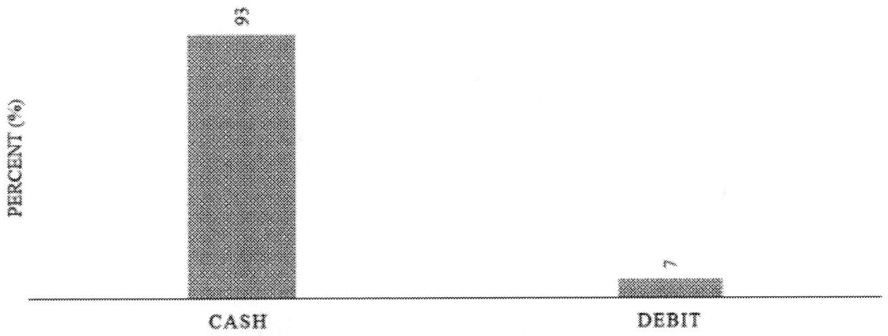

Fig. 17. Percentage of respondents' answers related to the method of payment when buying beef per week

3.5. Result and Post Purchase

The results showed that consumers will repurchase to the same place as shown in Figure 18. Only some of the consumers will not buy again at the current place.

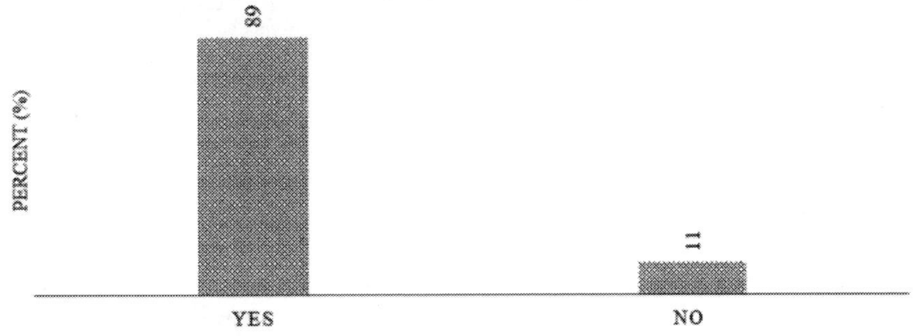

Fig. 18. Percentage of respondents' answers related to the desire to buy back beef at the current place

The reason respondents bought from this place was the quality of the meat in line with expectations. Another reason was cheap price. Expensive was the main reason for consumers not to repurchase at the current place.

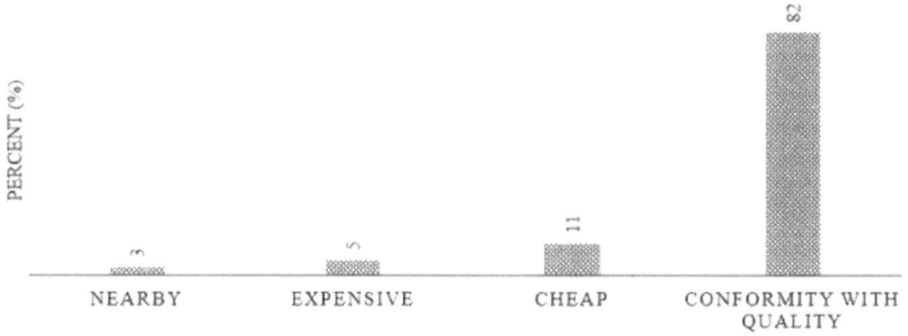

Fig. 19. Percentage of respondents' answers related to reasons to buy back beef at the current place

Respondents also willing to recommend the current place to family and friends. The main reason was that the taste of the meat matches the price with the quality.

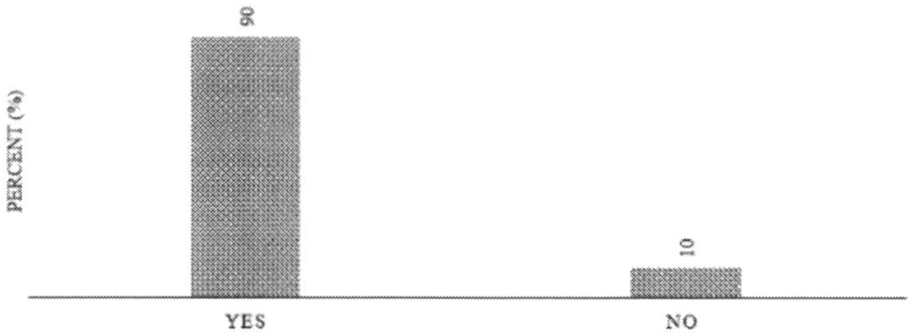

Fig. 20. Percentage of respondents' answers related to willingness to recommend to family/friends where to buy beef current time

In this study, it is also known that if the price of beef increases, the respondents will buy chicken meat as a substitute as shown in Figure 21.

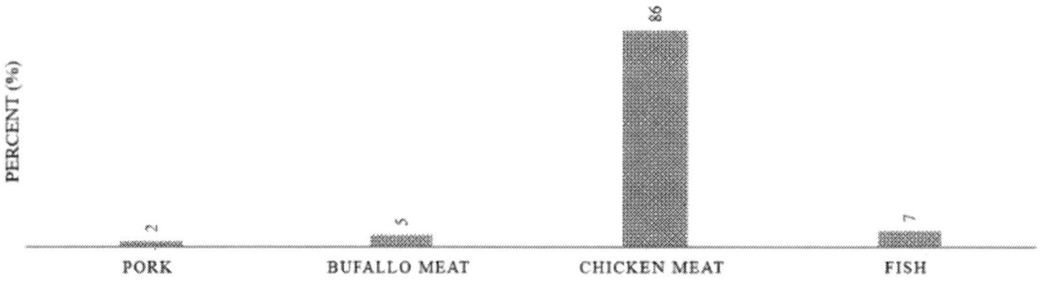

Fig. 21. Percentage of respondents' answers related to the type of meat that will be purchased if the price of beef increases

The second option was fish. The surprising thing was buffalo meat where only a small number of people choose it as a substitute for beef. This was contrary to the government's policy to import buffalo meat as an alternative to beef.

4 Conclusion

In the process of buying beef, there are 5 stages that are passed, namely need recognition, information search, alternative evaluation, purchase, post-purchase evaluation. In the need recognition stage, it was found that 54% of respondents in deciding whether to buy beef had planned and 42% were done based on the situation, while 4% were done suddenly. One of the respondents' sources of information was through promotions/advertising, 71% of respondents stated that they had seen beef promotions/advertising and 29% had never. Then, in conducting the evaluation, it was found that the thing to be evaluated was related to the part of the meat to be purchased. At the purchase stage 76% of respondents made purchases in traditional markets. Furthermore, in the post-purchase evaluation stage, it was found that 89% of respondents would repurchase to the same place and the other 11% would not. If the price of beef goes up, 86% of respondents will replace it with chicken. The remaining 11% replace it with fish, buffalo meat or pork.

References

1. Ministry of Agriculture. Beef Farming Commodity Outlook Book. Jakarta: Agricultural Data Center and Information System (2020)
2. Rusdiana, S. & Maesya, A. Economic Growth and Food Demand in Indonesia. Agriekonomika, 6, 1, 12-25 (2017)
3. Komalawati, R.W. Asmarantaka, R.Nurmalina, D.B. Hakim. Beef Price Volatility and Transmission in Indonesia: Case Studies in Jakarta, Bandung, Semarang and Surabaya, Buletin Ilmiah Litbang Perdagangan, 15, 1 (2021)
4. Sugiyono. Research Methods: Quantitative, Qualitative, and R & D. Alfabeta, Bandung (2016)
5. P. Kotler, G. Armstrong. Marketing Principles. 13th Edition. Volume 1, Erlangga, Jakarta (2016)
6. A. Usman, Hastang, V.S. Lestari. External Factors That Determine Beef Purchase Decisions at Pasar Terong. Journal of Animal Science and Industry, 3, 2 (2016)
7. J. Purwono, S. Sugyaningsih, M. Andryani. Beef Purchase Decision Analysis by Household Consumers, NeoBis Journal 8, 1 (2014)
8. W. Verbeke, I. Vackier. Profile and Effect of Consumer Involvement in Fresh Meat. Meat Science, 67, 159–168 (2004)
9. H. Asyrif, S.Susilowati, O. Rahayu P. Factors Affecting the Level of Consumer Preference and Decision Making in Choosing Beef at Traditional Markets, Sape District, Bima Regency. Dinamika Rekasatwa Journal 4, 1 (2021)
10. Merlino, VM., Borra, D., Girgenti, V., Dal Vecchio, A., Massaglia, S. Beef Meat Preferences of Consumers from Northwest Italy: Analysis of Choice Attributes. Meat Science, Vol 143, page 119-128 (2018)
11. A.P. Saksono, Mukson, S.I. Santoso. Analysis of Beef Consumer Behavior during the Pandemic Period in Banyumanik District. Agribusiness Paradigm 3, 2, 1-4 (2021)
12. A. Priyanti, I. Inounu. Price Behavior of Livestock Products on Religious Holidays. Agricultural Policy Analysis. 14, 2, 149-162 (2016)
13. R.S.Siregar. Analysis of Factors Affecting Consumer Behavior in Buying Beef (Case Study: Petisah Market, Medan Petisah District, Medan City), JASc 01,02 (2018)

Aquaculture and Its Impact of The Covid-19 Pandemic on The Fish Processing Industry: Case Study from Local Community

Ima Kusumanti[1*], *Muhammed* A. Oyinlola[2], *Muslikhah* Nanda[1], *Mugi* Pangestu Putu Hamka[1]

[1]Production and Management Technology of Aquaculture Study Program, College of Vocational Studies IPB University (Bogor Agricultural University), Jl. Kumbang No.14 Cilibende, Bogor, Indonesia

[2]Institute for the Oceans and Fisheries, University of British Columbia: Vancouver, BC, CA

Abstract. The Indonesian government has facilitated farmers through the role of agricultural extensionists (AE). Covid-19 pandemic has caused heavily impacts on the fisheries sector especially on the socio-economic conditions of the stakeholders, e.g., fishers, fish farmers, traders, as well as consumers. This impact on aquaculture is about its production is largely influenced by the demand from the food service sectors, processing factories, and export. Movement restrictions of fish farmers and less demand from consumers needs to maintaining the stocks of cultured commodities becomes more expensive as most of the products could not be harvested. From the snow ball sampling methods and research interview, the covid-19 pandemic makes the fish processing industry decrease more than 56%. It impacts 3 main activities, such as reducing demand for fish, low prices due to cancellation of shipment by buyer and lack of technical service provider. This decrease has a positive impact on the community to overcome the existing problem, diversifying the product and recognize need for new strategies to identify marketing opportunities that use technology.

1 Introduction

1.1 Background

The COVID-19 outbreak has disrupted the lives and livelihoods of everyone around the world. Covid-19 spreads very quickly, mutates, and changes the economy and livelihoods of the entire population. As a result, the export potential for many industries has declined significantly, including the fishing sector, particularly shrimp and lobster production (1). The Covid-19 pandemic is a world-changing event that has caused a severe economic downturn in the fishing sector that has changed the lives and livelihoods of countless workers around the world (1), but especially those related to small-scale fisheries and coastal fishing communities, including Indonesia.

The movement of fishing activities in Indonesia mainly includes capture, aquaculture, and fish processing products. The impact of the WFH system that is implemented to avoid the spread of Covid-19 has more or less affected the fishing industry. Most online media reported lower fish prices caused by lower levels of consumer demand from households to exporters. Therefore, in some areas, fishers have stopped fishing because they fear that their production will not be absorbed by the market. Fish market activity in certain areas over the past month has begun to decline due to lack of buyers, so many fish are in decline in the market.

The Covid-19 pandemic has become a major challenge for the national fisheries sector in Indonesia. The short and long-term impacts of the pandemic have the potential to significantly reduce the welfare of small fishermen and coastal communities who are very vulnerable due to declining demand for fishery products. This study aims to determine the impact of the Covid-19 pandemic on the production and distribution of fishery products in local communities, especially fish processing industry.

Indonesia which has three different time zones in its territory: Western Indonesia Time (WIB) — seven hours ahead (UTC+07:00); Central Indonesia Time (WITA) — eight hours ahead (UTC+08:00); and Eastern Indonesia Time (WIT) — nine hours ahead (UTC+09:00) makes it has different characteristics either in their fisheries resources and also the way to added value of their fish product. Comparative between three time zones is needed to have a look the way its changes and factors which affected its changes. In this study, it will discuss focused on the time zone Western Indonesia Time (WIB), how the pandemic impact of the fish processing industry.

Several study site in Western Indonesia has been selected, one of which is in Jakarta Capital and its surrounding due to several places known as the key player in fish processing activity, one of which is in Muara Angke. Based on the Governor's Decree No. 2293 of 1984 that Muara Angke is one of the centers for processing traditional fishery products, one of which is salted fish. Muara Angke fish processing, is one of the largest salted fish or dried fish producing areas in Jakarta. The center of the salted fish processing industry is in the salting block of Muara Angke in the Kapuk village, Penjaringan sub-district, North Jakarta. Muara angke is a very busy fishing village and also one of the largest in the capital area. This industry even supplies most of the fish needs for the people of Jakarta and its surroundings. In addition to the existence of traditional fishermen, Muara Angke is also a place to market fish from various regions on the island of Java.

Early when the Covid-19 pandemic hit, the amount of salted fish production was still stable and did not decrease. Nevertheless, with rules enforcement of restrictions on community a activities (PPKM) insulation in some cities, especially Jakarta makes the marketing of fishery products to be blocked. For the medium to large scale fish processing industry, this is something that needs to be adjusted quickly. However, like some household-scale processing industries, of course it will be different, where in this area there are tens to hundreds of small-scale salted fish processing. Seeing these existing conditions, the level of community vulnerability to pandemic conditions needs to be studied more deeply and in detail. In this study, it is the first step to determine the supporting factors so that the current capital assets, namely in the form of salted fish processing activities to be able to continue to survive so that the community can carry out livelihood strategies.

Most of the fish processing activity derives from the Jakarta Bay (JBE) ecosystem on the mainland and in the islands. Small pelagic fish are one of the major resources of the JBE [2], [4]. Apart from being a cheap source of nutrients, it can increase the income and welfare of coastal communities. Studies reported that the types of species found in JBE were *Rastrlliger brachysoma* (kembung), *Rastrelliger kanagurta* (banyar), *Selar crumenophthalmus* (selar), *Selaroides leptolesis* (selar kuning), *Atule mate* (selar hijau), *Megalaspis cordyla* (tetengkek), *Sardinella* sp. (tembang), *Amblygaster sirm* (siro), and *Stolephorus* sp. [5].

According to the quantitative data from Ministry of Marine Affairs and Fisheries (2020) [6], the recent fish stock decline linear with increasing population in Jakarta [7]. It can be indicated from the several species lost and the number declined. Development of Jakarta City, human activities as well as some domestic waste flow to JBE, give a negatively impact for the communities who dependent on the fisheries sector [10]. These environmental changes likely have the major impact of the fisheries resources in JBE as fish resources to support communities in their fish resources used

2. Method

2.1 Data Collection

This research was part of out in several stages to obtain an overview of the impact they felt during the pandemic condition, then linked it to the Sustainability Livelihood Framework on the activities of the fish processing industry players and make it comparable between several places in Indonesia. However, this research is still in the early stages where primary data is obtained using a questionnaire and secondary data is obtained from government institutions such as data from the Central Statistics Agency and the Ministry of Marine and Fisheries.

Recently, on the first stage, the survey was carried out using a questionnaire to collect the relevant information from different stakeholders in Indonesia. The information was collected through WhatsApp, email and discussion via zoom to the respondents.

To understand the resource use in fish processing activity, a ranking of the preferred species was conducted which result in ten different targeted species that are considered as the major economically important to the respondents in the study areas. Economically important in this term means "Resources that can be either consumed or sold'' where multiple answer was possible. According to their answers, some species were grouped based on three ranks; 1-highest, 2-medium, 3-lowest.

2.2 Data Analysis

Data Analysis was divided into two parts: statistical analysis and descriptive analysis. Quantitative answers were obtained in the data collection method. To achieve the significant difference in both study sites, a non-parametric test for independent sample was used. A cross-tabulation was used to show the relationship between two or more categorical variables by SPSS (Statistical Package for Social Sciences) [8] IBM series 20. Chi-square (X2): testing for goodness to fit used to test the hypotheses regarding statistical distribution on people perception of resources changes. Data on socioeconomic characteristics including location, community structure, occupation, resource use and livelihood will be explained

In this study, increased population, highly dependent on coastal resources are contributing to declining coastal resources stock. Due to these factors, the majority of the people that live in these coastal communities live below the poverty line with poor sanitation. This environmental crisis is likely to have major consequences for the livelihoods of those dependent upon these resources. To find out the livelihoods in both study areas, a framework about Sustainable Livelihood Approach (SLA) should be described.

2.3 Sustainable Livelihoods Approach (SLA)

The Sustainable Livelihood Approach (SLA) in the coastal context was developed in the frame of the Sustainable Coastal Livelihoods (SCL) Project South Asia funded by the UK Government's Department for International Development's (DFID) Policy Research Program. Krantz (2001) reported, the adoption of a livelihood approach within DFID resulted from the publication of the 1999 UK Government White Paper on International Development.

Apart from DFID, there are other organizations using SLA. They include the International NGOs CARE, OXFAM, Save the Children, ITDG (Intermediate Technology Development

Group), as well as bilateral organizations (DANIDA, SIDA) and multilateral organizations (UNDP, FAO, WFP, IFAD, World Bank).

DFID (1999) defined, ''A livelihood comprises the capabilities, assets and activities required for a means of living. A livelihood is sustainable when it can cope with and recover from stresses and shocks and maintain or enhance its capabilities and assets both now in the future, while not undermining the natural resources base''.

Sustainable Livelihoods Framework (SLF) is a way of looking at the complexity of people's lives and responding to that complexity (Figure 1), used to understand and examine the livelihood of the poor [8].

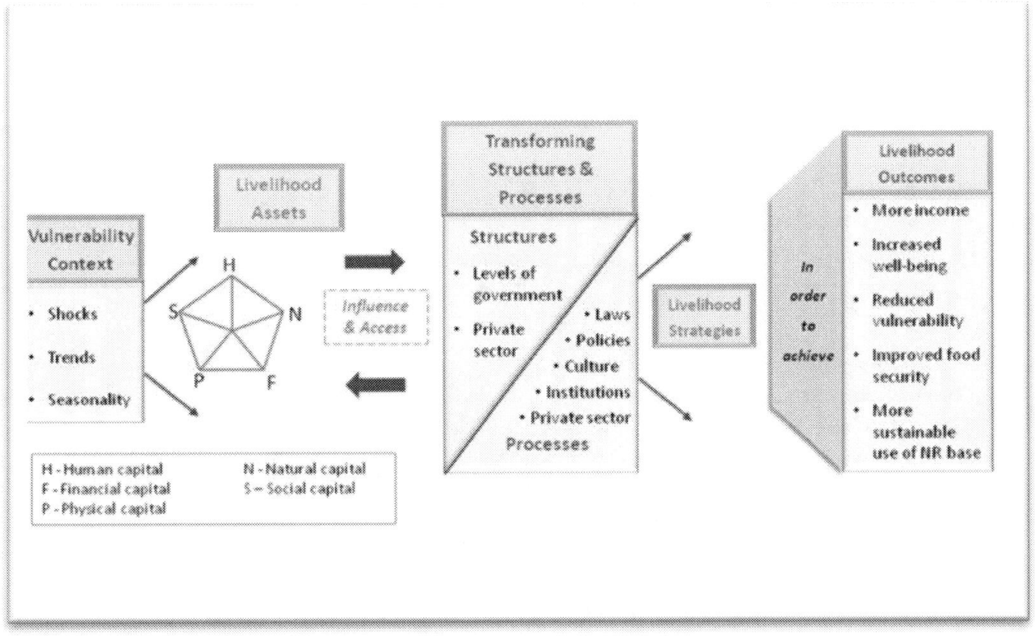

Fig. 1. Sustainable Livelihoods Framework

The SLF constructed based on five principal categories of livelihood assets (human, social, physical, financial, and natural) Carney (1998). In this study, comparison of livelihood assets between the islands and the mainland are based on the SLF with respect to the adaptive capacity of the fishermen.

3. Results and Discussions

3.1 Marine Living Resource Use

The most important species mentioned by the interviewed households differed between islands and mainland communities (Table 1). The mean ranks of species from the household survey on the islands and the mainland are presented below:

Table 1. List economically important species (in term of consumption and trade)

No.	Local name	Latin name	English name
1	Ekor kuning	*Caesio cuning*	Redbelly yellowtail fusilier
2	Tongkol	*Euthynnus affinis*	Mackerel tuna
3	Banyar	*Rastrelliger kanagurta*	Indian mackerel
4	Kerapu	*Epinephelus* sp./*Plectropomus* sp.	Grouper
5	Selar kuning	*Selaroides leptolepis*	Yellowstirpe scad
6	Baronang tompel	*Siganus guttatus*	Rabbit fish
7	Tembang	Clupeidae: *Sardinella albella, Sardinella brachysoma, Sardinella fimbriata*	Fringe scale sardine
8	Lemuru	*Sardinella lemuru, (Amblygaster leiogaster)*	Bali sardinela
9	Tenggiri	*Acanthocybium solandri*	Narrow - barred spanish mackerel
10	Kerang hijau	*Perna viridis*	Green mussel

In this study, economic importance was defined as use of these resources either for consumption or as a source of income. According [9] who wrote about the economic theory of natural resource utilization, economically means the resources can be processed in the fishing industry to achieve sustainable usage and economic values should represent the financial benefits, which are based on the demand of that community [12].

Ten categories of marine resources were listed related to economic importance. This information was gained from the respondents assigning ranks for the resources. Three ranks were used based on importance both for consumption and trade; 1-highest, 2-medium and 3-lowest. Observed differences in target species are due to four reasons; differences in fishing ground, types of fishing gear, market demand, and the availability of technology.

Large-Scale Social Restrictions (Pembatasan Sosial Skala Besar known as PSBB) in several cities in accordance with PP No. 21 of 2020 in the context of handling the Covid-19 outbreak in Indonesia which regulates restrictions on the movement of people, academic and workplace activities, religious activities, activities in public facilities, and distribution of goods to other cities. Another region. The implementation of this policy caused the Indonesian fisheries sector to experience a significant decline, including aquaculture, fishing and processing of fishery products. Countless workers, who normally work long hours in fishing operations have also had to stop working to prevent the spread of Covid-19.

The Covid-19 pandemic has impacted national fishery products, decreased market value of seafood, and consumer demand, including restaurants and hotels, resulting in the closure of many retail businesses, disruption of trade routes, and increased potential for pollution related to unsold fish products and excessive packaging for avoid the transmission of Covid-19 during transportation. The pandemic causes a 30-40% drop in export demand.

The results of the study found that there are several factors that make a difference to the sales of the medium-scale and traditional processed fish sector, namely, the source of raw materials, processing processes, and human resources. In terms of raw materials, local communities use high-grade fish raw materials freshness is very low. Of course, this affects the quality of the salted fish produced and ultimately affects the selling value. Then, in terms of fish processing, salted fish use the wet salting method, namely salting using a salt solution and soaking it for 12-24 hours.

3.2 Financial Capital

In general, the majority of fishermen both on the islands and the mainland depended on fishing as their main source of income (Figure 2). Fishermen tended to save their income every day to fulfill their daily needs. However, as the fish stock is decreasing most fishermen had two or more sources of income to increase their financial standing. These additional sources of income are grouped here into fish-related and non-fishing activities.

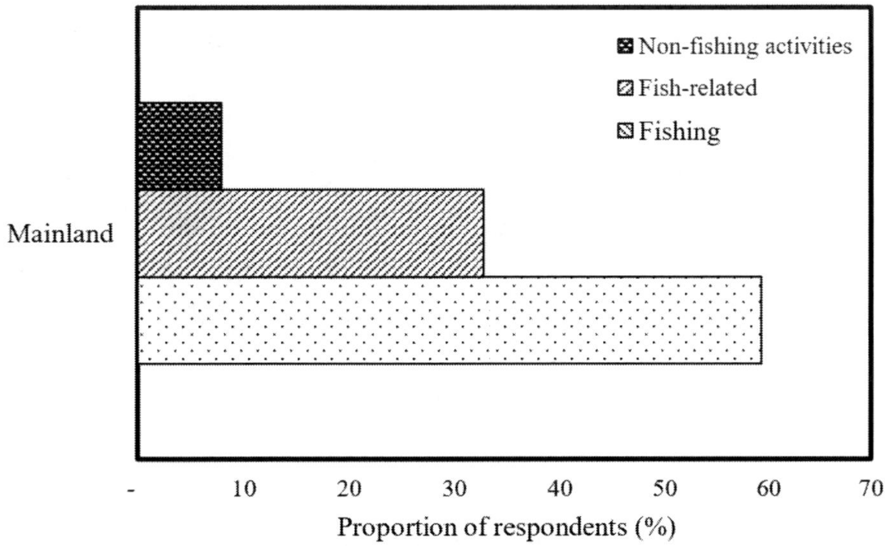

Fig. 2. Main sources of income of respondents on the mainland

There are two main sources of financial capital; available financial stocks (savings, credit) and regular inflows of money (pensions, other transfers, and remittances) [8]. According to the respondents, both on the islands and the mainland, the financial capital is a highly fluctuating and uncertain asset for developing their livelihoods. Fishing is a high-risk occupation [17], as it is seasonal [7] and subject to cyclical fluctuation in stock size [1] and highly unpredictable occurrence of stocks.

As a result, the fishermen cannot guarantee how much money they stand to earn in a day. Under this condition, they aim to save any extra income to build sufficient reserves. But in reality, the amount of money accumulated was barely sufficient to cover their daily needs. An example was given by one fisherman in Kamal Muara. During the dry season he was not able to go fishing regularly. This situation caused him to fish harder. Normally, he would earn up to IDR 70,000 a day, but in the dry season, that figure would be halved. Sometimes, he did not even make any money from fishing, and only caught enough fish to eat.

On the other hand, the effort of fishermen to fish harder can affect the financial assets they have. By aiming to catch more fish, fishermen in JBE try to move to other fishing grounds, which means they need more money to buy fuel. This may lead to the condition where the use of motors can be detrimental in both the dry and wet season on economic grounds. In that case the fishermen used more fuel, but did not catch enough fish to cover their cost.

Local communities mentioned pollution as a major cause of perceived changes. With a high number of tributaries flowing into Jakarta Bay and about 50 factories, the coastal waters of JBE are increasingly affected by human and natural impacts [4]. Pollution not only affects fisheries resources, it also can cause illness. It was reported that Escherichia coli was found in the water pollution [5] and it's as the cause of health effects in several diseases; eye, skin, respiratory and gastrointestinal. In the present study, respiratory diseases were mentioned most frequently as being caused by pollution, which indicates a high perception of air pollution.

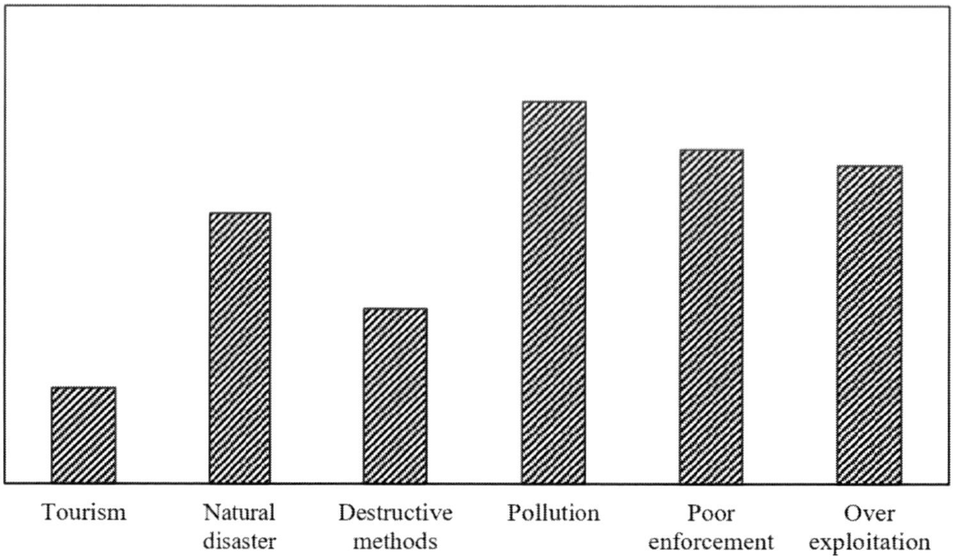

Fig. 3. Cause of perceived changes

Besides pollution, natural disasters and poor enforcement were mentioned to different degrees as perceived causes on the islands and mainland. Natural disasters are frequently experiencing on the mainland in the form of flooding, which occurs frequently and erosion increases along the coast. Already 25 years ago, it is reported that the coast of Jakarta Bay changes due to urban stress on the coastal environment [16]. Furthermore, the high number of respondents perceiving poor enforcement as a cause of environmental degradation on the mainland may be related to the fact that many here have felt direct impacts of corruption in their daily life. On the islands, people are less frequently involved with bureaucracy or the details of official life in the city, thus reducing their exposure to corruption.

4. Conclusion

Fish processing industry influenced the fishers' decision, as Indian mackerel is a valuable fish resource with a high economic value after value addition, which is done by the communities themselves. To the inhabitants of both areas, fishery is a way of life. It has a high cultural value rather than simply being a source of livelihood. During the pandemic, local community has a positive impact on the community to overcome the existing problem, diversifying the products in the smoked fish business in several other commodities such as milkfish, gourami and catfish smoked fish can be good a strategy to be adapted. They also recognize the need for new strategies to identify marketing opportunities that use technology.

References

1. Allison, E. H. and F. Ellis. (2001). The livelihoods approach and management of small-scale fisheries. *Marine Policy,* 25: 377-388.
2. Arifin, Z. (2004). Local millennium ecosystem assessment: condition and trend of the Greater Jakarta Bay Ecosystem. The Ministry of Environment, Republic Indonesia. Jakarta.
3. Baum, G., Kusumanti, I., Breckwoldt, A., Ferse, S. C., Glaser, M., Adrianto, L.& Kunzmann, A. (2016). Under pressure: Investigating marine resource-based livelihoods in Jakarta Bay and the Thousand Islands. *Marine pollution bulletin*, *110*(2), 778-789.
4. Breckwoldt, A., Dsikowitzky, L., Baum, G., Ferse, S. C., van der Wulp, S., Kusumanti, I., ... & Adrianto, L. (2016). A review of stressors, uses and management perspectives for the larger Jakarta Bay Area, Indonesia. *Mar. Pollut. Bull*, *110*, 790-794.
5. Cheung, et.al. (1990). Health effects of beach water pollution in Hongkong. *Epidemology and Infection*, 105 (01).
6. Cleary, D. F. R., Vantier, L., Vail, L., Manto, P., Voogd, N. J., Rachello-Dolmen, P. G., Tuti, Y., et al. (2008). Relating variation in species composition to environmental variables: a multi-taxon study in an Indonesian coral reef complex. *Aquatic Sciences*, *70*(4), 419-431.
7. Crawford, R. J. M. (1980). Seasonal pattern in South Africa's Western Cape purse-seine fishery. *Fish Biology*, 16: 649-664.
8. DFID. 1999. Sustainable livelihoods guidance sheets. London, UK: Department for International Development.
9. Gordon, H. S. (1954). The economic theory of a common property resource: The fishery. *Political Economy*, **62** (2): 124 – 142
10. Kusumanti, M., Ferse, S., & Glaser, M. (2021). Sustainable livelihoods frameworks in investigating household assets in Jakarta Bay, Indonesia. *Aquaculture, Aquarium, Conservation & Legislation*, *14*(1), 204-217.
11. MFMA. (2020). *Kelautan dan perikanan dalam angka* (Marine and fisheries in figure). Jakarta: Indonesia. Annual report Ministry of Marine Affairs and Fisheries Republic of Indonesia. 118 p.
12. Rönnbäck, P. (1999). The ecological basis for economic value for seafood production supported by mangrove ecosystems. *Ecological Economics*, **29:** 235 – 252.
13. Statistik Daerah Kepulauan Seribu. (2020). Badan Pusat Statistik Kabupaten Administrasi Kepulauan Seribu. (http://jakarta.bps.go.id).
14. Suman, A. (2011). Alternatif pengelolaam sumberdaya ikan di Teluk Jakarta. Sumberdaya ikan di perairan Teluk Jakarta dan alternatif pengelolaannya. Balai Penelitian Perikanan Laut. (Ed.): 222-241.
15. Taufik, M., Tuti, H., M. Fauzi. (2011). Keragaan sumberdaya ikan pelagis kecil di perairan Teluk Jakarta dan kepulauan Seribu. Sumberdaya ikan di perairan Teluk Jakarta dan alternatif pengelolaannya. Balai Penelitian Perikanan Laut. (Ed.): 179-195.

16. Verstapen, H. T. (1988). Old and new observations on coastal changes of Jakarta Bay: An example of trends in urban stress on coastal environments. *Coastal Research*, 4 (4): 573-587.
17. Poggie, J. J. And C. Gersuny. (1972). Risk and Ritual: An interpretation of fishermen's folklore in a New England community. *American Folklore*, 85 (355): 66-72.

The Role of Dipping Duck Hatching Eggs with Cherry Leaf Extract as Natural Sanitizers on Hatching Performance and Eggshell Bacterial Counts

Gilang Ayuningtyas[1], *Rina* Martini[1], and *Wina* Yulianti [1]

[1]College of Vocational Studies, IPB University (Bogor Agricultural University), Jl. Kumbang No. 14 Kampus IPB Cilibende 16128, Bogor, Indonesia

Abstract. The research aim is to study the role of dipping duck hatching eggs with cherry leaf extract as a natural sanitizer on hatching performance and eggshell bacterial counts. A number of 533 Alabio duck hatching eggs were used in this study. Hatching eggs produced from The College of Vocational Studies IPB University; were collected three times a day. Five hundred thirty-three hatching eggs were divided into five treatment groups. The first group was as control, the second was dipping into commercial disinfection, and the third and the fifth group were dipping into cherry leaf extract with concentrations 250 ppm, 500 ppm, and 750 ppm, respectively. The results showed eggshell bacterial counts from cherry leaf extract 750 ppm is not significantly different from commercial disinfection treatment. The bacterial count is lower than other treatments. The dipping treatment of 750 ppm cherry leaf extract in Alabio duck hatching eggs resulted in the lowest embryo mortality and the highest hatchability compared to other treatments. The highest hatchability percentage was in the cherry leaf extract treatment of 750 ppm (90.9%) and the lowest was in the control (75.2%).

1. Introduction

The development of poultry embryos requires an ideal environment, one of them is the cleanliness of the hatching eggs. Efforts to produce clean (physically and microbiologically) hatching eggs are carried out starting management on breeding farms. For example, providing nest boxes, collecting hatching eggs as often as possible (4-7 times per day), storing hatching eggs in a clean egg tray, and doing physical cleaning on dirty hatching eggs [1]. These actions do not guarantee the loss of microorganisms on the surface of the hatching eggshell or even those that have penetrated the interior of the egg. Hatch eggs contaminated with microorganisms cause a decrease in hatching performance such as increased embryo mortality and decreased hatchability. The correct egg sanitation process will play a role in reducing embryo mortality and correlate with hatchability. [2] stated that hatchability is directly related to embryonic mortality. An effective hatchery sanitation program is critical to achieving a high level of hatchability [3]. The fumigation process using formaldehyde gas has long been applied to the hatching egg sanitation process. This process plays a role in reducing and eliminating microbial contamination on the surface of the eggshell before the eggs are hatched using an incubator. However, formaldehyde gas which is made from the reaction of formalin and potassium permanganate (KMNO4) is a dangerous gas for both embryos and humans as operators of the hatching process. The dangers that formaldehyde can pose to humans include irritating, carcinogenic, and mutagenic.

Based on these conditions, it is necessary to research alternative materials for the process of sanitizing duck eggs as a substitute for the fumigation process of formaldehyde gas. Hatchery researchers report the use of natural sanitizing agents such as propolis [3], clove essential oil [2], garlic extract [4] gives a positive response to hatchability and has no negative impact on the development of poultry embryos. Cherry leaves (*Muntingia calabura*) contain several active compounds that can work as antibacterial. Flavonoids, saponins, and tannins contained in cherry leaves are antibacterial. This is in line with the research of [5], the results of the screening of phytochemical components in the ethanol extract of cherry leaves contained secondary metabolites: sterols, triterpenoids, flavonoids, alkaloids, saponins, glycosides, and tannins. Flavonoids and saponins have more concentrated intensity than other compounds. The results of the antibacterial activity test of 70% ethanol extract of cherry leaves showed antibacterial activity against *Bacillus subtilis* and *Shigella dysenteriae*, with a Minimum bactericidal concentration (MBC) against *Bacillus subtilis* of 6.25% and *Shigella dysenteriae* of 3.25% [6]. Cherry leaf ethanol extract is also known to have antibacterial activity against *E. Coli, P. aeruginosa, S. Typhimurium, S. aureus, B. subtilis, and C. Albicans* bacteria, and the highest inhibition zone diameter was shown in S. *aureus* [5].

Recent research has been carried out to use cherry leaves as an antiseptic in the animal husbandry sector, including as an antiseptic in the process of dipping the nipples of dairy cows [7] and cherry leaf extract as a sanitizing agent for hybrid duck eggshells [8]. The results of [8], ethanol extract of 20% cherry leaves showed significant results in suppressing embryo mortality and increasing hatchability in the hybrid duck egg hatching process. However, research on the application of a concentrated extract of cherry leaves as a sanitizing agent for Alabio duck hatching eggs and knowing its role on hatching performance has not been carried out. The process of sanitizing eggshells using natural sanitation ingredients such as concentrated extract of cherry leaves is expected to be a solution for the negative impact of using formaldehyde gas on the fumigation process of hatching eggs. This research was conducted to examine the role of concentrated extract of cherry leaves as a natural sanitizing agent in the cleaning process of duck eggshells and its effect on the hatching performance of duck eggs.

2. Materials and Methods

This research was carried out at the poultry laboratory of Livestock Management and Technology Study Program, The College of Vocational Studies Bogor Agricultural University. Hatching eggs produced from Alabio duck were collected three times a day and stored no longer than 24 hours in the room with a temperature of 18°C. Five hundred and thirty-three hatching eggs were divided into 5 groups and replicated four times. Each group represented a sanitation treatment based on cherry leaf extract at various concentrations in comparison to a chemical disinfectant and a control. The research treatment is demonstrated in Table 1.

Table 1. Research treatment

Treatment	Concentration	Method
Water (38-40 °C)/Control (P1)	Water (38-40°C)	Dipping
Chemical disinfectant (P2)	0,15% (1.5 ml per 1 liter water)	Dipping
Cherry Leaf Concentrated Extract (P3)	250 ppm (50 ml Cherry Leaf Concentrated Extract 5 000 ppm, 950 ml Aquabides)	Dipping
Cherry Leaf Concentrated Extract (P4)	500 ppm (100 ml Cherry Leaf Concentrated Extract 5 000 ppm, 900 ml Aquabides)	Dipping
Cherry Leaf Concentrated Extract (P5)	750 ppm (150 ml Cherry Leaf Concentrated Extract 5 000 ppm, 850 ml Aquabides)	Dipping

Note: Extract per 1 liter water

2.1 Incubation

The hatching process uses an automatic incubator with a capacity of 1000 eggs per day on both the setter and hatcher machines. The machine was cleaned using detergent, disinfected, and fumigated. The temperature and humidity in the incubator were set to 37-38 °C and 60-70%, respectively. Eggs to be incubated are given the identity of the egg number on the shell. Eggs were turned on from the 3rd day of incubation until the 25th day. Turning eggs is done automatically every one hour with a tilt angle of 45°. The rest treatment for hatching eggs was carried out from the 17th to the 25th day, 2 times a day (morning and evening) for 15 minutes each. From the 26th day to the 28th day, the frequency of rest is increased to three times a day (morning, afternoon, and evening). Hatch eggs were candled on the 7th and 25th days to see the condition and development of the embryos during the hatching process. Observations on the seventh day of the incubation period were carried out to determine embryo development, the number of fertile and infertile eggs. In addition to distinguishing between live and dead embryos. Egg Infertile and dead eggs were removed from the setter machine. Candling on day 25 or at the time of egg transfer was done to see if the embryo was still alive or dead. The live embryos would be transferred to the hatcher (egg transfer process). The hatching process took place on the hatcher machine until the 28th day.

2.2 Embryonic Death and Hatchability

Eggs that failed to hatch were broken out and examined macroscopically. This observation was done to estimate the development of the embryo and determine the age of death. Embryonic mortality was recorded on the 7th day and 25th day during the hatching process. The embryo mortality was expressed as the percentage of fertile eggs. Hatchability was expressed as percentages of setting and fertile eggs.

2.3 Duck Weight at Pull Duck

Duck embryos began to hatch on day 28 and after the hatching process was complete, the ducklings are left in the hatcher machine until the feathers dry. Then the ducklings would be weighed and recorded as hatching weight. Afterward, the ducklings were removed from the hatcher to be transferred to the duck starter house.

2.4 Statistical Analysis

Statistical analysis was carried out using Minitab 17 software with the experimental design model as follows:

$$Yij = \mu + \tau i + bj + \epsilon i$$

Information:
Yij : The observation record
μ : mean
Ʈi : the effect of treatment
bj : the effect of group ke-j
ϵij : the random error

Mean differences were separated by TUKEY HSD (honestly significant difference) test.

3 Results and discussion

The weight of hatching eggs of Alabio ducks ranges from 70.9 – 73.3 grams. The weight of Alabio duck eggs ranges from 59-65 grams/egg [9]. The duck egg weight which is proper for hatching is between 65-75 grams with a normal egg shape [10]. The eggs used in this study came from 1.5-year-old Alabio ducks. Alabio ducks reach sexual maturity at the age of 5-5.5 months with an egg production period of 2.5 to 3 years [9]. Based on this, the hatching eggs used to come from productive parent stock which is still in the standard age of rearing. Five sanitary treatments of duck hatching eggs and their effects on embryo mortality, hatchability, and hatching weight can be seen in Table 2

Table 2. Fertility, embryonic mortality, hatchability, and duck body weight

Treatment	Average egg weight (g)	Fertility (%)	Embryonic Mortality (%)	Hatchability (%)		Duck body weight (g)
				Egg set	Fertility	
P1	72,7	94,89	24,78	71,9	75,2	41,61
P2	70,9	95,92	18,20	78,6	81,8	42,85
P3	72,0	92,38	13,90	79,3	86,1	41,27
P4	71,0	92,29	11,14	82,0	88,9	40,93
P5	73,3	94,70	9,10	86,0	90,9	41,63

Note: P-value ≥ 0,05 ; not significantly different

The egg fertility rate shows a high percentage, ranging from 92.29-95.92%. This figure is following with previous research [11] that reported the fertility rate of Alabio duck eggs was 95.67%, and [12] which was 93.97%. The high fertility of duck eggs shows that maintenance management in breeding farms has been carried out optimally, including management of feed, mating, health, and also the age factor of the parent stock. The hatching eggs of the P1 group had the highest average embryo mortality of 23.4%, while the lowest embryonic mortality was in the P5 group. This is thought to be related to the level of cleanliness and the amount of microbial contamination in the hatching eggshells. The presence of microbes on the surface of the eggshell can penetrate the egg and can cause contamination of the embryo. This causes the death of the embryo in the hatching process. Based on [10], the dirty duck eggs will be easily contaminated with bacteria that enter through the pores of the shell and cause the death of the embryo. The microbes on eggshells of newly laid eggs can multiply rapidly when exposed to appropriate ambient conditions, and penetrate the eggshell through pores. This could lead to a dramatic reduction in hatching success [13]. The causes of high embryo mortality in the hatching process can be caused by bacterial and fungal contamination, poor egg storage conditions (storage room temperature), disease, genetics, damaged egg, and shell quality, and improper incubation conditions ([12]; [14]; [15]; 16]; [17]; [18]; [19]).

Previous research [20] reported the amount of TPC of duck egg shells sanitized with warm water was 4.19 log 10 CFU/egg, higher than sanitation using chemical disinfectants and concentrated extract of cherry leaf 750 ppm, 2.59 log 10 CFU/egg, and 2.97 logs 10 CFU/egg respectively. These results indicate that the eggshell sanitation process is not proper enough if it is only washed using warm water. Concentrated extracts of cherry leaves (250 ppm, 500 ppm, and 750 ppm) showed a lower embryonic mortality percentage than chemical disinfectants group treatment. These results are consistent with research results [8] that reported that dipping duck eggs using 20% cherry leaf infusion reduced embryo mortality by 43.42%. Cherry leaves (*Muntingia calabura*) contain several active compounds that can work as antibacterial. Flavonoids, saponins, and tannins contained in cherry leaves are antibacterial. This is in line with the research [5], the results of the screening of phytochemical components in the ethanol extract of cherry leaves contained secondary metabolites: sterols, triterpenoids, flavonoids, alkaloids, saponins, glycosides, and tannins. Flavonoids and saponins have more concentrated intensity than other compounds, while the result from [20] showed that the intensity of tannins is more concentrated than that of alkaloids and flavonoids. The mechanism of antibacterial tannin is to inhibit extracellular enzymes of microbes, taking over the substrate required for microbial growth. Other mechanisms work directly on metabolism by inhibiting the process of oxidation, resulting in the release of water and gas in eggs can be prevented [21]. Cherry leaf ethanol extract is also known to have antibacterial activity against *E. Coli, P. aeruginosa, S. Typhimurium, S. aureus, B. subtilis, and C. albicans* bacteria, and the highest inhibition zone diameter was shown in S. *aureus* [5]. The bacteria *E. Coli, S. Typhimurium,* and *S. aureus* were the dominant bacteria in the eggshell.

Hatchability as a percentage of hatchability based on setting eggs and fertile eggs showed no significant difference in each treatment, however, the average percentage of egg hatchability in the P5 treatment group was higher than the other treatments. This is inversely proportional to the embryo mortality percentage. Low embryonic mortality percentage will lead to high hatchability. The dipping method of hatching eggs with cherry leaf extract of 250 ppm, 500 ppm, and 750 ppm showed a higher average egg hatchability than the treatment with warm water and chemical disinfectants. The use of natural sanitizing agents in the hatching egg disinfection process has been widely practiced, [3] reported the use of 14% propolis extract gave higher hatchability than the group of eggs fumigated with formaldehyde. On the other hand, [22] and [23] found that there were no significant differences between the hatchability of oregano oil and formaldehyde fumigation groups. [24] reported that the hatch of fertile eggs in alcohol and control groups have been slightly lower than oregano vulgaris and they found a significant difference between oregano vulgaris and formaldehyde fumigation in the hatchability of fertile eggs. Also, [25] stated that the highest significant percentages for hatchability of fertile eggs were recorded in egg groups with cumin 04%. On the other hand, formaldehyde fumigation recorded the worst and highest significant percentage of embryonic mortality. [26] reported the use of essential oils (EOs) for egg sanitizer have significant reductions in the number of total aerobic mesophilic bacteria (up to 80.77%) and fungi (up to 69.33%) on shells, with hatchability rates significantly improved by up to 12.59%.

Based on [8], the treatment of 20% cherry leaf extract was able to increase egg hatchability by 12.01% compared to the control group (without being sanitized). Factors that affect hatchability include quality and handling of hatching eggs, egg fertility, and hatchery management (temperature, the humidity of the setter and hatcher machine, egg turning). In this hatching process, the temperature of the setter and hatcher machines tends to be stable as well as humidity. The setter and hatcher machine's temperature is set at 37-38 °C and with relative humidity at 60-70%. The hatching weight of ducks in each treatment was not significantly different, with an average hatching weight of 40.93 - 42.85 grams per bird. The weight of day-old duck (DOD) in this study was the following [27], which is a minimum of 37 grams per bird. Hatching weight is strongly influenced by hatching egg weight and hatchery management.

4. Conclusion

The treatment of dipping 750 ppm cherry leaf extract on hatching eggs of Alabio ducks as a natural sanitation agent resulted in the lowest embryo mortality percentage with the highest hatchability percentage compared to other treatments. The use of cherry leaf extract also did not affect the hatching weight of the ducks produced. Based on the results, concentrated extract of cherry leaves can be an alternative to natural egg sanitizers.

Reference

1. R.A. Ernst, F.A Bradley, U.K Abbott, R.M Craig, *Egg candling and breakout analysis*, ANR Publication 8134, California (2004). Available in http://anrcatalog.ucdavis.edu/pdf/8134.pdf.
2. G.D. S. Oliveira, V.M. dos Santos, S.T. Nascimento, J.C. Rodriguesy, Alternative sanitizers to paraformaldehyde for incubation of fertile eggs, J. Poultry Science, 1-6 (2019)
3. E.H.A Shahein, K. Sedeek Eman, Role of spraying hatching eggs with natural disinfectants on hatching characteristics and eggshell bacterial counts. Egypt. Poult. Sci. **34** (I), 211-228 (2014)
4. M. Baylan, G.C. Akpınar, S.D. Canogullari, T. Ayasan, The effects of using garlic extract for quail hatching egg disinfection on hatching results and performance, Braz. J.Poult. Sci., **20(2):** 343-350 (2018)
5. W.P.C. Buihan, R.O. Rubio, Jr DLV, Bioactive metabolite profiles and antimicrobial activity of ethanolic extracts from *Muntingia calabura* L. leaves and stems, Asian Pac. J. Trop Biome **6(8),** 682–685 (2016)
6. A.D. Prasetyo, H. Sasongko, Aktivitas antibakteri ekstrak etanol 70% daun kersen (Muntingia Calabura L.) terhadap bakteri Bacillus subtilis dan Shigella dysenteriae sebagai materi pembelajaran biologi SMA kelas X untuk Mencapai Kd 3.4 pada Kurikulum 2013, JUPEMASI-PBIO, **1**(1) (2014)
7. D.R. Prasetyanti, C. Budiarti, D.W. Harjanti, Efektifitas daun kersen (Muntinga calabura L.) dalam menurunkan jumlah bakteri dalam susu dan peradangan pada ambing sapi perah, J. Ilmu-Ilmu Peternakan, **19**(1), 1-16 (2016)
8. F.H. Al Hakim, MN Huda, GD Fitri, D Ambarwati, H Tistiana, Pengaruh ekstrak daun kersen terhadap daya tetas dan mortalitas telur itik hibrida, J. Ilmu-Ilmu Peternakan, **26** (2), 8 – 13 (2016)
9. Direktorat Jendral PKH, Itik alabio (2021), Available in: http://bibit.ditjenpkh.pertanian.go.id/content/itik-alabio

10. A.R. Setioko, *Penetasan telur itik di Indonesia.* Wartazoa.**7**, 40- 46 (1998)
11. D. Darmawati, Rukmiasih, R. Afnan, Daya tetas telur itik cihateup dan alabio, J. Ilmu Produksi dan Teknologi Hasil Peternakan, **4**(1), 257-263 (2016)
12. D.S. Kumalawati, T. Susanti, *Effect of breed on fertility, hatchability and embryo mortality of Alabio, Mojosari and Peking x White Mojosari crossbred duck eggs, in Proceedings*: Seminar Nasional Teknologi Peternakan dan Veteriner (2019)
13. M.I. Cook, S. R. Beissinger, G.A. Toranzos, J.W. Arendt, Incubation reduces microbial growth on eggshells and the opportunity for trans-shell infection. J. Ecology Letters. 8 (5), 532–537
14. E.E. Onbaşılar, Ö. Poyraz, E. Erdem, Effects of egg storage period on hatching egg quality, hatchability, chick quality and relatif growth in Pekin ducks. Arch Geflügelk. **71**,187-191 (2007)
15. A. Abudabos, The effect of broiler breeder strain and parent flock age on hatchability and fertile hatchability. Int.J. Poult Sci. 9,231-235 (2010)
16. C.W. van der Pol, I.A.M van Roovert-Reijrink, C.M. Maatjens H. van den Brand, R. Molenaar, Effect of relative humidity during incubation at a set eggshell temperature and brooding temperature post-hatch on embryonic mortality and chick quality. Poult Sci. **92**,2145-2155 (2013)
17. H. van den Brand, M.P. Sosef, A. Lourens, J. van Harn, Effects of floor eggs on hatchability and later life performance in broiler chickens. Poult Sci. **95**, 1025-1032 2016)
18. A. Addo, J.A. Hamidu, A.Y. Ansah, K. Adomako, Impact of egg storage duration and temperature on egg quality, fertility, hatchability and chick quality in Naked Neck Chickens. Int J Poult Sci. **17**,175-183 (2018)
19. A. Jabbar, A. Hameed, A. Yousaf, A. Riaz, Y.A. Ditta, The influence of hairline crack eggs on hatchery parameters and performance of chicks. World's Vet J. **9**,76-83 (2019)
20. G. Ayuningtyas, R. Martini, W. Yulianti, The potential Of cherry leaf extract (Mutinga calabura L) s a sanitary agent for eggshell in the alabio ducks hatching process, J. Sains Terapan, **10** (2), 50 - 61 (2020)
21. Y. Nurwantoro, B. Resmisari, Pengaruh perendaman jus daun sirih (Piper Betle LINN) terhadap jumlah bakteri pada telur itik, J. Indonesia Tropic Animal Agriculture, **3**, 156-160 (2004)
22. I. Yildirim, M. Ozcan, R. Yetisir, The use of oregano (Origanum vulgare L.) essential oil as alternative hatching egg disinfectant versus formaldehyde fumigation in quails (Coturnix coturnix japonica) eggs. Food Control. **15**,169-172 (2003)
23. G. Copur, M. Arslan, M. Duru, M. Baylan, S. Canogullari, E. Aksan, Use of oregano(Origanum onites L.) essential oil as hatching egg disinfectant. African, J.of Bio. **8**, 2531 -2538 (2010)
24. I. Yildirim, N. Ozcan, Use of oregano and cumin essential oils as disinfectants on hatching Quail eggs, J.Anim.Res.Inst.**11**, 61 -63 (2010)
25. Bekhet, M. Gamal, Impact of hatching egg disinfection on hatching characteristics and chick embryos, Indian J. of Animal Research, **55**, 353-358 (2021)
26. G.D.S. Oliveira, V.M. dos Santos, S.T. Nascimento, Essential oils as sanitisers for hatching eggs, World's Poultry Science Journal (2021).
27. Badan Standardisasi Nasional, Bibit niaga (final stock) itik Alabio meri umur sehari. SNI 7358:2008 (2008).

A Development of Web Application for Hydroponic Monitoring Systems

Hendi Hermawan[1,3,*], *Nur* Uddin[1,3], and *Teddy Mohamad* Darajat[2,3]

[1]Universitas Pembangunan Jaya, Department of Informatics, Tangerang Selatan, Indonesia
[2]Universitas Pembangunan Jaya, Department of Product Design, Tangerang Selatan, Indonesia
[3]Universitas Pembangunan Jaya, Center for Urban Studies, Tangerang Selatan, Indonesia

Abstract. A development of web application for a hydroponic monitoring system (HMS) is presented. The HMS is to collect data of hydroponic plants such as quality and quantity of nutrient solution. The HMS is integrated to internet such that the collected data is stored in a cloud server. The data can be accessed by users through a web application. This study is to develop the web application as an interface for accessing the HMS data in real time. The web development is done by implementing the user centred design (UCD) method. A study case of monitoring hydroponic temperatures is presented in this development. It resulted in a web application that presents real-time data of the temperatures in numerically as well as graphically. This web application provides an informative, attractive, and user-friendly interface of the HMS.

1 Introduction

Agriculture is an essential sector in human life for foods production. There are several methods in agriculture. Hydroponic is one of the popular agriculture methods especially in urban area. The hydroponics is a modern cultivation method without using soil as media but nutrient solution [1]. The solution contains substances needed for plant growth. Without using the soil, the hydroponic promises some advantages compared to traditional agriculture methods such as in cleanness, space requirement, maintenance, and productivity [2]. The less space requirement makes the hydroponic to be suitably applied in urban farming for food sustainability. Base on the way of feeding the nutrient to plants, the hydroponic can be classified into two types: dynamic and static hydroponics [3]. In the dynamic hydroponics, nutrient solution is distributed to the plants by flowing the solution through pipes and circulating it with the help of a pump. Meanwhile in the static hydroponic, the nutrient solution is only filled into a container without any flow or circulation. The solution remains in the container forever until harvesting.

The nutrient solution is the most important part of the hydroponic such that the quantity and quality have to be maintained periodically [4–6]. Maintaining the quality is to assure that the nutrient solution is available at a certain volume range. This maintenance can be simply done through a visual observation by human. However, maintaining the quality of nutrient

* Corresponding author: hendi.hermawan@upj.ac.id

solution cannot be done in a simply way as the nutrient quantity. The solution quality is determined by various parameters, such as: temperature, acidity, and concentration [7]. Those parameters cannot be precisely observed by using human senses but have to be measured using sensors such as temperature sensors, pH sensors, and total dissolved solids (TDS) sensors.

One of the mandatory maintenances in hydroponics is maintaining the nutrient solutions periodically. Conducting this maintenance manually is definitely inefficient. In order to assist the hydroponic maintenance, an electronic system known as the hydroponic monitoring system (HMS) has been developed. The HMS contains of a microcontroller and sensors as the main components. The HMS does automatic and periodic measurements of quality and quantity of the nutrient solution. The traditional HMS store the measurement data in a memory at local device or local network. By the advance of computer and internet technologies, the HMS is integrated to internet such that allows to store the measurement data into a web server. As stored in the webserver, the monitoring can be done by accessing the data anytime and from anywhere through the internet. This gives a flexibility to do a remote observation to the hydroponics. Moreover, the system can be used to expands the monitoring coverage as long as the internet is available. The HMS that is integrated to the internet is named as the internet of things based hydroponic monitoring system (IoT-based HMS).

Several studies on the development of IoT-based HMS have been conducted. An IoT-based HMS was presented in [7]. This system was developed by including several sensors, i.e.: temperature sensor, pH sensor, TDS sensor, and liquid level sensor. The system utilizes Arduino Uno and the ESP8266 WiFi module as the microcontroller and the WiFi communication, respectively. Meanwhile, a Raspberry Pi 2B is utilized in building a web server. Data of the monitoring system were presented a web page. Another IoT-based HMS was developed in [8] and called as the iHydroIoT. This system applied more sensors than the system in [7], such that it was able temperature, CO2, pH, volume, and TDS of the nutrient solution. Moreover, the iHydroIoT was also equipped with light intensity sensor, humidity sensor, and air temperature sensor. The iHydro applied an Arduino Uno as the microcontroller. Webserver of the iHydro was built using a Raspberry Pi 2. A BLE (Bluetooth Low Energy) module was installed on the Arduino Uno such that the microcontroller was able to send data to the webserver via Bluetooth communication. In order to make the data being accessible from anywhere, the webserver was connected to internet through a WiFi communication.

An advanced IoT-based HMS by involving artificial intelligence was presented in [9]. The system was equipped by a bunch of sensors to measure many parameters such as air temperature, air pressure, altitude, air humidity, light intensity, ultraviolet, CO2, nutrient temperature, pH, dissolved oxygen, electricity conductivity (EC), and (TDS). This monitoring system was also equipped with two cameras to get vision data of the plants. The system utilized an Arduino Mega as the microcontroller that collects the data from the sensors, while the vision data was collected using a Raspberry Pi 3B+. The vision data was applied in a deep learning algorithm to predict the harvest time. Previously, a study of integrating IoT-based HMS and machine learning to provide appropriate control action for hydroponic was presented in [10]. Prototype of the system was applied in tomato plant growth.

The previous research works show how the advanced computer technology could contribute in the agriculture sector in order to improve the food production in quality as well as in quantity. We have been working on developing a prototype of IoT-based HMS to collect data of hydroponic. A study case presents a monitoring hydroponic temperature. Development of the prototype includes two parts: hardware and software. The hardware development has been done and presented in [11]. This paper presents the second part development, which is the software development. This is to build a web application as user interface of the hydroponic monitoring system. Presentation of this paper is organized as follows. Section I provides background and motivation of the work. Section II presents the method of by explaining how the measurement, the web application development method, and the realization of constructing the web application. Section III presents the resulted web application. Finally, the Section IV concludes the work.

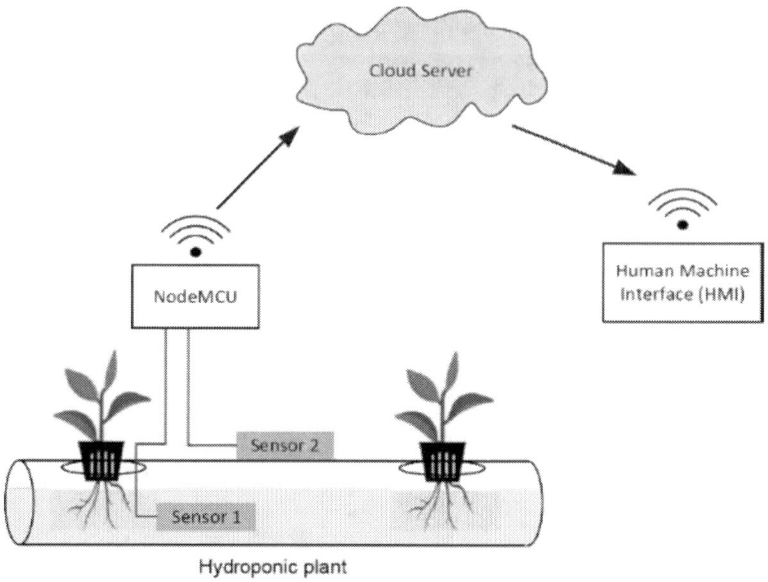

Fig. 1. Diagram of the hydroponic monitoring system [11].

2 Method

2.1 IoT-based hydroponic monitoring system

A design of IoT-base hydroponic monitoring system was presented in [11]. The monitoring was concerned on the solution temperature and the air temperature. Diagram of the monitoring system is shown in Fig. 1. The system has four main parts: sensors, NodeMCU, cloud server, and human machine interface (HMI). The sensors are to measure the temperatures, where the measurement data is sent to the NodeMCU. The NodeMCU is a microcontroller with a built-in WiFi module. The NodeMCU is to collects and processes the measurement data, and then send the processed data to the cloud server through a WiFi communication. The data are saved in the server and can be accessed by users. The users access the data from their devices such as PCs, laptops, or smartphones. These devices communicate to the cloud server through internet connection using WiFi or wired connection. The HMI is applied as an interface for the user in accessing the data.

The Fig. 1 shows that the sensors include Sensor 1 and Sensor 2. The Sensor 1 measures temperature of the nutrient solution, while the Sensor 2 measures the air temperature surrounding the plant. The Sensor 1 is placed by dipping the sensor in the nutrient solution such that a water-proof temperature sensor is required. The Dallas temperature sensor DS18B20 is one of the choices and selected to be applied as the Sensor 1. This sensor is a low cost and waterproof temperature sensor that is able to measure temperature in range of -55 to 125 °C with accuracy ±0.5 °C. The Sensor 2 is placed in the air, beside the plant. Due to the hydroponics is uncovered, the DS18B20 is also applied as the Sensor 2 for anticipating water contamination to the temperature sensor.

Penelitian_LogData_01

Date	Time	Index	Temperature 1 (C)	Temperature 2 (C)
01/07/2021	23:44:31	8659	29.15	29.19
01/07/2021	23:45:04	8660	29.15	29.19
01/07/2021	23:45:38	8661	29.21	29.19
01/07/2021	23:46:11	8662	29.15	29.19
01/07/2021	23:46:44	8663	29.15	29.12
01/07/2021	23:47:18	8664	29.21	29.19
01/07/2021	23:47:53	8665	29.15	29.12
01/07/2021	23:48:26	8666	29.15	29.12
01/07/2021	23:48:59	8667	29.15	29.12
01/07/2021	23:49:35	8668	29.15	29.12
01/07/2021	23:50:09	8669	29.21	29.12
01/07/2021	23:50:42	8670	29.21	29.12
01/07/2021	23:51:15	8671	29.21	29.06
01/07/2021	23:51:49	8672	29.21	29.06
01/07/2021	23:52:22	8673	29.09	29.06

Fig. 2. Temperature measurement data in Google sheet [11]

The cloud server applied in this system was the Google Sheets. The experimental results in [11] show that the developed system was working well in monitoring the hydroponics temperatures. The temperatures data was saved in the Google Sheets and accessible from anywhere through internet. A sample of the temperatures data presented in the Google sheet is shown in Fig. 2.

2.2 Development method of web application

The Fig. 2 shows measurement data of the IoT-based HMS. It is shown that the data is presented in a Microsoft Excel format which is not a quite user friendly as interface for the monitoring system. A more user-friendly interface is required for better and more convenient in monitoring the temperatures. Therefore, a web application is developed for providing a better user-interface. This web application is developed by implementing the user cantered design (UCD) method. This method allows to understand the user needs by exploring the user point of view in detail. The UCD method is guided by a belief that end users may influence in the web design through participating in the design process. It is a multi-stage problem-solving process, where the designer is not only evaluating and predicting of how the consumers will use the product but also validating based on the user behaviours through real-world experiments [12].

The International Organization for Standardization (ISO) defines a standard of requirements and recommendations for human-centred design principles and activities throughout the life cycle of computer-based interactive systems which is documented as the ISO 9241-210. Implementation of the UCD includes several steps as shown by a flow chart in Fig. 3 and explained as follows [13, 14]:

a) Plan for UCD
   This is the initial step for determining which development approaches will be used. The determination is done by considering the business case for usability and budget, timeline, resource, expertise, and other constraints into account [14].

b) Specify the context of use
This stage is to specify use context of the developed application. The specification is done through user analysis and task analysis. The user analysis is to examine and learn characteristics of the users [15]. In this case, the users should be persons who have experience in hydroponics. It is necessary to pay attention to how to care for plants to thrive and produce optimal results. The required temperature of each plant is unique. Finding optimal temperature of a plant requires a temperatures observation for a long period. This observation will be a tedious job. A task analysis process entails determining the user's objectives and the system functionality that is required. As a result, users require a method that simplifies the temperature observation of the plant, and in this case is the hydroponic plants.

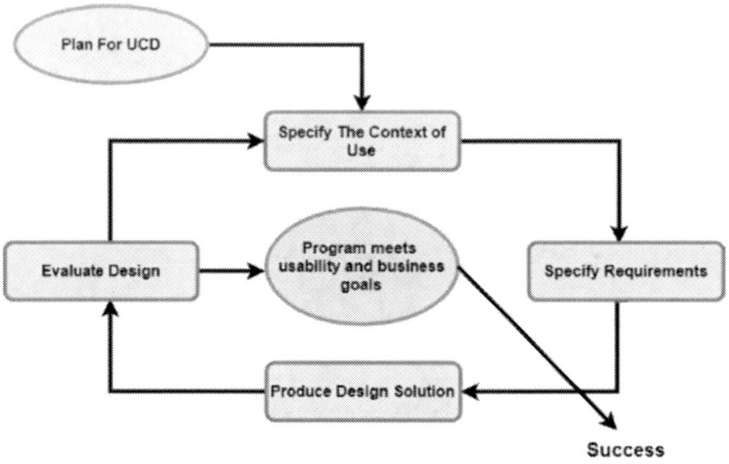

Fig. 3. Flow chart of User-Centered Design (UCD) Method [13]

c) Specify requirements
This stage is accomplished by bringing end-users into the design process and communicating with potential end-users and other key stakeholders to aid with required usability testing. Participatory design with end-users is a critical component of UCD and should be carried out throughout the system design and development process [16]. In this study, users will be able to monitor the temperature conditions of hydroponic plants via the internet. Users can view the minimum and highest temperatures, the average daily temperature, and a chart of the average temperature. Additionally, there is a table consist of date, time, and data temperature that is updated every 30 seconds.

d) Product design solution
The evaluation results are used at this stage to guide the system's design and execution. This principle emphasizes the critical importance of user-centred evaluation in guiding and enhancing design across the technology lifecycle. In this study, we build a web application according to the needs.

e) Evaluate design
This stage involves comparing the iterative interface design to the ISO UCD procedure's requirements. The end-user is included in this iterative process. This includes active user involvement in evaluation and design throughout the development process, as well as evaluation of use in real-world user situations and contexts.

2.3 Realization

The web development is realized using the Google Data Studio. The Google Data Studio provides a simple way in interacting with charts as well as filtering other charts [17]. Therefore, building a hydroponics temperatures data visualization can be done easily and quickly.

Based on the requirements specification, the web is developed to present the measured temperature data in real-time. Data flow in this web development is shown in Fig 4. The measured data is available in the Google Sheet, where the sample data are shown in the Fig. 2. The data include the date, time, temperature 1 (nutrient temperature), and temperature 2 (ambient temperature). The data model which is the Google Sheet, is connected to Google Data Studio to generate data source. The data source is a component to build a monitoring report on a web application.

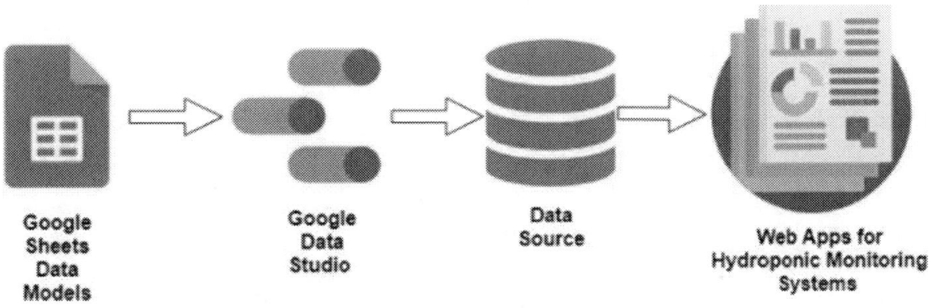

Fig. 4. Data flow in the web application.

3 Results

The web development results in a web that visualizes the hydroponic temperatures data as shown in Fig. 5. The web includes the title of the web application, a table containing the temperature data per 30 seconds, and gauges displaying the minimum, average, and maximum values of temperature 1 (nutrient temperature) and temperature 2 (ambient temperature). The web was designed by selecting colours that are soothing to the eyes.

According to UCD method shown in Fig. 3, evaluations of the developed web are required to assure that the developed product meets to the user needs. Evaluation was done together with the user. It was found a weakness as follows. The visualized sample data in the Fig. 5 shows that the minimum value of temperature 1 was 26.02. This was the minimum value of temperature 1 in the period of the 1st to 14th August 2021. However, it was no information when the minimum temperature was happened. It could be found in the table but it is very difficult due to the large amount of data in the table.

Based on this evaluation results, the user needs information not only the temperatures data but also the time. Therefore, the developed web was improved by adding a time series graph of the temperatures. It resulted in a web application version 2 as shown in Fig. 6. This web provides a chart that presents time series data of the temperature 1 and the temperature 2 for a certain time period. Using this chart, the users can easily observe the daily hydroponics temperatures such as: finding the maximum or minimum temperature, and estimate the average temperature. The user satisfies with the web application version 2, and therefore it is applied as the HMI of the hydroponics temperatures monitoring system.

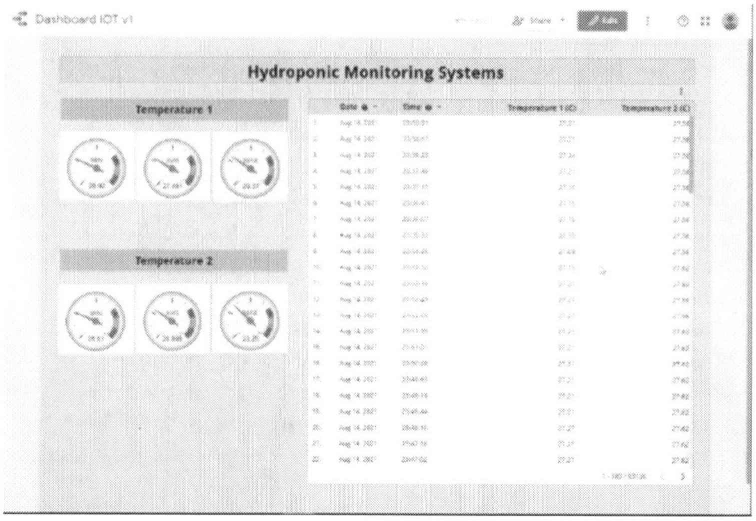

Fig. 5. Web application version 1.

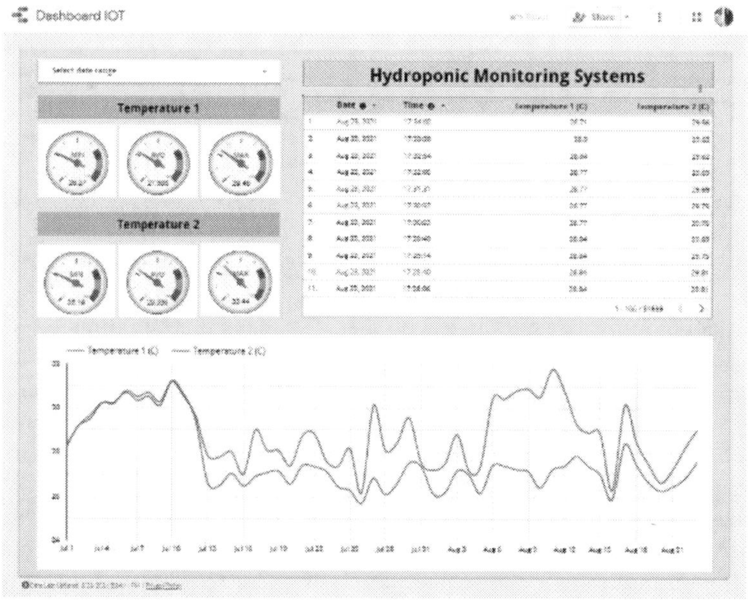

Fig. 6. Web application version 2.

4 Conclusion

Web Application for an IoT-based hydroponic monitoring systems has been developed. The web application was developed by using the Google Data Studio and implementing the user-centered design method. The resulted web application provides a user-friendly interface for the monitoring system. The web application displays the values of temperature 1 (nutrient temperature) and temperature 2 (ambient temperature) in real-time. Both temperatures data are displayed in numerically as well as graphically.

Acknowledgment

The authors acknowledge financial support from Universitas Pembangunan Jaya through Internal Grant No. 008/PER-P2M/UPJ/11.20

References

1. J.E. Son, H.J. Kim, T.I. Ahn, in *Plant factory* (Elsevier, 2020), pp. 273–283
2. F. Hidayanti, F. Rahmah, A. Sahro, International Journal of Advanced Science and Technology **29**, 5157 (2020)
3. C.L. Chang, G.F. Hong, *Design and IMPLEMENTATION of a human-machine-interface based hydroponic nutrient solution irrigation system*, in *2014 Montreal, Quebec Canada July 13–July 16, 2014* (American Society of Agricultural and Biological Engineers, 2014), p. 1
4. A.F.M. Filho, C.A.V. de Azevedo, M.R. de Queiroz Almeida Azevedo, J.D. Fernandes, C.R. da Silva, P.D. Fernandes, Australian Journal of Crop Science **12**, 572 (2018)
5. W.J. Cho, H.J. Kim, D.H. Jung, D.W. Kim, T.I. Ahn, J.E. Son, Computers and electronics in agriculture **146**, 51 (2018)
6. W.J. Cho, H.J. Kim, D.H. Jung, C.I. Kang, G.L. Choi, J.E. Son, Transactions of the ASABE **60**, 1083 (2017)
7. P.N. Crisnapati, I.N.K. Wardana, I.K.A.A. Aryanto, A. Hermawan, *Hommons: Hydroponic management and monitoring system for an IOT based NFT farm using web technology*, in *2017 5th International Conference on Cyber and IT Service Management (CITSM)* (IEEE, 2017), pp. 1–6
8. G. Marques, D. Aleixo, R. Pitarma, *Enhanced hydroponic agriculture environmental monitoring: An internet of things approach*, in *International Conference on Computational Science* (Springer, 2019), pp. 658–669
9. S. Park, J. Kim, Electronics **10**, 1400 (2021)
10. M. Mehra, S. Saxena, S. Sankaranarayanan, R.J. Tom, M. Veeramanikandan, Computers and electronics in agriculture **155**, 473 (2018)
11. N. Uddin, H. Hermawan, T.M. Darajat, S. Marwanto, *Internet-Based Temperature Monitoring System for Hydroponic*, in *The 3rd International Conference on Agricultural Technology and Engineering, and Environmental Sciences (3rd ICATES)* (2021)
12. E.S. Poole, Translational behavioral medicine **3**, 402 (2013)
13. S. Wardhana, M.K. Sabariah, V. Effendy, D.S. Kusumo, *User interface design model for parental control application on mobile smartphone using user centered design method*, in *2017 5th International Conference on Information and Communication Technology (ICoIC7)* (IEEE, 2017), pp. 1–6
14. N. Bevan, I. Curson, *Planning and implementing user-centred design*, in *CHI 98 conference summary on Human factors in computing systems* (1998), pp. 111–112
15. T.T. Hewett, R. Baecker, S. Card, T. Carey, J. Gasen, M. Mantei, G. Perlman, G. Strong, W. Verplank, *ACM SIGCHI curricula for human-computer interaction* (ACM, 1992)
16. I.O. for Standardization, *Ergonomics of Human-system Interaction: Part 210: Human-centred Design for Interactive Systems* (ISO, 2010)
17. G. Kemp, G. White, *Google Data Studio for Beginners: Start Making Your Data Actionable* (Apress, 2021)

Marketing Development of Beef Cattle, Sheep, Goat and Derivative Products through the Application of Digital Marketing in Facing the Impact of Covid-19 Pandemic

Muh Faturokhman[1], *Liisa Firhani* Rahmasari[1*], and *Fariz Am* Kurniawan[†2]

[1]Agribusiness Management Department, Collage of Vocational Studies IPB University, Jalan Kumbang No.14 Bogor 16151, West Java, Indonesia
[2] Livestock Management and Engineering Department, Collage of Vocational Studies IPB University, Jalan Kumbang No.14 Bogor 16151, West Java, Indonesia

Abstract. Livestock business especially beef cattle, sheep and goat, experienced the marketing problems during the pandemic of Covid-19, which resulted a decrease in farmers income. Although there was a decline in purchasing power, data showed that during the pandemic, the demand of processed livestock products increased 3-4 times compared to before the outbreak of pandemic. This condition due to the changes in consumer preferences who wanted the product that have quality standards of health and food safety, and indicated a great opportunity for increasing the sales of processed products for beef cattle, sheep and goats. To support the level of sales of both primary and processed products, it was necessary to change the promotion and marketing strategy. Utilization of digital marketing platforms as a new philosophy and modern business practice related to the marketing products, services and information that can be carried out by small-scale farmers. The digital platform which developed for digital marketing of livestock products is Vilstock.id. The application used the SDLC (*System Development Life Cycle*) Waterfall method, while the comparative analysis of the value of implementing digital marketing platforms before and after using was the Return Cost Ratio (R/C ratio) a comparison between revenues and costs. The result of the analysis showed that there was an increasing income of SME farmers as indicated by the value of the R/C ratio from 1.014 (before) to 1.072 (after). This shows an increase in product sales by 5.4% within 4 weeks.

1 Introduction

The Covid-19 pandemic has impacted several sectors. Starting from economic sector, education and also agricultural include the agribusiness sector of livestock, fisheries, plantation and other sectors. The impacts experienced by the agricultural sector include: (a) farming activities are disrupted, especially in the food crops and horticulture sectors, as well as animal husbandry and fisheries because of an appeal to limit activities outside the home so that the cultivation and farming are less than optimal and have an impact on farmer's

* Corresponding author: liisa@apps.ipb.ac.id

income receipts and breeder, (b) difficulties in conducted marketing of agricultural products, (c) the decline in people's purchasing power and the ability of the households to access affordable food also has an impact on the economic resilience of farmers and breeders. However, compared to the industrial sector, tourism services, and others, the agricultural sector on a large scale is considered as one of the most stable sectors in facing the crisis due to the COVID-19 pandemic. Based on the data of Badan Pusat Statistik Indonesia [1], it is said that the agricultural sector is able to absorb 35.41 million workers. This should be used as a momentum to realize food self-sufficiency and make the agricultural sector on a broad scale which includes livestock, fisheries, plantations, forestry as well as agro-industry as a priority sector for development, this condition also related to the sustainable development goals (SDGs) program.

Livestock businesses, especially beef cattle, sheep and goats, during the Covid-19 pandemic were also affected, which resulted in a decrease in farmer's income and obstacles in marketing livestock products, especially beef cattle. In fact, livestock business as one of the agricultural sub-sectors providing protein has an important role and contributes greatly to national income receipts. The decline in sales of livestock products (beef cattle, sheep and goats) was due to the declining purchasing power of the people due to the large number of workers being laid off (termination of employment) and the low economic turnover of the community since the Covid-19 pandemic broke out. This condition much more alike with what happen in US. The research model used the dynamic system demonstrate the likely effects of a pandemic on the USA's food system and it reveals that a severe pandemic like Ebola for example with greater than a 25 % reduction in labor availability can create significant and widespread food shortages [2]. With low purchasing power, people tend to look for beef cattle substitutes with food sources that have relatively the same protein level as beef, goat or sheep meat. This causes the level of demand for beef cattle and meat to decrease relatively. However, another problem that also affects the low level of sales is the reluctance of consumers to have direct contact with sellers, especially if transactions are carried out in the market because they avoid to get exposure to COVID-19. This is evidenced by the low level of household consumption which only grew 2.84% in the first quarter of 2019-2020 compared to the first quarter of 2018-2019 which reached 5.02% [1].

Consumers of beef, goat and lamb on average are people from upper-middle income, so that with the Covid-19 pandemic situation, this consumer group still has relatively purchasing power to buy meat and its derivative products, so it is estimated that the demand level, although in aggregate, will still decline in the positive range. Data shows that during the COVID-19 pandemic, the level of demand for processed products increased 3-4 times compared to before the pandemic [3]. This shows that the opportunity to sell processed products from beef cattle is still wide open to be increased. The increase in demand for processed products is due to changes in consumer preferences which tend required products that meet quality standards of health and food safety.

Marketing strategies that can be optimized to increase sales are promotional strategies. One form of change in marketing strategy (promotion) is through digital marketing platforms. Electronic marketing through digital marketing can be viewed as a new philosophy and modern business practice involved with the marketing of goods, services, information and ideas [4]. Digital marketing is a supporter of marketing strategies because it is considered very effective in selling and marketing products and is considered an option to meet purchasing needs quickly [5]. Marketing activities involved in the digital marketing series are part of marketing management.

During the COVID-19 pandemic, consumer characteristics and preferences were different from before. The old consumer behavior has begun to be disrupted by the presence of network technology. This is done by consumers to avoid the risk of being exposed to covid19. According to Kotler et.al., [6], consumer decisions nowadays are strongly influenced by advertising, public opinion, personal knowledge, and experience. Advertisements, public opinion, and personal knowledge are now formed from online interactions sourced from websites, social media, and business platforms such as e-commerce that disrupt media such as newspapers, billboards and magazines. The digital marketing system is considered capable of responding to the needs in the era of the COVID-19 pandemic, so that companies (SMEs, cooperatives and farmer/livestock groups) can survive the disruption caused by the Covid-19 pandemic [7].

Marketing is one of the spearheads of companies in reaching consumers in order to get revenue for a business. In the conditions of the Covid-19 pandemic, many business sectors, especially beef cattle agribusiness, experienced difficulties in marketing agricultural products. Large-scale social restrictions which is a policy to limit the movement and physical space of people, in fact also have an impact on the flow of goods and the sustainability of the business sector. This is one of the factors that underlies research studies on the development of marketing for beef cattle and their derivatives through the application of digital marketing in the face of the COVID-19 pandemic which is expected to be one of the contribution efforts of universities in conducting collaborative efforts with the business world (SMEs, Cooperatives and Farmer/Livestock Group).

Demand for cattle, sheep and goats from consumers, both in the form of livestock (primary) and derivative products, must be balanced with efforts to implement the latest technology. How to compete with feedloters and meet the demand for qurban cattle (cows), for example, cannot be done traditionally, especially in the midst of the COVID-19 pandemic, but must be directed at marketing development that require a managerial role, besides that the provision of sales services that must fulfill the protocol standards for handling covid-19 . This is important to do, so that consumers become comfortable in the transaction process. Digital marketing that will be carried out includes creating consumer value, capturing consumer value, and maintaining consumer value [8]. Therefore, in implementing digital marketing, it is necessary to use digital marketing platforms such as websites, social media, and supporting applications. Therefore, these are the aims of our research:
1) Assisting in solving marketing problems for beef cattle (cows, sheep and goats) and derivative products through digital marketing applications for SME Animal Husbandry companies
2) Performing a comparison analysis of the value of using digital marketing platforms for SME and Animal Husbandry companies before and during the COVID-19 pandemic

2 Materials and Method

2.1 Digital Application System

One of the goals of this research is to produce a digital marketing application prototype that is easy to use by business actors in the field of agribusiness, especially SMEs, Cooperatives and Farmer/Livestock Groups that cultivate cattle, sheep and goats who want to sell online. Technically making digital marketing designs using the waterfall method.
The software life flow approach is sequential or sequential starting from analysis, coding, testing and support stages. The stages of the Waterfall method are described as follows:
1) Software Requirements Analysis in this software requirements analysis aims to analyze all requirements including the documents and interfaces needed to determine the software solution that will be used as the process of computerizing the system.
2) Design at this stage, a design according to system requirements will be made regarding the database design, software architecture and User Interface that will be created. The use of

the Unified Modeling Language (UML) is intended to explain in more detail in the program design and database design. The UML that will be used is the Activity Diagram.
3) System Development (Code Generation). At this stage, the design implementation is made into a software program. At this stage a new system is created using Visual Studio 2010 as the programming language, MSQL to create a database. ODBC Connector as database connection, XAMMP as database server liaison and Crystal Report 8.5 as report generation tools.

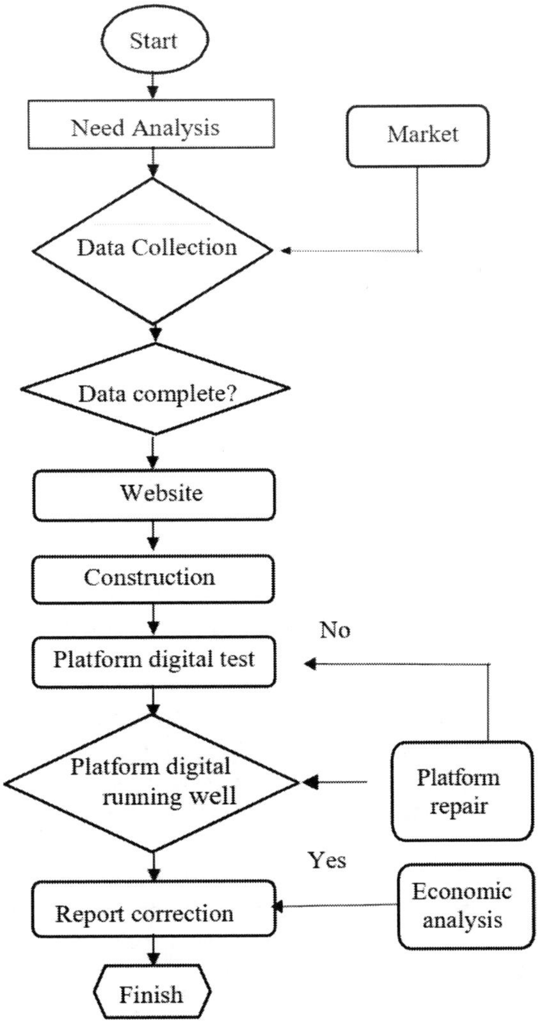

Fig. 1. Research Framework Flowchart

1) Testing program testing using black boxes by producing in accordance with previously designed expectations. The use of black boxes in application testing will provide an explanation of the suitability of expectations in program making.
2) Support the process or maintenance, The author seeks to develop a system that has been made related to the software and hardware that will be used. The stages in this research can be seen in Figure 1.

This digital marketing application design research uses the SDLC (System Development Life

Cycle) Waterfall method. This model takes a systematic and sequential approach starting from the level of system requirements and then moving on to the analysis, design, coding, testing, and maintenance stages. It is called a waterfall because the stages that are passed must wait for the completion of the previous stage and run sequentially, for example the coding stage must wait for the design stage to be completed [9].

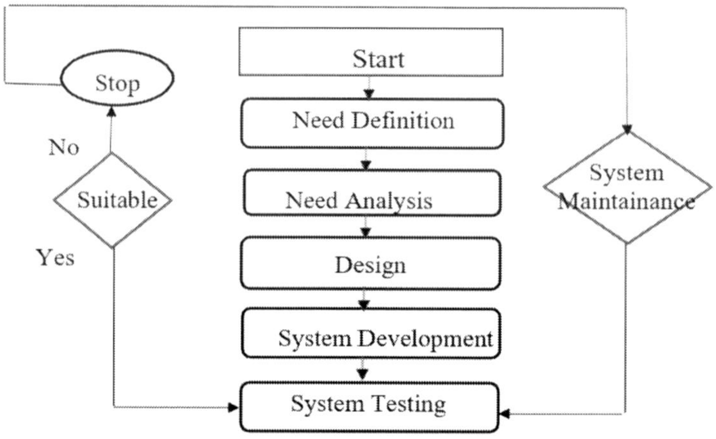

Fig. 2. SDLC (*System Development Life Cycle*) Waterfall Method

2.2 Economic Approaches

2.2.1 Profit-Lost Analysis (L/R)

Profit and loss analysis is a report that describes the results of operations in a certain cycle which contains the amount and source of income as well as the amount and type of costs incurred [10]. Profit and loss analysis is carried out by analyzing all revenues and expenses incurred during one year to find out the actual amount of profit and loss that the company earns. The income statement has components, namely sales, variable costs, fixed costs, gross profit and before tax, net income, interest and taxes.

This analysis makes it easy to determine the amount of cash flow obtained by the company and can calculate the company's minimum sales either quantity or value for money to find out the break even point. The income statement is also used to estimate taxes to be included in cash flow. The preparation of the income statement needs to be done as an illustration of the company's performance in achieving its goals, whether the planned business can provide benefits or vice versa by carrying out two activities, namely:

a) Calculating Total Cost (TC) is the expenditure of production costs to produce goods to be marketed by adding Total Variable Cost (TVC) and Total Fixed Cost (TFC)

b) Calculating Total Revenue (TR) as company revenue originating from marketing the main or side products sold

Through the income statement, the company obtains information about the business being run, whether it is profitable or not and the amount of net profit earned by the company. This analysis can make it easier to determine the amount of cash per year obtained by the company. Profit and loss analysis can be calculated using the following formula:

Total Revenue	: $TR = P \times Q$
Description	: TR = Total Revenue (IDR)
	: P = Output Price (IDR)
	: Q = Output Quantity (Unit)
Total Cost	: $TC = TFC + TVC$

Description : TC = Total Cost (IDR)
: TFC = Total Fixed Cost (IDR)
: TVC = Total Variable Cost (IDR)

(TR – TC) > 0 the company gains profit and (TR – TC) < 0 the company lost

a) R/C ratio Analysis

R/C ratio (Return Cost Ratio) is a comparison between revenues and costs. The analysis of the R/C ratio can be calculated using the following formula:

$$R/C \text{ ratio} = TR/TC$$

Description:
Total Revenue (TR) = The company revenues/ incomes (IDR)
Total Cost (TC) = The company expenditures (IDR)

The R/C ratio is used to assess the efficiency of a business. Efficiency is a figure of the best comparison between an effort and the results achieved. The efficiency of a business is determined by the size of the results obtained from the business, as well as the size of the costs required to obtain these results. If the R/C ratio >1 the business is said to be efficient, the R/C ratio = 1 the business is at the break-even point, and the R/C ratio <1 the business is said to be inefficient and lost [11].

3 Result and Discussion

3.1 Economic Small and Medium Business Partner

This research involves two Small and Medium Enterprises (SMEs) whose products are used in digital marketing, namely: UKM Mitra Tani Farm and UKM Happy Farm.
(1). Mitra Tani Farm UKM: Mitra Tani Farm is an SME engaged in the cultivation and trading of sheep. In addition, MT Farm entered the processing of derivative products into various canned products. Mitra Tani Farm or better known as MT Farm was established in 2004 with its initial business only focusing on the sale of sacrificial animals, over time and increasing market needs, MT Farm built a cage with a capacity of 1,000 goats, sheep and 250 cows. Currently MT Farm also serves aqiqah packages with an average of 4-5 aqiqah packages per day. The development of derivative products is also continuously carried out by producing around 500 canned products per day. The types of canned products produced include: lamb rendang, beef rendang, lamb tongseng, lamb fried rice seasoning, kebuli rice, lamb curry, lamb curry, lamb soup and tengkleng lamb. The average sales of canned products are 800-900 canned products per month.
2) Happy Farm UKM: Happy farm is a sheep breeding and cultivation farm located in Caringin District, Bogor. Happy farm was founded 4 years ago in Caringin Bogor, and currently has a nursery development branch in Tenjo District, Bogor. In addition to fattening sheep, Happy Farm also sells breeds and breeds of sheep from crosses between New Zealand sheep and superior local sheep. Another business carried out by Happy Farm is running *Aqiqah* packages. The products sold on the digital marketing platform are breeding sheep and sheep breeds.

The method that the author uses to make the Vilstock.id website is the System Development Life Cycle (SDLC) waterfall model proposed by Pressman [12]. The following is a discussion of the stages used for the development of the Radal Smart application.
(1). Analysis

At this stage several things related to the scope of work that will be needed to make decisions in making this system are described.

a) Problem Identification
At this stage, the identification of the problem is carried out, the following is a description of it:

a. Ruminant livestock producers do not have knowledge of website building

b. Customers often have difficulty finding information about existing products and their prices.

c. The online product purchase service is not yet available so that the sales system is still going on conventionally where customers buy goods by coming directly to the cage or shop directly.

b) Current System Analysis
The current system works manually. It is felt that there are still many obstacles because there are so many outputs that must be printed and there are many other shortcomings. And in this system the customer still has to come directly. Therefore, as a solution to the problems above, the authors propose some changes that will be made with an integrated system using a web-based interface and can be accessed via a web browser via the internet.

c) Analysis of the Proposed System
From the previous system problems, a new system is needed that is developed with better information technology so that it can facilitate data processing and tracing any information that is done by the system because the application to be submitted is web-based or online.

In this analysis stage the author describes the ideal online ordering system or the proposed system for Vilstock.id with the following steps:
a. To start the ordering process and purchase transactions, the user can select the product he wants to buy.
b. If the user has selected the product, then click the buy button.
c. If the user wants to continue shopping or buy other products, the user can click the click back button on the selected product.
d. Then click the shopping data menu, to see the shopping list that we have selected.
e. If you want to delete the shopping list that we have selected, please click cancel.
f. Next, click the menu fill in personal data, then fill in the form provided clearly such as name, address, telephone number, e-mail. Then click finish shopping.
g. Then the order form or invoice will be displayed as evidence that the customer has made a transaction.
h. User orders will be processed by Vilstock.id and the ordered items will be sent immediately. The proposed system can be seen in Figure 3.

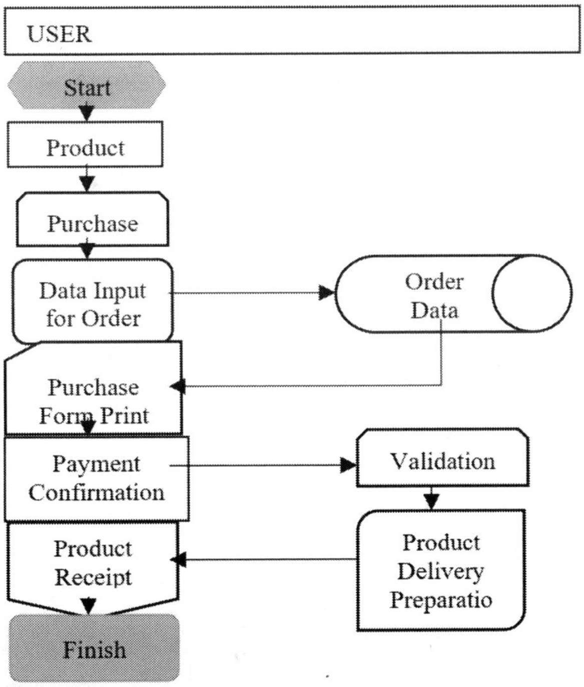

Fig. 3. Proposed System flowchart at *Vilstock.id* website

According to figure 3. Proposed System flowchart is implemented in e-commerce form with the order flowchart system that can be seen in figure 4.

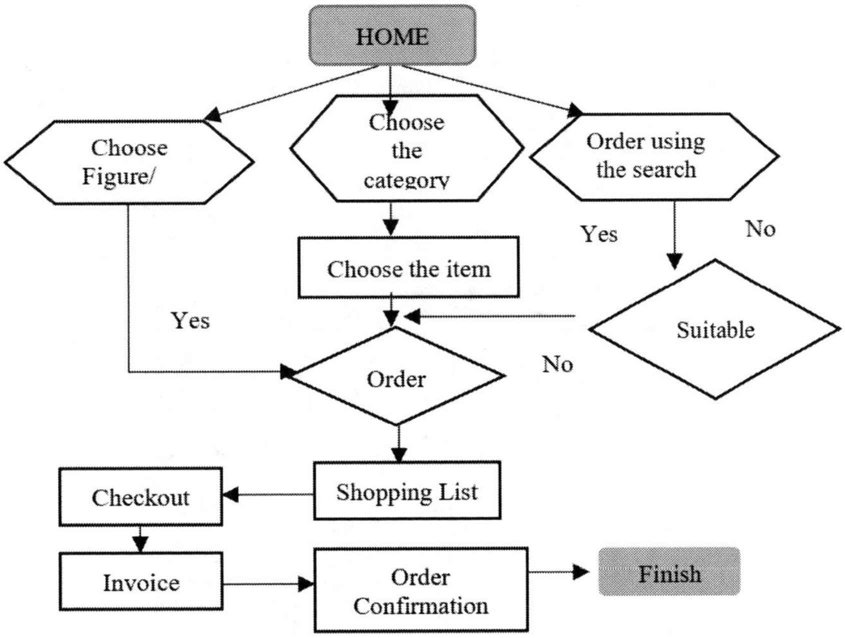

Fig.4. Order Flowchart System at *Vilstock.id*

3.2 Website Design Results

The results of the website design are the final result after going through all the stages in making the website. The design result consist of the main page display, shopping page, cart page, shipping page, payment page, and admin page.

3.2.1 Vilstock.id Platform Performance

Digital Marketing is marketing that uses platform found on the internet in carrying out activities to reach target consumers. Basically, digital marketing is marketing that uses digital platform on the internet. In the formation of a digital marketing strategy, vilstock.id utilizes 3 platforms such as Instagram, Facebook, and Website which then uses internet advertising using social media and website. This social media is widely used as a medium for online promotion of goods/ services through various application features, namely photos and descriptions. Instagram and Facebook play a role in providing information regarding what products are sold in vilstock.id. The following is a display of social media Instagram and Facebook of Vilstock.id.

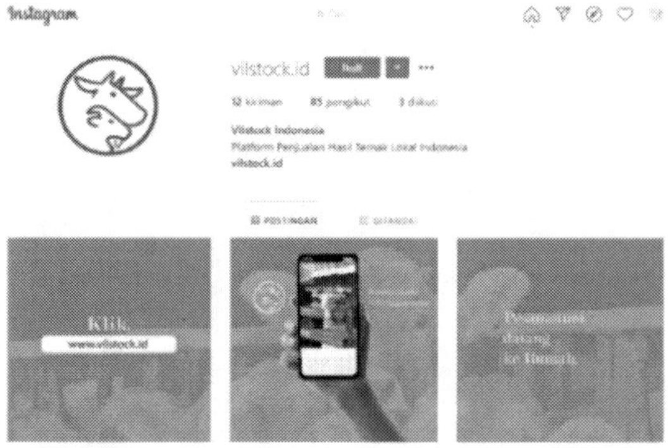

Fig. 5. Vilstock.id Instagram social media display

Social media and website must be connected to form a good digital marketing ecosystem. So that in doing internet advertising the results obtained will be more optimal and an increase in sales will be achieved. In doing digital marketing, vilstock.id through 2 ways, first organic way through Search Engine Optimizing (SEO) and second the paid way through Facebook ads that are linked to the website. Here's an example of how to display SEO and paid ads from vilstock.id

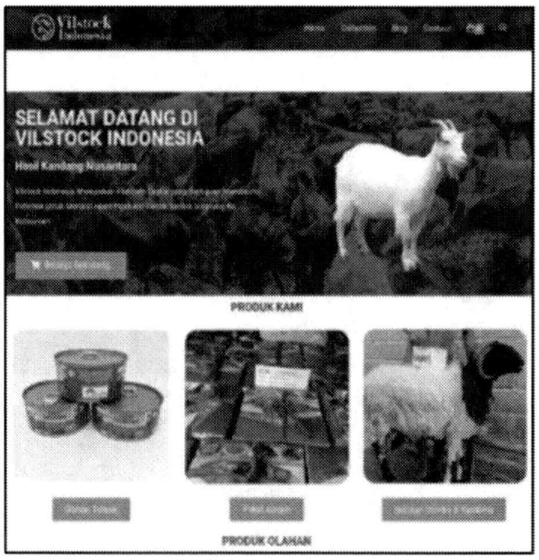

Fig.6. Vilstock.id front page display

The advertisement was made aimed to consumers that have an interest in consuming processed meat especially whom live in Bogor. The advertising reach is set at 153-400 people per day for one month. For one month, the website is always shared to various existing platform with the aim of maximizing traffic to the website. Then the result obtained are seen in figure 7.

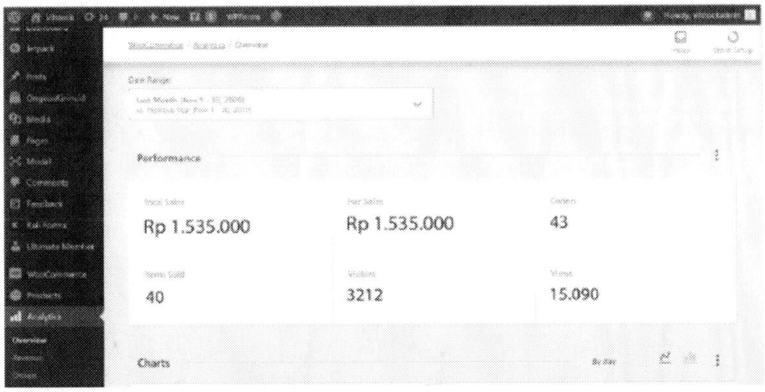

Fig.7. D*igital marketing* result display

Figure 7 shows that ads with a full reach of 12000 people per month there are about 26% of people who see the ad and then decide to come to the website to visit. The average visit from each person is 4-5 times. Repeated visits will potentially trigger the desire to buy the promoted product. The results of the promotion resulted in a conversion value of 1.2%, which means that each IDR 1,000 of promotional costs incurred will generate an income of IDR 1,200.

3.2.2 Income Analysis

Income analysis is used to determine the income of farmer from the sale of products from SME Mitra Tani Farm. SME Mitra Tani Farm's revenue is done by adding up the receipts from 9 products sold on vilstock.id then the average revenue is reduced by the average total cost incurred. For the time dimension used in this income analysis, revenue is for 3 months, namely September, October and November according to the length of the research period.

3.2.2.1 Breeder SMEs Acceptance

Revenue in this study is the result of multiplying the total sales of 9 Mitra Tani SME's products with the selling price set for each product. Sales of MT Farm SEME products through two events, namely conventional sales (offline) and online sales. The result of offline sales during September to October reached 1,172 products with the most sold product being beef rendang. While online sales reached 379 products with the most sold product was also beef rendang that reached 172 pieces products. Thus, the result of online sales are still greater than online sales. In addition, it's turn out that in the two types of sales, the product that is most favored by customers both offline and online is beef rendang. This can seen from the the sales of beef rendang which reached 403 cans or about 24% of total sales. The prices set for 9 item products are IDR 40,000 and some are IDR 35,000 per can with an average price of IDR 37,222 per each can. With total sales reached 1,679 item products, the total revenue for there month is reached IDR 58,785,000.

3.2.2.2 Production cost

Production cost consist of fixed cost and variable costs. Fixed costs include: depreciation cost, taxes, levies and other costs. While the variable cost consist of the cost of raw and supporting materials, marketing cost, labor cost, and transportation cost.

The average cost per can is IDR 4,000. The average tax and retribution cost are IDR 4000 and other cost are IDR 1,000. Based on the variable cost component, the highest cost in producing canned products is the cost of raw and supporting materials. The average cost of raw and supporting material reaches IDR17,555,56 or about 74% of the total variable cost. With the average fixed cost per can IDR 9,000 and the average variable cost of IDR 23,566 so the total cost per can is about IDR 32,636 with the average selling price per can is about IDR 33,500 then the profit earned per can is IDR 4,566.44. If the total sales per 3 month reached 1,679 item products, then the total revenue will reach IDR 58,785,000. If the average total cost per can is Rp. 32,636, then sales are 1,679 products, so the total cost is IDR 54,828,677.78, so the profit obtained is about IDR 3,956,322.22.

3.2.2.3 R/C Ratio Analysis

Based on the results of the study, it was found that the feasibility level of selling processed food products on raw materials for beef cattle (cattle, sheep and goats) after selling them through digital marketing vilstock.id showed that R/C = 58,785,000 / 54,828,677,78, then R/C = 1.072. Thus, the production of canned products is feasible because the R/C > 1. Whereas before sales were made through digital marketing the R/C ratio reached 1.014. So that with digital marketing, it can increase sales by 0.058 within only 4 weeks.

3.2.2.4 Analysis of Digital Marketing's Contribution to Sales

Product sales through digital marketing vilstock are expected to increase SME sales results. Since the launch of the vilstock.id platform in late October or early November, 40 cans have been sold. Of the 9 products on the platform, shredded beef is the product that consumers buy the most. The number of shredded beef sold was 10 cans or 33% of the total product sold. In November, MT Fam's off-line sales reached 735 products, with sales through vilstock in the same month totaling 40 products. So, by marketing through the vilstock platform, sales of 40 product items increased or an increase of 5.4%. Thus, sales through digital marketing can increase product sales. An increase of 5.4% is quite good because the platform has only been introduced to the public for a month. To increase sales through vilstock, it takes longer time and more intensive promotions so that more people are familiar with the vilstock.id platform and make vilstock.id an option for online shopping.

4 Conclusion

Digital marketing also helps solve problems in marketing beef cattle products (cattle, sheep and goats) and products at partner companies (SMEs, Cooperatives and Farmers/Livestock Groups). This can be seen from the comparative analysis of the value of using digital marketing platforms in partner companies (SMEs, Cooperatives and Farmer/Livestock Groups) before and during the COVID-19 (New Normal) pandemic. sheep and goats), namely R/C = 58,785,000 / 54,828,677,78, then R/C = 1,072, whereas before the implementation of digital marketing the R/C ratio reached 1,014. Thus, the production of canned products is feasible because R/C > 1. In addition, the successful use of digital marketing platforms as a means of marketing beef cattle and derivative products during the Covid 19 (New Normal) pandemic shows that there has been an increase in product sales of 5.4% in a period of 4 weeks.

The Vilstock.id website is expected to continue as a marketing strategy development for beef cattle products (cattle, sheep, goats) and their derivative products by SMEs by developing content to reach more consumers. In addition, the development of livestock derivative products also needs to be added to be more varied and provide consumers with a more diverse choice of products, so that consumer interest will be higher and SMEs will gain the customer satisfaction and loyalty. This will certainly have an impact on increasing the sales volume of SME products in the future.

-This research was financially supported by Collage of Vocational Studies IPB University-

References

1. Badan Pusat Statistik. *Pergerakan Harga Produk Peternakan*. Website BPS Pusat. (2020).
2. A. G.Huff, W.E.Beyeler, N. S. Kelley, J. A.McNitt. *How resilient is the United States' food system to pandemics?*. Journal of environmental studies and sciences, 5(3) p337–347.Tersedia link: https://doi.org/10.1007/s13412-015-0275-3(2015).
3. L.Aprilia, L Cyrillia, Burhanudin. *Analisis Strategi Pemasaran Ternak Berbasis E-Commerce di PT. X.* Jurnal Ilmu Produksi dan teknologi Hasil Peternakan. IPB. Bogor (2018).
4. H. El-Gohary, *E-Marketing - A literature Review from a Small Businesses perspective.* International Journal of Business and Social Sciences. Vol. 1 No 1. ijbssnet.com/journals/20.pdf (2020).
5. D.Chaffey, F.Ellis-Chadwick. *Digital Marketing : Strategy, Implementation and Practice.* Pearson. UK. (2019)
6. P. Kotler, K. Keller, M. Brady, M. Goodman, T.Hansen. *Marketing Management : 4th European Edition.* Pearson.UK (2019)
7. H.M. Taiminen, H. Karjaluoto, *The Usage of digital marketing channels in SMEs*, Journal of Small Business and Enterprise Development, Vol. 22 No. 4, pp. 633-651.(2015).
8. M. Faturokhman, M. Sarma, M. Najib, *Analisis Saluran Distribusi Daging Sapi di DKI Jakarta. Jurnal Sains Terapan*. Diploma IPB Bogor (2016).
9. A.S. Rosa, M. Salahudin. *Rekayasa Perangkat Lunak Terstruktur dan Berorientasi Objek.*Informatika Bandung. Bandung (2014).
10. Kasmir. *Manajemen Perbankan*. PT Raja Grafindo Persada. Jakarta (2012)
11. Soekartawi. *Agribisnis: Teori dan Aplikasinya*, Cetakan ke- tujuh, PT. Raja Grafindo Persada, Jakarta (2003).
12. R.S Pressman. *Software Engineering : A Practitioner's Approach.* Mc.Graw-Hill Companies Inc. New York (2001).

Sensory Properties and antioxidant activity of *Chrysanthemum* Flower tea bags with lemon peels and mint leaves

Ai Imas Faidoh Fatimah[1*], *Anita* Ristianingrum[2], and *Leni* Lidya[2]

[1]Food Quality Assurance, College of Vocational Studies IPB University, Indonesia
[2]Agribusiness Management, College of Vocational Studies IPB University, Indonesia

Abstract. The Covid-19 pandemic has had an impact on many business sectors including the cut-flower of Chrysanthemum. Sales of cut-flowers have decreased and the farmer suffering huge losses. It is necessary to develop a Chrysanthemum product to increase its economic value. This study aims to determine the sensory properties and antioxidant activity of Chrysanthemum tea in bags with the addition of lemon peels and mint leaves. The treatments in this research consisted of 6 formulations of Chrysanthemum: lemon peels: and mint leaves; FI (80%: 0%: 0%); F2 (80%: 20%: 0%); F3 (80%: 0%: 20%); F4 (80%: 10%: 0%); F5 (80%: 0%: 10%) and F6 (80%: 0%: 0%). Sensory properties were determined by hedonic tests (color, aroma, and taste of tea brewed water) with 30 panelists. Antioxidant activity was determined by the DPPH method and polyphenol content by the spectrophotometric method. The results of the hedonic test showed that F2 was the most preferred on the color and aroma parameters, while F2 and F4 had the same average value for the taste parameter. The highest antioxidant activity value was F2 with an IC_{50} value was 134,38 ppm and polyphenol content was 81.42 ppm.

1 Introduction

The pandemic of Covid-19 affected all aspects of people's lives, such as social, economic, health, and psychological. The impact of this pandemic on the society in Indonesia is not much different from the society in other countries that are also facing this pandemic [1]. The policy to control the Covid-19 pandemic in Indonesia through the Large-Scale Social Restriction Policy has an impact on the public activity and movement both nationally and internationally, for example, limiting events such as weddings and birthday parties, and another celebration event. The farmers of Chrysanthemum flowers are among those affected by the restriction policy which limited celebration events causes the demand for this flower to be reduced or even non-existent. Farmers have tried to get the benefits of Chrysanthemum by processing it into herbal tea. *C. morifolium* has been reported to contain numerous bioactive compounds such as phenolics, which possess antioxidant activity include 5-*O*-caffeoylquinic acid (chlorogenic acid), 3-*O*-caffeoylquinic acid,4- *O*-caffeoylquinic acid,

* Corresponding author: aiimasfaidohfatimah@apps.ipb.ac.id

3,4-di- caffeoylquinic acid, 3,5-di-*O*-caffeoylquinic acid, 4,5- di-*O*-caffeoylquinic acid, caffeic acid, luteolin, luteolin-7-*O*-glucopyranoside, apigenin, apigenin-7-*O*-glucoside, acacetin-7-*O*-rutinoside, acacetin [2].

Some Chrysanthemum tea products of the Indonesian market are sold in simple forms of bulky flower buds and packaged simply using low-grade packaging with unattractive design and uncompleted labels. This causes the product to be less attractive to consumers. In addition, in the bulky form, the different amounts of each brewing process can affect the quality of tea. Therefore, development is needed to increase consumer interest in chrysanthemum tea so that it can increase its economic value. The development of chrysanthemum tea products can be pursued by changing the bulk tea that is packaged in plastic into tea bags. In addition, other materials can be added to improve the organoleptic and functional properties.

Tea bags are widely known by people in Indonesia, but are commonly used for tea (*Camellia sinensis*) products and have been used widely used for Chrysanthemum tea. Tea bag innovation is carried out by packaging the tea in special paper or nylon packaging to shorten the tea brewing process and after brewing the tea dregs can be separated immediately. The main advantage of tea bags is their convenience because each tea bag contains about 2 grams of loose tea leaves which are ideal for making a cup of tea. Other advantages of teabags are also related to their easy availability, the small size of the bags used to pack the right amount of ingredients, which makes them practical compared to regular teas, the ease of mixing different ingredients (herbs) with various health benefits [3].

Chrysanthemum flower plants are popular in Indonesia as ornamental flowers and have not been optimally used as herbal tea beverages products. Improvement quality of Chrysanthemum tea can be carried out by combining tea with new materials or ingredients that can enrich the organoleptic quality of tea. Ingredients that potentially can be added to Chrysanthemum tea are lemon peels and mint leaves. Lemon peels have a bright color and contain chemical compounds that are beneficial for the body as antioxidants. Orange peels contain different compounds, depending on the variety, so the aroma is different. However, the dominant compound is limonene ($C_{10}H_{16}$) that varies for each citrus variety. Lemon is a plant that has benefits as natural antioxidants because it contains vitamin C, citric acid, essential oils, bioflavonoids, polyphenols, coumarins, flavonoids, and volatile oils in the skin such as limonene (± 70%), -terpinene, -pinene, -pinene, as well as coumarins, and polyphenols [4]. Mint leaves are also widely used for drinks. Mint leaves have a fresh effect and distinctive aroma due to a large number of menthol compounds [5]. The aqueous extract and EO of the mint plant are potent natural antioxidants [6]. This study aims to determine the sensory properties and antioxidant activity of Chrysanthemum tea in bags with the addition of lemon peels and mint leaves.

2 Material and Method

2.1 Materials

Chrysanthemum tea was obtained from PT Rahmat Purnama Farm, Sukaresmi Cianjur. Fresh lemon and mint leaves were purchased from the local market in Bogor. Dried chrysanthemum was grounded by using a blender to a particle size of 20-mesh. Lemon peels were cut into small pieces and dried at 60-70ºC for 7 h, and dried lemon peels were grounded by using a blender to a particle size of 20-mesh. Mint leaves were dried at 60-70ºC for 2-3 h, were also grounded by using a blender to a particle size of 20-mesh. All materials were mixed according to the formulation and put into bags of 2 grams/bag. The formulation of chrysanthemum tea can be seen in Table 1. As the sample for analysis, tea bags were brewed in hot water as much as 200 ml for 5 minutes.

Table 1. Formulation of Chrysanthemum tea, Lemon peels, and Mint leaves.

Material	Percentage (%)					
	F1	F2	F3	F4	F5	F6
Chrysanthemum tea	100	80	80	90	90	80
Lemon Peels	0	20	0	10	0	10
Mint leaves	0	0	20	0	10	10

2.2 Methods of Analysis

2.2.1 Sensory analysis

Sensory tests are performed by the hedonic testing method on color, aroma, and taste parameters. The test was conducted by 30 untrained panelists with ages ranging from 17 to 55 years old. The panelists were asked to evaluate (aroma, taste, and color) and give scores to each sample of Chrysanthemum tea bag based on the panelist's preference. The rating scale used for the hedonic test is a 5-point hedonic scale, 1: Dislike very much, 2: Dislike slightly, 3: Neutral, 4: Like 5: Like very much. The data of this research were analyzed by Analysis of Variance (ANOVA) and further test using the Duncan Multiple Range Test (DMRT).

2.2.2 Determination of antioxidant activity

Antioxidant activity testing was carried out on 6 formulations of Chrysanthemum tea with the addition of lemon peels and mint leaves. Tea bag based on each formulation were brewed at 90°C then moved up and down for 5 minutes and the tea bags were removed. The tea solutions were then diluted with methanol for different concentrations from 20 ppm, 40 ppm, 60 ppm, 80 ppm. and 100 ppm. The antioxidant activity test of the sample was carried out by taking 0.2 ml of the test solution added by 3.8 ml of a solution of 1.1-diphenyl-2-picrylhydrazyl (DPPH) in methanol. The solution was homogenized with vortex and then incubated in a dark place for 30 minutes. Then the absorption was measured by Ultraviolet-visible (UV-Vis) spectrophotometer at a wavelength of 518 nm. As a blank, the same treatment was carried out with the sample was replaced with methanol [7]. The absorbance data was calculated as an antioxidant activity with the following equation:

$$AA\ (\%) = \frac{A-B}{A} x\ 100$$

Note :
AA = Inhibition activity
A = Absorbance of control
B = Absorbance of sample

DPPH scavenging activity of the sample was made into a regression curve to obtain the regression equation used to measure IC_{50} that refers to the concentration of the sample which is required to scavenge 50% DPPH free radicals. The formulation treatment thatgave the highest IC_{50} result was followed by a polyphenol content test to determine the number of polyphenol compounds in Chrysanthemum tea bags.

3 Results and Discussion

3.1 Sensory Properties

3.1.1. Aroma hedonic test

Aroma is one of the critical aspects of tea quality which can determine acceptance or rejection of a tea before it is tasted [8]. The aroma of a food and beverage product can be determined by the sense of the nose through the smell caused by the presence of volatile compounds. The average value of hedonic aroma taste for tea bags with the formulation of Chrysanthemum, lemon peels, and mint leaves is shown in Table 2.

Table 2. The hedonic score of Aroma of Chrysanthemum tea with the addition of lemon peels and mint leaves

No.	Treatment	Average Score of Aroma
1	F1	3.17[ab]
2	F2	3.47[b]
3	F3	3.05[a]
4	F4	3.37[ab]
5	F5	3.10[ab]
6	F6	3.13[ab]

Note: Different notation showed significant differences (α=0.05)

Based on the result of Analysis of Variance (ANOVA), the formulation of Chrysanthemum, lemon peels, and mint leaves in F1, F4, F5, and F6 have no significant effect on the panelist's preference on aroma parameters because the addition of lemon peels and mint leaves to Chrysanthemum tea is 10%. Meanwhile, treatment of F2 and F3 is significantly different because the addition of lemon peels is 20% and contributes a strong aroma to the Chrysanthemum tea bags. Lemon peels have complex volatile essential oils that give a distinctive aroma [9]. Twenty-six volatile compounds in the lemon peels oils were identified by GC-MS and can be grouped into four main chemical families: (i) monoterpenes (20 compounds); (ii) sesquiterpenes (3 compounds); (iii) aldehydes (2 compounds), and (iv) esters (1 compound) [10]. The constituents of citrus essential oils include limonene, β-myrcene, α-pinene, p-cymene, β-pinene, and terpinolene are the main aromatic compounds [11].

According to Table 2, it was found that the higher proportion of mint leaves, the lowest score of panelist's preference. Mint leaves contain menthol compounds and essential oils that give the tea a fresh aroma when brewed. In fresh peppermint leaves, 36 components were identified in total that accounting for 93.84% of the total volatiles. In the fresh mint leaves, the principal components were menthol (44.39%), menthone (15.36%), menthofuran (10.27%), 1.8-cineole (5.81%), menthyl acetate (4.78%), neoisomenthol (2.37%), and limonene (1.87%) and drying of the peppermint leaves at ambient air as well as hot air temperatures of 50°C, 60°C, and 70°C can be recommended for high oil yield and resulted in a better-quality product in terms of menthofuran, neoisomenthol, and 1.8-cineole [12]. Peppermint leaves hold a resilient sweetish odor and a spicy flavor with a freshening after mastication [13]. There are 58 types of volatile compounds contained in

Chrysanthemum flowers, including -Humulene, Ledene oxide-(I), cis-Z-α- Bisabolene epoxide, 3.4-Dihydro-2.2-dimethyl-2H- 1-benzopyran, trans-limonene oxide, and limonene. *C. morifolium* Ramat flower is rich in flavonoids luteolin-7-glucoside and quercitrin as well as volatiles β-humulene and ledene oxide-(I), this B- Humulene has been described as having a woody-spicy, dry, clove-like tenacious odor and somewhat bitter taste [14]. This result shows that mint leaves are not suitable to mix with Chrysanthemum tea which each material has a unique and distinctive aroma.

3.1.2. Taste hedonic test

Besides aroma, the taste of tea is an important factor in determining the quality of tea. The taste of tea is related to its ingredients or composition. Based on the analysis of variance on the taste parameters, the significance value of the sample was 0.295 or more than α (0.05), this result showed that there is no significant effect of formulation on the taste of tea bags. The average taste score of hedonic taste is shown in Table 3 with a value ranging from 2.85 (dislike slightly) to 3.2 (neither like nor dislike).

Table 3. Hedonic score of taste of Chrysanthemum tea with the addition of lemon peels and mint leaves

No.	Treatment	Average Score of Taste
1	F1	3.10
2	F2	3.22
3	F3	2.98
4	F4	3.22
5	F5	3.03
6	F6	2.85

Table 3 showed the formulation of F2 has the same hedonic taste score with F4 indicating that the panelist prefers the taste of Chrysanthemum tea with lemon peels than mint leaves. The lowest score on hedonic taste was observed in sample F6. Based on the researchby Shofiati, 2014 [15], Lemon peels was added to the red dragon fruit peels tea bags and the result was the addition of 40% lemon peels gives the highest taste value and the addition of lemon peels is very influential on the taste of the red dragon fruit peels tea bags. The bitter taste of orange peels is due to the presence of hesperidin and naringin [16]. The addition of peppermint extract provides freshness and has a distinctive minty taste when brewed due to the menthol content on peppermints but the higher concentrations above 4% will cause a bitter taste and cause a spicy taste when brewed [17]. The astringent and bitter taste in beverages with the addition of mint leaves is caused by compounds catechins and alkaloid compounds found in the leaf. The content of tannins in a material determines qualities related to color, taste, and steeping aroma because tannins can coagulate protein to produce an astringent taste in steeping water [18].

3.1.3 Color hedonic test

The consumers' preference for food and beverage products is also determined by color. The average hedonic score for color is presented in Table 4 with the score range was 2.67-3.55 means the preference of panelists is from slightly dislike to like. The formulation of F2 has the highest score for the panelist's preferences 3.55 and F6 has the lowest score of 2.67. Panelists prefer chrysanthemum tea with light yellow lemon peel compared to brown chrysanthemum tea which is influenced by the addition of mint leaves. The color hedonic score of Chrysanthemum tea with the addition of lemon peels and mint leaves can be seen in Table 4.

Table 4. The color hedonic score of Chrysanthemum tea with the addition of lemon peels and mint leaves

No.	Treatment	Average Score of Color
1	F1	3.27[b]
2	F2	3.55[b]
3	F3	3.22[b]
4	F4	3.35[b]
5	F5	3.28[b]
6	F6	2.67[a]

Note: Different notation showed significant differences (α=0.05)

The original chrysanthemum tea has a slightly tea yellow color. Based on Table 4, the color of the formulation F2 and F4 have a higher average score (3.55 and 3.35) than other formulations. These data indicate that the addition of lemon peels to chrysanthemum tea resulted in the color of the tea that was preferred by the panelists. The addition of lemon peels gives yellowness to the chrysanthemum tea, with the higher amount of addition, the color is yellow turns to orange. Chrysanthemum flowers are known for their beautiful flowers with different colors including yellow, orange, orange-red, which are mainly contributed by carotenoids that are responsible for the deep yellow and orange coloration in the chrysanthemum flower [19]. Citrus peels are subdivided into the epicarp or flavedo and mesocarp or albedo. The flavedo is the colored peripheral surface of the peel while the albedo is the white soft middle layer of the peel [20].

The statistical analysis results of sensory tests (α=0.05) for color parameters showed that there was no significant difference of color based on the panelist's preference between the formulation F1 until F5 but there were significant differences between those formulations with F6. Lemon peels have carotenoid pigments that contribute to lightness and yellowness in tea [21]. The brewed tea of formulations F3 and F5 has a slightly brown color with the higher the percentage of mint leaves addition causes the color of the tea brew to be darker. Mint leaves contain chlorophyll 3.21-8.49 g chlorophyll-a; 0.66 - 1.79 g chlorophyll b; and 10.3 mg carotene/100 g [22]. Drying in the processing of mint leaves powder and brewing with hot water causes a change in the color of the mint leaves. The longer the drying time, the more the chlorophyll pigment will fade because of the heating. Based on the research of Mg2+ ions contained in the chlorophyll of mint and stevia leaves will be released and replaced by hydrogen ions (H+) and form pheophytin compounds which are light green in color and will cause the green color of the mint and stevia leaves to turn brown so that the color of the resulting filtrate tends to be dark [23]. At the next level, the change of the C atom with a hydrogen atom causes pheophytin to be converted into pyro pheophytin which has a brown color [24]. However, the combination of lemon peels and mint leaves in F6 has the lowest preference value because the color of tea that was produced has a less bright brown ten to dark color.

3.2 Antioxidant activity test

Nowadays, consumers are increasingly aware of the importance of maintaining health, therefore they prefer to consume healthy foods and beverages such as products that contain antioxidants. Antioxidants are compounds that can inhibit oxidation reactions by binding to free radicals and highly reactive molecules. Free radicals can damage macromolecules leading to cause many diseases. The antioxidant activity test in this study was conducted to determine the formula of Chrysanthemum tea bags with high antioxidant activity as an advantage for this tea product. Antioxidant activity test using the 1.1-diphenyl-2-

picrylhydrazyl (DPPH) method which is easy, fast, and sensitive to test the antioxidant activity of plant extracts. The changing color of DPPH solution from purple to yellow determines the antioxidant activity in the sample against DPPH measured by UV-Vis spectrophotometer to obtain the value of IC_{50} as the antioxidant concentration required to obtain 50% radical inhibition [25]. The percentage of inhibition activity was measured for each formulation and the result is shown in Table 5.

Table 5. Percentage of inhibition of Chrysanthemum tea with the lemon peels and mint leaves

Concentration	% Inhibition Activity of Antioxidant					
(ppm)	F1	F2	F3	F4	F5	F6
20	9.27	7.20	11.24	10.65	10.45	0.99
40	22.29	10.26	18.44	16.17	10.85	12.72
60	24.65	24.85	21.79	26.53	21.30	16.47
80	34.62	38.36	26.82	27.22	31.16	19.63
100	43.79	47.73	33.93	46.84	41.32	26.63

The inhibition activity of each formulation on five varying concentrations (20, 40, 60, 80, and 100 ppm) was demonstrated in Table 5 that shows the inhibition activity of each extract of the formulation was increased with the increasing concentration of the sample. The 100 ppm of Chrysanthemum tea bags showed the best antioxidant activity. Based on Table 5 The best formulation that contributes the highest percentage of inhibition is F2 (80% Chrysanthemum tea and 20% lemon peels). Citrus peels are described as a rich source of unique phenolic compounds to citrus, especially the characteristic flavanone glycosides (mainly naringin, hesperidin, narirutin, and neo hesperidin). Citrus peels are a good source of phenolic compounds that can be extracted and employed as natural antioxidants to prevent oxidation of some foods or may be utilized in designing functional foods, Huge amounts of flavanones and many polymethoxylated flavones which are very rare in other plants are contained in citrus peels [26].

IC_{50} value is the concentration of samples that are required to scavenge 50% DPPH free radicals. The IC_{50} value of each sample formulation is calculated by using the regression equation of the log dose inhibition curve (Table 6).

Table 6. IC_{50} value of Chrysanthemum tea with the lemon peels and mint leaves

Treatment	IC_{50} (ppm)
F1	166,42
F2	134,38
F3	431,54
F4	183,34
F5	218,27
F6	550,26

The lower value IC_{50} refers to the higher antioxidant activities. Based on Table 6 can be known that the formulation of F2 is the treatment with the highest antioxidant activities based on DPPH assay and F6 is the lowest antioxidant activity. For further information, to determine the content of polyphenols that support the antioxidant activity of formula F2, a polyphenol test was carried out by spectrophotometric methods with results showing that the content of polyphenols of F2 was 81.42 ppm. The citrus peels are rich in nutrients and contain many phytochemicals that can act as antioxidants and protect cells against free radical damage, e.g. polyphenols, carotenoids. Phytochemicals are known to display their health-protective effects in diverse ways with strong potential for use in drug production or as food supplements [27].

4. Conclusion

This study aims to determine the sensory properties and antioxidant activity of Chrysanthemum tea in bags with the addition of lemon peels and mint leaves. The results of the hedonic test showed that the formulation F2 was the most preferred by panelists on the color and aroma parameters, while F2 and F4 had the same average value for the taste parameter. The highest antioxidant activity value was F2 with an IC_{50} value was 134,38 ppm and polyphenol content was 81.42 ppm

5. References

1. M.C. Widyasari, H. Soesanto, IJEBAR, **5**, 238 (2021)
2. J. Gong, B. Chu, L. Gong, Z. Fang, X. Zhang, S. Qiu, J. Wang, Y. Xiang, G. Xiao, H. Yuan, and F. Zheng, Antioxidants, **8**, 325 (2019)
3. P. Bassi, V. Kumar, S. Kumar, S. Kaur, Y. Gat, I. Majid, JFPE, **43**, 1 (2020)
4. S. Mujdalipah, S.L. Brilianty, L. Yosita, Mardiani, Edufortech. **5**, 15, (2020)
5. M.W. Apriliyanti, M. Ardiansyah, M.E. A. Santoso, *Antioxidant activity and sensory properties in packaged beverages with melinjo peels, mint leaves, and stevia leaves formulations,* IOP Conf. Ser.: Earth Environ. Sci, 7 November 2020, Jember, Indonesia (2020)
6. L.G. Riachi, C.A. De Maria, Food Chem, **176**, 72 (2015)
7. K. Arumsari, S. Aminah, Nurrahman, JGP, **9**, 128 (2019)
8. V.S.P. Chaturvedula, I. Prakash, JMPR, **5**, 2110 (2011)
9. E. Palazzolo, VA. Laudicina, M.A. Germanà, Curr. Org. Chem, **17,** 3042 (2013)
10. M.G. Aguilar-Hernández, P. Sánchez-Bravo, F. Hernández, Á.A. Carbonell-Barrachina, J.J. Pastor-Pérez, P. Legua, Foods, **9**, 241 (2020)
11. MF. Ramadan, *Fruit Oils: Chemistry and Functionality*, (Springer, Egypt, 2019)
12. M. Beigi, M. Torki-Harchegani, A.G. Pirbalouti, Int. J. Food Prop, **21**, 267 (2018)
13. P. Nayak, T. Kumar, A.K. Gupta, and NU. Joshi, J. pharmacogn. phytochem, **9,** 1519 (2020)
14. Q. Sun, S. Hua1, J. Ye1, X. Zheng, Y. Liang, Afr. j. biotechnol. **9**, 3817 (2010)
15. A. Shofiati, M.A.M. Andriani, and C. Anam, Jurnal Teknosains Pangan, **3**, 5 (2014)
16. K. Manjarres-Pinzon, M. Cortes-Rodriguez, E. Rodríguez-Sandoval, Braz. J. Chem. Eng. **30**, 667 (2013)
17. T. Anggraini, D. Silvy, SD. Ismanto, F. Azhar, JLI, **4**, 79 (2014)

18. D.A. Apriliyani, S. Prabawa, B. Yudhistira, Agrointek, **15**, 876 (2021)
19. D. Ramful, T. Bahorunb, E. Bourdonc, E. Tarnus c, O. I. Aruoma, Toxicology, **278**, 75 (2010)
20. P. Nayak, T. Kumar, A.K. Gupta and N.U. Joshi, JPP, 9, 3 (2020)
21. S. Ullas P, Namita, K. P Singh, A. Kundu, S. Panwar, G. Krishnan, G. Kumar. Indian J. Agric. Sci. 88, 393 (2018)
22. W.B. Sunarharum, A.N. Yudawati, and N.E. Asih, *Effect of different brewing techniques and addition of lemon peel (Citrus limon) on physico-chemical characteristics and organoleptic of cascara tea,* IOP Conf. Ser.: Earth Environ. Sci. 733 012086 (2021)
23. E. Hely, M.A. Zaini, A. Alamsyah, Jurnal Agrotek UMMat, 5, 1 (2018)
24. E. Straumite, Z. Kruma, R. Galoburda, Agron. Res. 13, 1104 2015
25. J. Flieger, M. Flieger, *Molecules* **25,** 1 (2020)
26. M.H.A. Abd El-ghfar1, H.M. Ibrahim, I.M. Hassan, A.A.A Fattah, MH. Mahmoud, *Int.J.Curr.Microbiol.App.Sci,* **5**, 777 (2016)
27. D. Suja, G. Bupesh, N. Rajendiran, V. Mohan, P. Ramasamy, N.S. Muthiah, A.A. Elizabeth, K. Meenakumari, K. Prabu, Int J Pharmacogn Chinese Med, **1**, 2 (2017).

Utilization of Cheesy Jackfruit (*Artocarpus heterophyllus* Lam.) Sw. Embutido

Noeme C Mosura[1], Rosevic *M Esbieto*[1]

[1] Iloilo Science And Technology University, Miagao, Republic of the Philippines

Abstract. This study aimed to utilize Jackfruit into Cheesy Embutido. This research was conducted to determine the level of consumer's acceptability of Jackfruit Cheesy Embutido. Embutido is a Filipino-style meatloaf made with a festive mixture of any type of meat, carrots, and raisins wrapped around slices of eggs, cheese, and sausage. A set of steps were conducted in this research, such as experimentation, evaluation, and data gathering and analysis. Results revealed that there was no significant difference in the level of acceptability of the jackfruit Embutido as to appearance (μ=405, p=0.375), Aroma (μ=405, p=0.370), Texture (μ=350, p=0.122), Taste (μ=338, p=0.089) and General Acceptability (μ=395.50, p=0.386).

1 Introduction

Local consumers' love of meat products in the Philippines saw a rise in meat consumption, specifically pork meat. According to Statista Research Department, in 2021, [1] pork was the most consumed type of meat in the Philippines, at approximately 15 kilograms per person per annum. As a meat-loving nation, meat dishes are commonly served even at any party, event, or fiesta [2]. Our country has a rich food culture since it has numerous indigenous and foreign influences, and one example is the embutido. Embutido is a Filipino-style meatloaf made with a festive mixture of any type of meat, carrots, and raisins wrapped around slices of eggs, cheese, and sausage [3]. The Filipino style Embutido can be traced to when Spain colonized the Philippines for three and a half centuries beginning in the middle of the 15th century. The Spaniards introduced different sausage recipes during that time, including chorizo, longaniza, and embutido.

On the other hand, jackfruit scientifically known as Artocarpus heterophyllus, is an exotic fruit grown in tropical regions of the world and has increased in popularity in recent years. It was locally grown in the country, wherein in 2020, the volume produced in the Philippines amounted to 40.2 metric tons. This fruit is part of the Moraceae plant family and has a spiky outer skin and is green or yellow. It also has a distinctive sweet flavor and can be used to make a wide variety of dishes. Likewise, it has a comparable texture to meat. Moreover, it is beneficial as it provides a moderate amount of calories and lots of fiber, vitamins, minerals, and antioxidants.

Numerous people are astonished at just how much Filipinos eat in a day. Breakfast, lunch, and dinner are all typically big meals. Filipinos are fond of experimenting with different kinds of dishes. Hence, with the abundance of jackfruit in the community, the researchers were challenged to utilize this fruit as a meat substitute to enhance the sensory characteristic of the food and create a cost-effective recipe. Likewise, this research was conducted to enhance the product's nutrient values and give an additional income source to the local jackfruit growers.

The primary purpose of this study was to utilize Jackfruit into Cheesy Embutido. This research was conducted to determine the level of consumer's acceptability of Jackfruit Cheesy Embutido. Specifically, this study aimed to:

1. Evaluate and compare the sensory qualities of Jackfruit Cheesy Embutido in two formulations in terms of appearance, aroma, texture, and taste after processing, during, and after storage.
2. Evaluate the perceptibility of Jackfruit Cheesy Embutido of the most preferred formulation.

2 Literature Review

The luscious embutido is one of the many recipes we acquired from the Spaniards. The word embutido is the generic term for cured sausages in Spain, Portugal, and some countries in South America. Embutido is a Filipino-style meatloaf with ground pork as its main ingredient. It is a dish generally served at special events like fiestas.

An embutido is a type of sausage found in Spain and Portugal. It contains hashed meat, generally pork, seasoned with aromatic herbs or spices that are served, which is wrapped in the skin of the pig's intestines. Mass-produced embutido are often wrapped in a type of artificial but edible skin. There are several types of sausages, including cured chorizo, salami, cured loin, and sobrassada or cooked morcilla or androlla. The Filipino embutido frequently contains raisins. Embutido is a type of meatloaf prepared in a Filipino style. Though a well-known dish for a holiday, embutido could be enjoyed every day without any hassle. Several meat processing companies now produce this meatloaf for commercial purposes.

Jackfruit is a tree that could be 30 to 70 ft. (9-21 m.) tall. The leaves are oblong, oval, or elliptical with 4 to 6 inches in length, leathery glossy, and deep green. It is the largest fruit globally, capable of reaching 100 pounds. This fruit is a large tropical Asian tree related to the breadfruit that yields fine-grained yellow wood and immense fruits containing an edible pulp and nutritious seeds and its fruit [4]. Many parts of this plant, including the bark, roots, leaves, fruit, and seeds, have medicinal properties. Jackfruit has many benefits, so the researchers utilize it to make embutido to improve its nutritive value.

Based on the study of Elevitvh and Manner [5], as cited in the preparation of jackfruit chips is effortless and can quickly be done. The bulbs are cut into 4cm x 2cm slices. The cut slices were blanched in water for 10 minutes. Then, the slices are weighed and immersed in 0.1% KMS for 15 minutes using 2 kg. of solution per kilogram of bulb material. After sieving the water, the slices are dried using a mechanical dryer to assure less oil absorption in the subsequent frying. The slices are fried in oil and stirred in a narrow wooden rod maintaining the temperature at 70 °C for 1 hour and 60 °C for another 6 hours. When the slices are light-yellow, the chips are taken from the saucepan. These prepared chips are mixed with some salt and spices in a bowl. Then processed chips can e packed in various packaging materials or containers. Considering moisture content (%), weight gained (%), quality aspect, and sensory attributes like crispiness, color, flavor, and overall acceptability, metal foil pouches were found most suitable for packaging jackfruit chips.

The prepared chips can be stored at ambient conditions in "metalex" foil for two months without loss of organoleptic quality.

Based on Amit and Ambarish study [6], the ripe jackfruit contains a good amount of fermentable sugar, which may be explained for the commercial production of vinegar and wine. It was reported that the maximum alcohol content in jackfruit wine was 10%, with a sugar utilization of 14% of total sugar solids. These early results show promise for using this fruit for commercial wine production. A certain maturity level and ripeness of jackfruit are essential for the production of jackfruit wine.

The jackfruit wine may protect against antioxidant and DNA damage and could become a valuable source of antioxidant-rich nutraceuticals. Additionally, the wine could be a commercially valuable by-product for jackfruit growers.

3 Methodology

3.1 Research Design

The experimental method used in this study was a collection of research designs that used manipulations and controlled testing to understand casual processes. Generally, one or more variables are manipulated to determine their effect on a dependent variable. According to McLeod [7], an experiment is an investigation in which a hypothesis is scientifically tested. The independent variable (the cause) is manipulated in an experiment, while the dependent variable (the effect) is measured, and any extraneous variables are controlled.

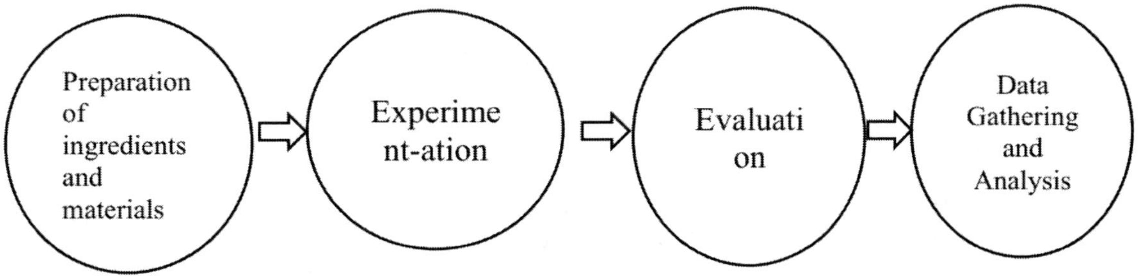

Fig. 1. Process of producing Jackfruit Cheesy Embutido.

3.2 Evaluators

Using purposive sampling, this study was evaluated by fifteen (15) target consumers and fifteen (15) Culinary Instructors during the second semester of Academic Year 2020-2021. The expertise of these people was well-considered, knowing that they could greatly help in achieving the purpose of this experimentation.

3.3 Materials and Instrumentation

The main tools used in making Jackfruit Cheesy Embutido are a steamer, mixing bowl, knives, chopping board, measuring equipment, LPG cooking range, foil, and freezer.

The ingredients used in the two formulations were as follows:

The ingredients of Formulation A (100 % Jackfruit) were (1) 100 grams chopped jackfruit, (2) 8 pieces Vienna sausage, (3) 100 grams bread crumbs, (4) 2 pieces raw eggs, (5) 100 grams chopped carrots, (6) 50 grams chopped bell pepper, (7) 50 grams chopped onions, (8) 50 grams chopped garlic, (9) 50 grams pickle relish, (10) 100 grams grated cheese, (11) 100 grams raisins, (12) ½ tablespoon salt, (13) ½ tablespoon pepper, (14) ½ tablespoon white sugar, (15) 4 pieces hotdog cut half lengthwise, (16) 2 pieces boiled eggs.

Moreover, the ingredients of Formulation B (50% meat and 50 % Jackfruit) were (1) 50 grams chopped jackfruit, (2) 50 grams ground meat, (3) 8 pieces Vienna sausage, (4) 100 grams bread crumbs, (5) 2 pieces raw eggs, (6) 100 grams chopped carrots, (7) 50 grams chopped bell pepper, (8) 50 grams chopped onions, (9) 50 grams chopped garlic, (10) 50 grams pickle relish, (11) 100 grams grated cheese, (12) 100 grams raisins, (13) ½ tablespoon salt, (14) ½ tablespoon pepper, (15) ½ tablespoon white sugar, (16) 4 pieces hotdog cut half lengthwise, (16) 2 pieces boiled eggs.

The instrument used was the sensory evaluation score sheet of the Five-point Likert scale for the characteristics and the nine-point Hedonic scale for general acceptability.

3.4 Data Gathering Procedure

The experimental study was composed of the following: Phase I is preparing materials, tools, equipment, and ingredients. Phase II is the preparation of the product, and Phase III is the evaluation of the finished product.

Phase I: The preparation of the materials and ingredients: (1) Gather all the materials and ingredients needed for cooking, (2) Place the ground meat and chopped jackfruit in a large container, (3) Add bread crumbs and raw eggs and mix well, (4) Put in the carrots, bell pepper, onion, garlic, pickle relish, and grated cheese and mix thoroughly, (5) Add raisins, salt, sugar, and pepper, then, mix well, (6) Place the meat mixture in an aluminum foil and fatten it, (7) Put the slice sliced hotdog, sliced boiled eggs, and sliced cheese alternately on the middle of the flat mixture, (8) Roll the foil to form a cylinder, then, lock the edges of the foil.

Phase II: Cooking procedure: (1) Place the product inside the steamer, (2) Cook for 1 hour, (3) Let it cool, (4) Unwrapped, slice, and serve.

Phase III: Evaluation of the finished product will be conducted after the product has been cooled down. (1) Prepare the evaluation sheet (2) Present the finished product to the evaluators (3) Gather the data (4) Compute and record the results.

3.5 Data Processing Techniques

The data gathered were tallied, tabulated, and analyzed. The gathered data were computed using different statistical tools. The mean and standard deviation were used for descriptive analysis. For inferential analysis, the Mann-Whitney U test was used to test if there is a significant difference between the two formulations of the Jackfruit Cheesy Embutido in terms of appearance, aroma, taste, texture and general acceptability as perceived by the evaluators. The Mann-Whitney U test was used because the assumption for normality of the date for the parametric test was not satisfied.

4 Results

The data shows the level of acceptability of jackfruit in terms of appearance, aroma, texture and taste as perceivedby the evaluators in two formulations.

Table 1: Mean Scores of the samples in terms of Appearance

Appearance	μ	SD	Remarks
Formulation A	4.80	0.407	Extremely Embutido Appearance
Formulation B	4.70	0.466	Extremely Embutido Appearance

As to Appearance, Formulation A was perceived as Extremely Embutido Appearance (μ = 4.80, SD = 0.407). Likewise, Formulation B was also perceived with Extremely Embutido Appearance (μ 4.70, SD = 0.466).

Table 2: Mean Scores of the samples in terms of Aroma

Aroma	μ	SD	Remarks
Formulation A	2.37	1.326	Extremely Jackfruit Aroma
Formulation B	2.07	1.202	Slightly Jackfruit Aroma

As to Aroma, Formulation A was perceived as Slightly Jackfruit Aroma (μ =2.37, SD =1.326). Likewise, Formulation B was also perceived with Slightly Jackfruit Aroma (μ = 2.07, S_D = 1.202).

Table 3: Mean Scores of the samples in terms of Texture

Texture	μ	SD	Remarks
Formulation A	3.93	0.980	Moderately Smooth
Formulation B	3.53	1.008	Moderately Smooth

As to Texture, Formulation A was perceived as Moderately Smooth (μ = 3.93, SD = 0.980). Likewise, Formulation B was also perceived as Moderately Smooth (μ = 3.53, SD = 1.008).

Table 4. Mean Scores of the samples in terms of Taste

Taste	μ	Sd	Remarks
Formulation A	2.87	1.036	Extremely Jackfruit Flavor
Formulation B	2.30	1.179	Slightly Jackfruit Flavor

As to Taste, Formulation A was perceived as Moderately Jackfruit Flavor (μ =2.87, SD = 1.036). Likewise, Formulation B was also perceived with Slightly Jackfruit Flavor (μ = 2.30, SD = 1.179).

Table 5. Comparison of the General Acceptability in Two Formulations

General Acceptability	μ	Sd	Remarks
Formulation A	8.10	0.740	Liked Moderately
Formulation B	8.27	0.759	Liked Extremely

Table 5 shows the result of the General Acceptability. The evaluation outcome revealed that formulation A was perceived as liked moderately (μ = 8.10, SD = 0.740) while formulation B was perceived as liked extremely (μ =8.27, SD = 0.759). Formulation B, which comprises 50% pork meat and 50% jackfruit, is the most acceptable type of embutido mixture. The data revealed that jackfruit could be used in making embutido as a new product that provides good nutrition.

Table 6. The difference between Formulation A and Formulation B in terms of Appearance, Aroma, Texture, Taste, and General Acceptability.

Variable	μ rank	∑rank	MWU Value	P	Remarks
Appearance					
Formulation A	29	870	405	0.375	NS
Formulation B	32	960			
Aroma					
Formulation A	28.57	857	392	0.370	NS
Formulation B	32.43	973			
Texture					
Formulation A	27.17	815	350	0.122	NS
Formulation B	33.83	1015			
Taste					
Formulation A	26.78	803.50	338.50	0.089	NS
Formulation B	34.22	1026.50			
General Acceptability					
Formulation A	26.68	860.50	395.50	0.386	NS
Formulation B	32.32	969.50			

The table shows the Mann-Whitney U Test result on the level of perception of the evaluators of Jackfruit Embutido in two formulations as to appearance, aroma, texture, taste, and general acceptability. Results revealed that there was no significant difference in the level of acceptability of the jackfruit Embutido as to appearance (μ=405, p=0.375), Aroma (μ=405, p=0.370), Texture (μ=350, p=0.122), Taste (μ=338, p=0.089) and General Acceptability (μ=395.50, p=0.386).

It further implies that the formulations of the Cheesy Jackfruit Embutido contributed to the variation of its appearance, aroma, texture and taste

5 Conclusion

As to appearance, the evaluators described both Formulations A and B has extremely Embutido Appearance. As to aroma, both Formulations A and B were described as Slightly Jackfruit Aroma. As to texture, the evaluators described both Formulations A and B as Moderately Smooth. As to taste, the evaluators described Formulation A as Extremely Jackfruit Flavor, while Formulation B was described as Slightly Jackfruit Flavor. As to General Acceptability, Formulation A was described by the evaluators as Liked Moderately, while Formulation A was Liked Extremely. Hence, Formulation B with 50% pork meat and 50% jackfruit is highly liked by the consumers.

Based on the study's findings, the following conclusions were made on Cheesy Jackfruit Embutido: it was generally accepted as it is Liked Moderately for Formulation A and Liked Extremely for Formulation B. Hence, jackfruit can be a good ingredient in enhancing the Embutido recipe. Also, utilizing this fruit as a meat substitute is cost-effective and provides more nutrients into the finished product.

References

1. Statista, (2021) Per capita meat consumption in the Philippines 2021, Statista Research Department.Retrieved from: <https://www.statista.com/statistics/756518/philippines-meat-consumption-per-capita-by- type/>

2. Rosemary, (2021) *10 Interesting Facts You Need To Know About Food in The Philippines*Retrieved from https://www.authenticfoodquest.com/10-facts-food-in-the-philippines/

3. Embutido (Filipino cuisine) "Embutido" (Filipino Meat Loaf). Saveur. Retrieved 11 December 2018. Retrieved from. https://en.wikipedia.org/wiki/Embutido_(Filipino_cuisine)

4. Jackfrut Retrieved from <Merriam Webster Dictionary>, (2018)

5. Elevitvh and Manner, Retrieved from <https://www.esearchgate.net/publication/328354285_Jackfruit_Artocarpus_hetero_phyllis_and_Breadfruit_A_altilis_Phytochemistry_Pharmacology_Commercial_Uses_and _Perspectives_for_Human_Nourishment>, (2006)

6. Amit and Ambarish. Jackfruit wine. Retrieved from <https://onlinelibrary.wiley.com/doi/full/10.1111/j.1541-43337.2012.00210.x\>, (2010)

7. McLeod, S. Experimental Method. Retrieved from <https://www.coursehero.com/file/19532256/Experimental-Method/>, (2012)

Contribution of Contract Farming to Improve Smallholder Seed Multipliers Access to the Market In Rwanda

Jean Pierre NDUWIMANA[1*]

[1]Rwanda Agriculture and Animal Resources Development Board, Huye - Rwanda

Abstract. This study assesses the effectiveness of contract farming in improving smallholder maize seed to have multipliers income so that the improvement strategies can be established. Contract farming is seen as linking and maintaining smallholder maize seed multipliers on the seed market and ensures sustainable maize seed supply to the Rwandan farmers after the decision of the government to pull out from bull buying of maize seeds. The results show that Contract farming could contribute to increasing maize seed production. They were reducing transaction costs and marketing risks due to different factors, such as the experience of smallholder multipliers involved. The study finds various factors that can hinder the performance of contract farming, such as limited quantity, low availability of pre-basic seeds essential for certification, pricing mechanism excluding smallholder multipliers, long process, and bureaucracy in seed payment and distribution.

1 Introduction

The Rwandan economy is highly dependent on the agricultural sector, which accounts for around 33% of national GDP, and almost 50% of all exports come from agriculture. However, only 2% of agriculture is exported. The sector also includes 90% of the total workforce [1]. Rwandan agricultural sector is dominated by small-scale subsistence farming as part of old farming practices and dependence on rain to farm. It faces a set of restrictions, including mainly its reliance on climatic conditions. Weak linkages to the markets and low level of productivity are mostly due to the low use of required inputs such as seeds and fertilizers [2].

Maize is one of the staple foods, and it is consumed in different traditional food preparation. The consumption of maize has been increasing, and it is becoming an important cash crop for smallholder farmers. This condition put maize in major crops in the country and it is ranked second to the sorghum among the important cereals and third in all crops. The maize crop covers 100,000 ha, which means 10% of the total cultivation land and is grown in all country's ecologies with an average of production 1.2 tons per ha [3]. According to the national seed policy of 2007 of Rwanda, the government through the

* Corresponding author: jp.nduwimana@gmail.com

Ministry of Agriculture(MINAGRI) intends to encourage the private sector actors in seed production and marketing activities. It can gradually withdraw from this trend and focus its efforts on coordination, regulation, quality control, and other key activities such as agriculture services delivery that cannot be carried out by the private sector [4].

To help and maintain smallholder seed multipliers and ensure the sustainability of seed supply to the Rwandan farmers, the government set different interventions such as organizing training sessions on seed processing, subside of processing materials, and help in the smallholder seed multipliers cooperative formation. Contract farming is one of these interventions to address this challenge of limited capacities of smallholders seed multipliers. Contract farming is an agreement between farmers, processing, and or marketing firms to produce and supply agricultural products. It is under forwarding agreements and predetermined prices for smallholders. Contract farming is expected to reduce market uncertainty, improve access to inputs, knowledge, services, and provide higher income [5].

This intervention still requires in-depth research to determine the capacities of smallholder seed multipliers, opportunities or alternatives, needs, trust, and willingness of smallholders towards seed companies. This research will also help understand the level of government facilitation that can lead to the type of contract farming necessary to ensure the farmers' availability and sustainable supply of maize seeds. This study focuses on maize seeds due to their importance as a staple food in Rwandan traditional food preparation. It also serves as a cash crop for small farmers since it is currently grown in all Rwandan ecologies. However, productivity is still very low due to the low adoption of new technologies and limited use of improved maize seeds, impacting annual production. As mentioned [3], maize grain must be imported in 2006 to supplement domestic production to satisfy local market demand. This study assessed the effectiveness of contract farming in improving smallholder maize seed of multiplier income to improve strategies.

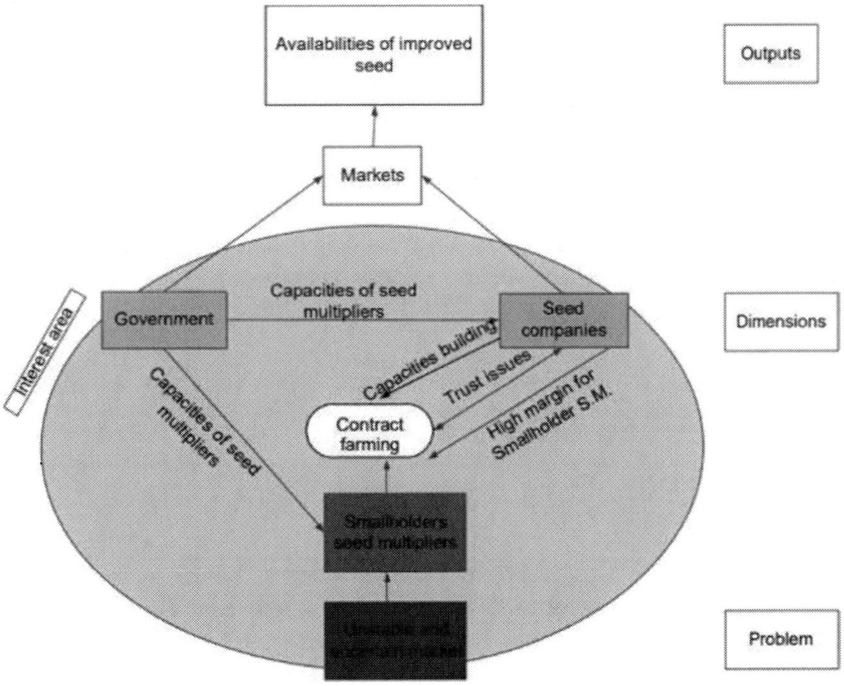

Fig. 1 Research framework

2 Literature Review

Contract farming is an arrangement between farmers and processing and or marketing firms to produce and supply agricultural products under forwarding agreements, often at pre-set prices [6]. [5] define that contract farming is an agreement between farmers, processing, and marketing firms to produce and supply agricultural products. Under forwarding agreements, it is frequently predetermined prices for smallholders. Contract farming is expected to reduce market uncertainty, improve access to inputs, knowledge, and services, and provide higher income. For this study, contract farming will be defined as an agreement between individual seed multipliers and seed companies to produce and supply maize seeds. For smallholders, Contract Farming will be the way to reduce uncertainty in the market.

The arrangement usually involves the buyer in providing certain production support through, for example, the delivery of inputs and the facility of technical advice. From this agreement, the farmer undertakes to supply specific produce in quantities and according to quality standards within the agreed deadlines. On the other hand, the company agrees to support the farmer's production and buy the produce at a predetermined price. According to [7], concerns of exploitation in the forms of large agribusiness firms using contracts to take advantage of cheap labor and transfer production risk to farmers. Despite these critics, it is also proven that contract farming generates high incomes and many other benefits for smallholder farmers. However, contract farming can answer the problems of market disappointment, the provision of credit, and agricultural inputs because the contracts often involve the provision of seed, fertilizer, and technical assistance on post payments and a defined price at harvest [7].

Contract farming is a form of "vertical coordination aimed primarily at correcting the market failure associated with spot markets due to imperfect information" [6]. The strict management and coordination of the agriculture value chain due to Market competition, consumer demands, technology development, government policies, and product traceability require a high level of organization, leading to vertical coordination in agriculture chains [5]. According to [8], contracting between producers and processing or marketing firms remains a practical method to strengthen vertical coordination in the agriculture value chain. The motives for undertaking CF vary for buyers and farmers, but all parties want to reduce market uncertainty and transaction costs. Farmers are motivated by access to markets for their product, technical assistance, and pre-financing for inputs by the contracting company on practical price. Farmers are also motivated by the new technology and innovations from the firms, which are adapted to their needs and help them satisfy market requirements. For the firms, their motivation is to protect a consistent and predictable supply of the needed product that meets their specifications in terms of quality and quantity.

3 Methodology

3.1 Desk Study

Secondary data acquired from different sources such as books, scientific journals, official reports, and the Internet were used to find information from the current literature review related to this topic. These data were very important to confirm the foundation of this research and the results. It also helps identify the stakeholders' roles, decision-making, and leading power in the seed value chain.

3.2 Survey

The survey was carried out in four districts, Muhanga, Ruhango, Kamonyi, and Rwamagana, where the Smallholder seed multipliers were the main target group for the study. A sample of 40 smallholder maize seed multipliers was randomly selected from a list of all maize seed multipliers working in 4 Districts. To understand the different 16 capacities ofsmallholder seed multipliers to produce the required quantity and quality seed and to get from smallholders' perspective the factors that can make contract farming successfuland what can be done to increase trust between them and seeds companies. Given the current trend of the COVID-19 pandemic, the researcher couldn't travel to the field for data collection. A substitute data collector with a bachelor's degree in statistics applied to economies was hired to collect quantitative data. To ensure the validity and accuracy of quantitative data, the commissioner has made an employee available to assist the data collector to the extent possible. The researcher conducted a meeting with both to discuss and clarify the questionnaire and to follow up through a telephone call. The data was filled in using an online form so that the researcher could verify and cross-check the data collected every day.

3.3 Interview

Key informants from government institutions (RAB) and other actors involved (seed company and individual seed multipliers) were purposively selected according to their role and position were consulted and interviewed at the time of research. To contribute to high income to smallholder seed multipliers and how contract farming can increase trust between actors. Given the pandemic period, to maintain the validity and accuracy of the qualitative data. A semi-structured online interview was conducted with highly focused questions. According to the interviewee's consent, some interviews were recorded as 40 Smallholder seed multipliers and 6 Key Informants (1 RAB, 3 Seed companies, and two individual seed multipliers).

4 Results

4.1 Stakeholders' role, decision making, and leading power in the Maize seedvalue chain

The maize seed sector has different stakeholders playing a crucial role in this sector, from production, processing, and distribution to last users. In this sector, there are supporters and influencers. All these stakeholders have different levels of decision and power in the coordination chain. The organization of the chain, position, and different functions of stakeholders is presented on the chain map (Figure 2).

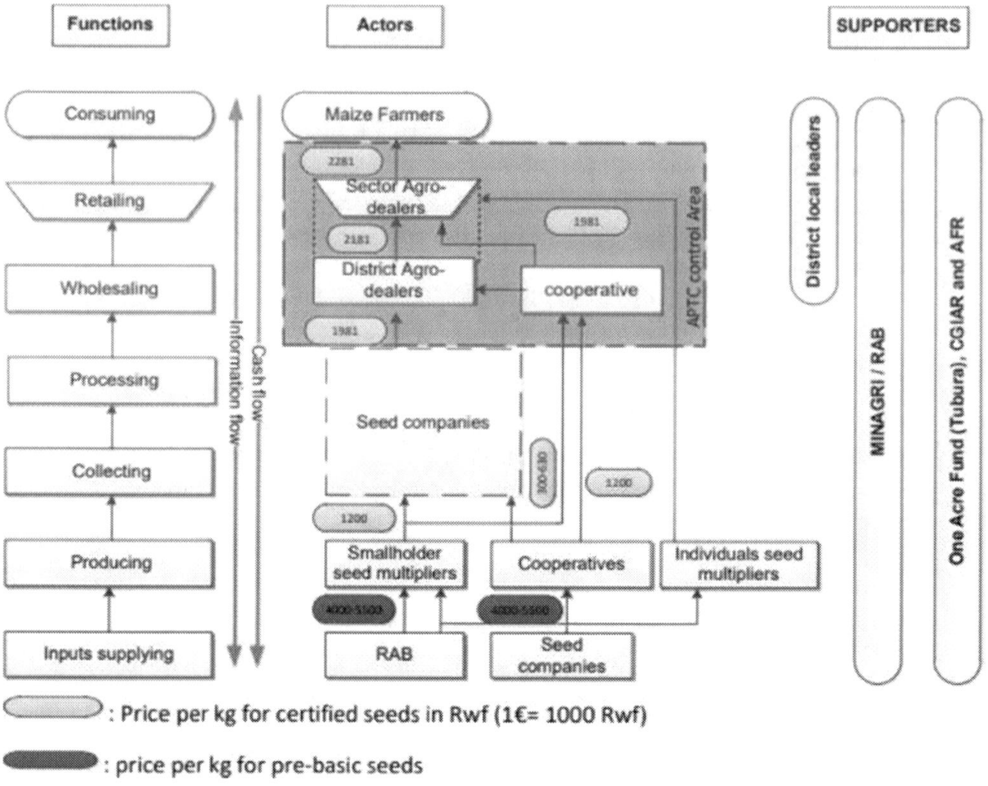

Fig. 2 Maize seed new value chain map

4.2 Success and failure factors that can affect contract farming

The survey shows that 40% of smallholder multipliers are motivated by the price in the first place, 47.5% the time of payment in the second place, and 30% are motivated by the consistency of the buyer on the third time. The results also show that 70% of the respondents do not have yet contract with their buyers. They only communicate via meeting and delivery notes after supplying their products. Most of them, 97.5% are using amicably and local leaders as a way of mediation. Only 2.5% plan to enforce their contract using courts, local leaders, and amicably as mediation.

4.3 Contribution of contract farming to the higher income of smallholder seed multipliers

The survey results show that 72.5% of smallholder seed multipliers are planning to sell their hybrid maize seeds through seeds companies. 20% of respondents jointhe cooperatives to sell their maize seeds, and only 7.5% have a plan of selling themselves their seeds direct to the open markets. The statistical test was used to verify whether there is a difference in production per ha between multipliers selling to companies, multipliers selling to cooperatives, and multipliers selling themselves to the open markets.

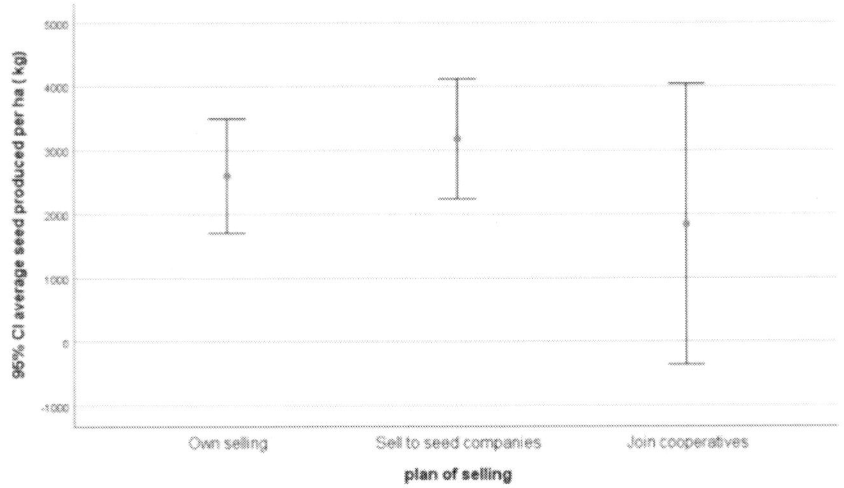

Fig. 3 Average seed produced per ha

The statistical test was also used to verify whether there is a difference in average total cost used for seed production per ha, between multipliers selling to companies, multipliers selling to cooperatives, and multipliers selling themselves to the open markets. After a one-way ANOVA test, the resultant shows a difference in average total cost used for seed production per ha, between multipliers selling themselves to the open markets and multipliers selling to companies, and between multipliers selling to the cooperative. However, there was no difference in the average total cost used for seed production per ha between multipliers sold to seed companies and multipliers sold to cooperatives.

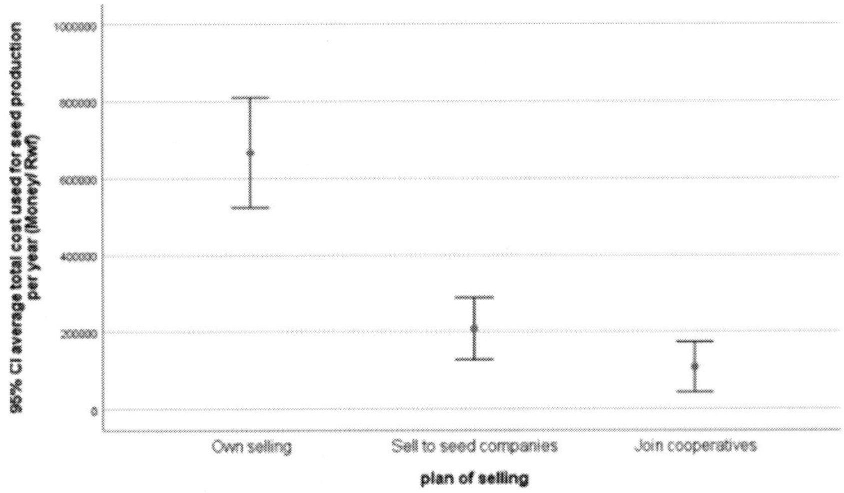

Fig. 4 Average total cost for seed production

The statistical test was also used to verify whether there is a difference in selling price between smallholder multipliers according to the buyer partners. There is a significant difference in selling price between multipliers selling their seeds to seeds companies and multipliers selling their seeds through cooperative, but they were no significant difference in selling price between multipliers selling themselves to the open markets and multipliers selling to the seed companies.

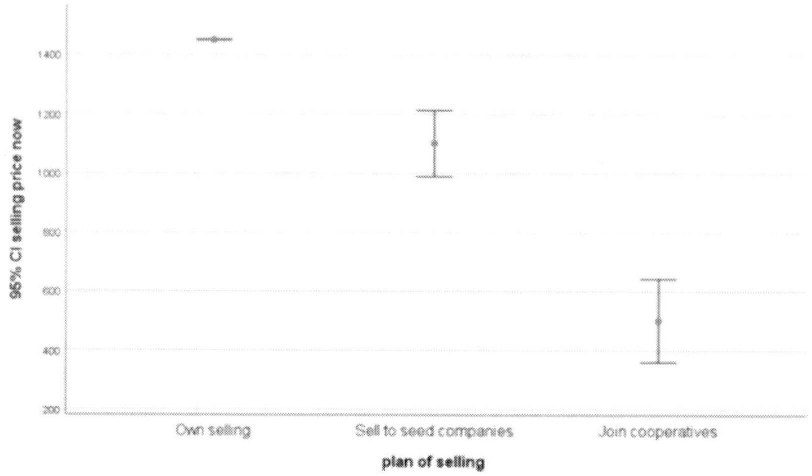

Fig. 5 Selling price

This difference in selling price inspires verification also where there is a difference in margin rate between multipliers according to the buyer partners. After a statistical test, the result shows a difference between multipliers selling to seeds companies, multipliers selling through cooperative, and multipliers selling themselves shown by p-value (0.000).

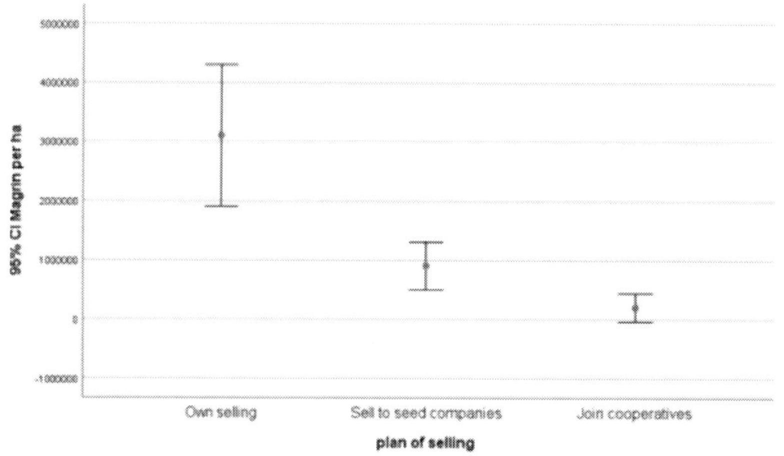

Fig. 6 Margin rate

5. Conclusion

Hybrid maize seed is also a highly valuable commodity with specific knowledge and investment for production. Strict control that prevents the high level of maize seed trafficking and smuggling to all bordering countries, will highly contribute to the success of contract farming. On the other hand, the study finds various factors that can hinder the performance of contract farming, such as limited quantity and low availability of pre-basic seeds essential for certified/ commercial seed multiplication, pricing mechanism excluding smallholder multipliers, long process, and bureaucracy in seed payment and distribution, and finally accessibility of smallholder seedmultipliers. All these factors can be the high risks of failure for contract farming.

References

1. Nelson, M. M. W. E. S. U. C.,. Rwanda Early Generation Seed Study, Feed the Future, Kigali (2016).
2. Giertz, A. G. G. M. M. R. R. D. a. S. K.,. Rwanda Agriculture Sector Risk Management, World Bank Group. Rwanda (2015).
3. NGABITSINZE, J. C., Analysis of Economic Efficiency of Maize Production inHuye District in Rwanda. Int. J. Agric. Innov. Res., 3(3) (2014).
4. MINAGRI. National Seed Policy, Ministry of Agriculture andAnimal Resources. Kigali, Rwanda (2007).
5. Holtaland, G. Contract farming in Ethipia: concept and practice. Ruth Daviesed. : AgriProFocus. Arnhem, The Netherlands. (2017).
6. Wainaina, P. J. J. O. J. N. Impact of contarct farming on smallholder poultry farmers' incom in Kenya. Nairobi, Departement of Agriculture Economiecs of University of Nairobi. (2012).
7. Nham, P. T. contract farming and its impact on income and livelhood for small-scale farmers: case study in Vietnam. J. Agrib.Rur. Dev. ,4(26), pp. 147-166 (2012).
8. FAO, Contract farming for inclusive market access. Carlos A. da Silva ed. Rome: FAO (2013).

Hypoglycaemic effect of *Bawang Dayak* extracts (*Eleutherine palmifolia* (L.) Merr.) on Sprague Dawley rats

Andi Early Febrinda[1*], *Nancy Dewi* Yuliana[2], *Tutik* Wresdiyati[3], *Made* Astawan[2]

[1] Department of Food Quality Assurance Supervisor, College of Vocational Studies, IPB University, Indonesia
[2] Department of Food Science and Technology, Faculty of Agricultural Technology, IPB University, Indonesia
[3] Department of Anatomy, Physiology, and Pharmacology, Faculty of Veterinary Medicine, IPB University, Indonesia

Abstract. The Dayak tribe in Kalimantan has long used *bawang dayak* as a traditional cure for various degenerative diseases. Previous research showed that *bawang dayak* bulb has an active compound that can inhibit the enzyme alpha-glucosidase, known as alpha-glucosidase inhibitor (AGI). Compounds that have such ability has a potential in controlling blood glucose level in diabetes mellitus. This research aims to study the effective dosage of aqueous extract and ethanolic extract of *bawang dayak* bulb as an AGI through an oral glucose tolerance test in male Sprague Dawley rats. Experimental rats were divided into 14 groups, 7 for the ethanolic extract test and 7 for the aqueous extract test. The extracts were applied in 5 dosage levels in each test, with acarbose as a drug control and distilled water as a positive control. Data were plotted to determine the area under the curve (AUC). The effective dosage was the lowest dosage that gave a significantly smaller AUC value than the AUC value of the positive control group. The effective dosage for *bawang dayak* aqueous extract and *bawang dayak* ethanolic extract was 200 mg/kg and 100 mg/kg, respectively.

1 Introduction

Diabetes mellitus (DM) is a degenerative disease that affects many of the world's population. In 2015 the number of people with diabetes globally was 422 million and was estimated to reach 642 million people in 2040. Prevalence is rising higher in low and middle-income countries. In 2015 the percentage of adults with diabetes was 8.5%, which means 1 in 11 adults live with diabetes. Diabetes became one of the largest health expenditures in the world in 2013, amounting to 612 billion dollars or 11% of total health spending. Diabetes is one of the global health emergencies in the 21st century. There are 318 million adults who have impaired glucose tolerance and are at high risk of developing diabetes [1]. Persons with impaired glucose tolerance (IGT) – 15.6% of the U.S. population

*Corresponding author: earlyfebrinda@apps.ipb.ac.id

– are at high risk for developing diabetes. In addition, IGT is an important risk factor for a number of other adverse health conditions and mortality. IGT is defined on the basis of an abnormal oral glucose tolerance test (OGTT). Persons without diabetes but with an OGTT 2-hour value of 140-199 mg/dl are considered to have IGT [2]. The person who develops pre-diabetes will become diabetes in several months or years, if he/she goes untreated appropriately and adequately.

One mechanism that can be used to prevent an increase in blood glucose levels is through inhibition of glucose absorption in the intestine. This mechanism involves a substance that can inhibit the action of enzymes that break down starch into glucose, known as alpha-glucosidase inhibitors (AGI). AGI works in the intestinal tract to prevent disaccharides from breaking down into glucose, thereby decreasing glucose absorption. Many studies have been carried out on various medicinal plants in vitro to determine the potential of these plants as AGI agents. *Bawang Dayak* (*Eleutherine palmifolia* (L.) Merr.) is an indigenous plant of Kalimantan Island, Indonesia. Dayak people have long used this plant as a traditional medicine for several diseases. It is known by various other names such as *bawang tiwai, bawang hantu, bawang sabrang, babawangan beureum, bawang siyem, bawang sayup, bawang kapal* and *bawang lubak*. *Bawang dayak* plants have good adaptations and can grow in various climates and soil types. In addition, this plant can also be reproduced and harvested in a short time so that it is easily developed for industrial scale [3]. *Bawang dayak* aqueous and ethanolic extracts have antioxidant and alpha-glucosidase inhibitor activities that have been proven *in vitro* [4]. The methanolic extract of *Eleutherine americana* bulb contains eleutherinoside A, a naphthalene derivative, which has strong AGI activity with IC_{50} 0.5 mM [5]. The existence of eleutherinoside A in *Eleutherine palmifolia* ethanolic extract also has been confirmed [6].

This research aims to verify the AGI activity of *bawang dayak* extracts work *in vivo* and determine the effective dosage of *bawang dayak* aqueous extract and *bawang dayak* ethanolic extract to decrease intestinal glucose uptake Oral Glucose Tolerance Test (OGTT) in male *Sprague Dawley* rats. This study will give pre-clinical proof and reveal the potential of *bawang dayak* extracts as a local resource for developing the phytopharmaceutical products in controlling and preventing hyperglycaemia in people with Impaired Glucose Tolerance (IGT).

2 Methods

2.1 Plant material and extraction process

Bawang dayak bulbs (*Eleutherine palmifolia* (L.) Merr.) were collected from Air Hitam Village, Samarinda City, East Kalimantan. The bulbs were identified and authenticated by Dr Joeni Setijo Rahajoe from Herbarium Bogoriense, Indonesian Institute of Sciences. *Bawang dayak* extraction was carried out by maceration using water and ethanol solvents according to the procedure explained by Febrinda *et al.* [4]. Fresh *bawang dayak* bulbs were added with water (or ethanol) with a ratio of 1:4 (b:v) and crushed using a blender. The solutions then were soaked for thirty minutes in an ultra-sonic bath and followed by incubation in a shaker incubator at room temperature for two hours. Then the solutions were centrifugated, and the dissolved phase was collected to be filtered using Whatman filter paper No. 1. The extract solutions obtained were frozen and dried using a freeze dryer for 48 hours, followed by pounding the extracts to become a dry powder named *bawang dayak* aqueous extract (BD aqueous extract) and *bawang dayak* ethanolic extract (BD ethanolic extract).

2.2 Experimental animal and oral glucose tolerance test

This research was conducted under the Guide for The Care and Use of Laboratory Animals [7]. 70 male Sprague Dawley rats aged around eight weeks weighing 180 to 200 g were obtained from the Food and Drug Supervisory Agency, Jakarta. The experimental animals were maintained under standard laboratory conditions (26 °C, 12 hours of light/ dark cycles) and fed and watered *ad libitum*. The animals were divided into 14 groups: 7 groups for BD aqueous extract test and the other seven groups for BD ethanolic extract test. The animals fasted for 10 hours, from 10:00 p.m. to 8:00 a.m. The following day, the fasting blood glucose levels of the experimental animals were measured immediately using a glucometer (Accu-Chek Active, Roche, Germany). The animals were grouped and given the following treatment: (1) positive control group (**PC**) treated with distilled water; (2) drug control group (**DC**) treated with 5 mg/kg BW of acarbose solution; (3) treatment group (**A1/E1**) treated with BD aqueous extract or BD ethanolic extract at a dose of 100 mg/kg BW; (4) treatment group (**A2/E2**) treated with BD aqueous extract or BD ethanolic extract at a dose of 200 mg/kg BW; (5) treatment group (**A3/E3**) treated with BD aqueous extract or BD ethanolic extract at a dose of 300 mg/kg BW; (6) treatment group (**A4/E4**) treated with BD aqueous extract or BD ethanolic extract at a dose of 400 mg/kg BW; and (7) treatment group (**A5/E5**) treated with BD aqueous extract or BD ethanolic extract at a dose of 500 mg/kg BW. Thirty minutes after the administration of treatment solution, the animals received sucrose solution with a concentration of 25% at a dose of 2 g/kg BW by oral. The solution was administered through gastric gavage. Blood glucose levels were collected and measured on the 30th, 60th, 90th, and 120th minutes after the sucrose administration. The data obtained were plotted to form a curve, and the area under the curve (AUC) was calculated for each experimental group.

2.3 Statistical analysis

The AUC values of all experimental groups were analyzed using Analysis of variance (Anova). If the Anova result showed significance, further statistical *post hoc* test would be carried out using DMRT. Data were displayed as Mean ± SD.

3 Results

3.1 OGTT result of BD aqueous extract

The curves of OGTT result on BD aqueous extract showed that the acarbose and all treatment groups had lower curves than the positive control (Fig. 1). The area measurement under the curve showed that the entire treatment group had a smaller AUC than the positive control (Fig. 2). However, the dose groups with a significantly smaller AUC than the positive control were 200, 300 and 500 mg/kg BW. These groups had AUCs, which were not significantly different from the acarbose group. Based on DMRT *post hoc* test at α 0.01, the lowest dose that gave significantly lower AUC than the positive control was 200 mg/kg BW (Fig. 2).

https://doi.org/10.1051/e3sconf/202234800029

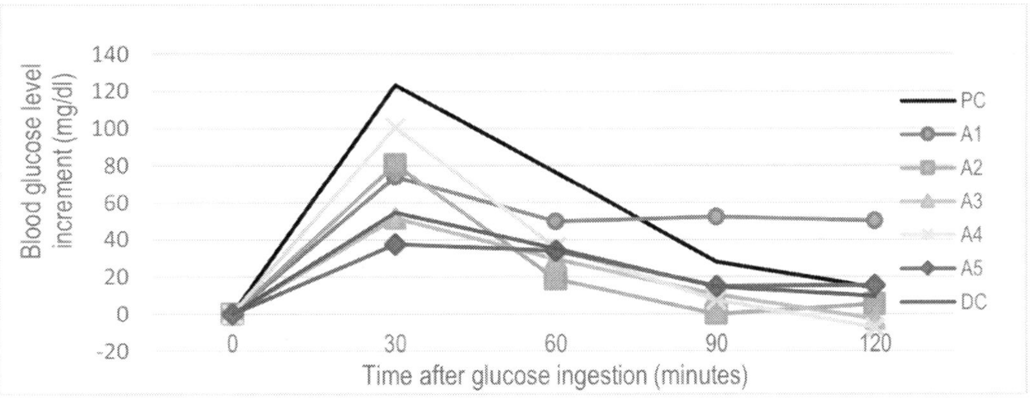

Fig. 1. OGTT curves for BD aqueous extract. PC= positive control group, treated with distilled water. A1= treated with BD aqueous extract dose 100 mg/kg BW. A2= treated with BD aqueous extract dose 200 mg/kg BW. A3= treated with BD aqueous extract dose 300 mg/kg BW. A4= treated with BD aqueous extract dose 400 mg/kg BW. A5= treated with BD aqueous extract dose 500 mg/kg BW. DC= drug control group, treated with acarbose 5 mg/kg BW.

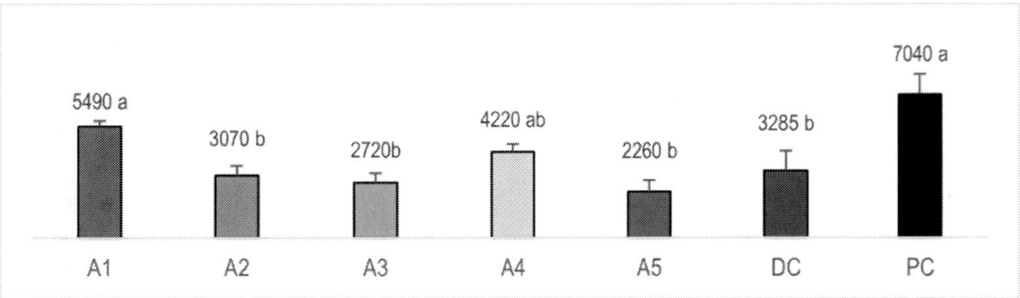

Fig. 2. The area under the curve (AUC) of BD aqueous extract. A1= treated with BD aqueous extract dose 100 mg/kg BW. A2= treated with BD aqueous extract dose 200 mg/kg BW. A3= treated with BD aqueous extract dose 300 mg/kg BW. A4= treated with BD aqueous extract dose 400 mg/kg BW. A5= treated with BD aqueous extract dose 500 mg/kg BW. DC = drug control group, treated with acarbose 5 mg/kg BW. PC= positive control group, treated with distilled water. Values not sharing common superscript differ significantly at $p<0.01$

3.2 OGTT result of BD ethanolic extract

The curves of OGTT result on BD ethanolic extract showed that the acarbose and all treatment groups had lower curves than the positive control (Fig. 3). The area measurement under the curve showed that the entire treatment group had a smaller AUC than the positive control (Fig. 4). However, the dose groups with a significantly smaller AUC than the positive control were 100, 300 and 500 mg/kg BW. These groups had AUCs, which were not significantly different from the acarbose group. Based on DMRT *post hoc* test at α 0.01, the lowest dose that gave significantly lower AUC than the positive control was 100 mg/kg BW

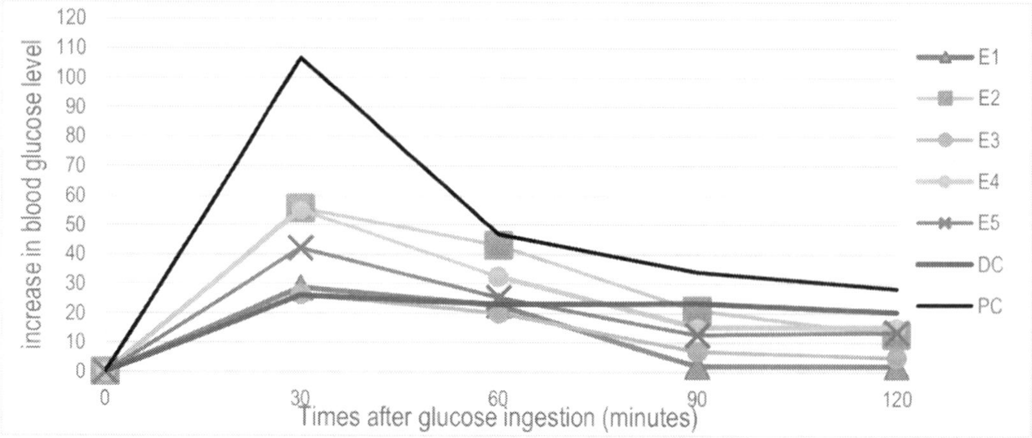

Fig. 3. OGTT curves for BD ethanolic extract. E1= treated with BD ethanolic extract dose 100 mg/kg BW. E2= treated with BD ethanolic extract dose 200 mg/kg BW. E3= treated with BD ethanolic extract dose 300 mg/kg BW. E4= treated with BD ethanolic extract dose 400 mg/kg BW. E5= treated with BD ethanolic extract dose 500 mg/kg BW. DC = drug control group, treated with acarbose 5 mg/kg BW. PC= positive control group, treated with distilled water.

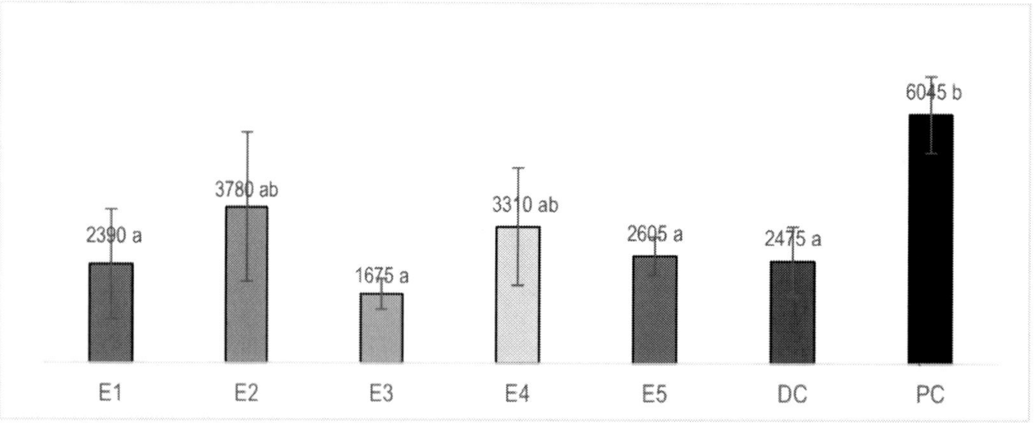

Fig. 4. Area under the curves of BD ethanolic extract. E1= treated with BD ethanolic extract dose 100 mg/kg BW. E2= treated with BD ethanolic extract dose 200 mg/kg BW. E3= treated with BD ethanolic extract dose 300 mg/kg BW. E4= treated with BD ethanolic extract dose 400 mg/kg BW. E5= treated with BD ethanolic extract dose 500 mg/kg BW. DC = drug control group, treated with acarbose 5 mg/kg BW. PC= positive control group, treated with distilled water. Values not sharing common superscript differ significantly at $p<0.01$

4 Discussion

The ability of *bawang dayak* extracts in inhibiting alpha glucosidase enzyme in the intestine was compared to acarbose as a well-known diabetes mellitus medicine. Both extracts showed that they had no significantly different ability from acarbose, although in different dose levels. BD ethanolic extract had a more remarkable ability to reduce blood glucose levels shortly after sucrose consumption than BD aqueous extract (Fig. 1 and Fig. 3). The administration of BD ethanolic extract at a dose of 100 mg/kg BW significantly gave a lower AUC value compared to positive control, while for the aqueous extract a dose of 200 mg/kg BW was needed. This is consistent with *in vitro* research carried out previously where BD ethanolic extract had significantly lower IC_{50} of alpha glucosidase inhibition (241 ppm) than the aqueous extract (505 ppm) and not significantly different from acarbose (288 ppm) [6].

A naphthalene derivated compound named Eleutherinoside A in methanolic extract of *Eleutherine americana* bulb, could act as an alpha-glucosidase inhibitor (IC_{50} of 0.5 mM) [5]. This compound, along with two other naphthalene derivates (eleuthoside B, and eleutherol), was also found in *bawang dayak* bulb from Air Hitam Village, Samarinda (*Eleutherine palmifoli*a (L.) Merr.) that was used in this research. The presence of those compounds was shown by an NMR analysis on *Eleutherine palmifolia* ethanolic extract [4]. A result of phytochemical screening analysis on *Eleutherine palmifolia* extract also showed that phytochemical compounds presence in BD ethanolic extract was more abundant than those found in the aqueous extract. There were strong positive indications of alkaloid and triterpenoid compounds in BD aqueous extract. In contrast, there were strong positive indications for triterpenoids, flavonoids, and phenolic compounds [6]. Those compounds also had AGI activities which were stated by several research reports [8-12].

Due to the higher presence of Eleutheinoside A and phytochemical compounds in BD ethanolic extract than in the aqueous extract, the ethanolic extract had a higher ability than the aqueous extract in inhibiting alfa glucosidase enzyme activity. Those resulted in a greater decrease of glucose absorption in the intestinal tract of the experimental animal. This result was consistent with the OGTT experiment result of another Kalimantan indigenous plant, *pasak bumi* root (*E. longifolia*), which showed that the methanolic extract had a higher hypoglycaemic effect than the water infusion due to its phytochemical contents [13].

Synthetic AGI, such as acarbose, has long been used to manage type 2 diabetes mellitus. In addition to slowing down the absorption of glucose by the intestine, various diabetes drugs with the mechanism of action of AGI are reported to have side effects in the form of gastrointestinal discomfort [14,15]. Although it has a significant effect on glycemic control and insulin levels, synthetic AGI such as acarbose does not significantly affect blood lipids. On the contrary, bawang dayak extract significantly improved blood lipid profile and antioxidant status in Sprague Dawley alloxan-induced diabetic rats. Those experimental results on *bawang dayak* revealed the potential of this Kalimantan indigenous plant to be a local resource for developing phytopharmaceutical products for treating diabetes mellitus and preventing its cardiovascular complication.

5 Conclusion

The AGI activity of *bawang dayak* extracts was effective *in vivo* in reducing glucose uptake in the intestine of Sprague Dawley male rats. The effective dose of BD aqueous extract and ethanolic extract to inhibit the action of the alpha-glucosidase enzyme determined through OGTT in male Sprague Dawley rats were 200 and 100 mg/kg BW respectively. The combination of antioxidant capacity and the ability to inhibit the activity of alpha glucosidase enzyme in *bawang dayak* extracts showed that this plant had great potential as a source of the phytopharmaceutical product in controlling and preventing hyperglycaemia. Therefore, it is necessary to further assess the acute and chronic toxicity of bawang dayak extract to determine its safety for long-term usage.

References

1. International Diabetes Federation, IDF diabetes atlas 7th edition (Brussels, 2015).
2. NHANES, Oral glucose tolerance test procedure manual (CDC, Washington, 2007)
3. R.Y Galingging, JPPTP **10**, 76 (2007)
4. A.E. Febrinda, N.D Yuliana, E. Ridwan, T. Wresdiyati, M. Astawan, IFRJ, **21,** 1405 (2014)
5. T. Ieyama , M.D.P.T. Gunawan-Puteri, J. Kawabata, *Food Chem* **128**, 308 (2011)
6. A.E. Febrinda, M. Astawan, T. Wresdiyati, N.D. Yuliana, JTIP **24**, 161 (2013)
7. National Research Council, Guide for the care and use of laboratory animals 8th ed. (The National Academies Press, Washington DC, 2011)
8. S.S. Lee, H.C. Lin, C.K. Chen, Phytochem. **69**, 2347 (2008)
9. K. Tadera, Y. Minami, K. Takamatsu, T. Matsuoka, J. Nutr. Sci. Vitaminol **52**, 149 (2006)
10. M.B. Patel, S.M. Mishra, J. Funct. Foods **4,** 79 (2012)
11. Y.C. Lai, C.K. Chen, S.F. Tsai, S.S. Lee, Phytochem **74,** 206 (2012)
12. H. Wang, Y.J. Du, H.C. Song, Food Chem. **123**, 6 (2010)
13. Fransisca, G.E. Kalangi, D.C. Saptasari, P. Hendra, J. Pharm. Sci. Community **15**, 1 (2018)
14. J. Feng, X.W. Yang, R.F. Wang. Phytochem. **72,** 242 (2011)
15. F.A. Van de Laar, P.L.B.J. Lucassen, R.P. Akkermans, E.H. Van de Lisdonk, G.E.H.M. Rutten, C. Van Weel, Cochrane Database Syst. Rev. **2** (2005)

How can blended learning contribute to the development of dairy professionals within the global dairy sector?

Nicolien van der Horst[1]

[1]Van Hall Larenstein University of applied sciences, Agora 1, 8934 CJ Leeuwarden, The Netherlands

Abstract. This research focuses on the goal to determine the effect of blended learning in the global dairy sector and to what extent this method has an added value. It was carried out by a 4th-year thesis student from the Animal Husbandry study at the Van Hall Larenstein University in the Netherlands. This research is substantiated by information gathered while participating in creating blended learning modules with Triple Dairy,through literature research, interviews with ten stakeholders in the transfer of knowledge, and two different surveys with Dutch farmers and African trainers of the EARNED project. The research goal is to determine how blended learning is most effective in the global dairy sector and what added value this method has. This research aims to deliver a report, article, and consultation that clarifies how blended learning can contribute to the development of dairy professionals in the global dairy sector.

1 Introduction

The Teaching method of *blended learning* is the central topic of this research. *Blended learning* uses a mix of *didactic methods* in the learning process, frequently shaped by combining confront to confront instruction and another popular instruction frame: e-learning [1]. Based on the demand for knowledge, innovation, and *practical skills*, this research explores how *blended improve dairy professionals* in the *global dairy sector*. *There* is a need for knowledge. Triple Dairy believes that the blended learning method has thebest outreach and highest impact. Triple Dairy wants to know how blended learning can create the highest value in the international dairy sector. Blended learning is already appliedin formal education and has a positive learning effect [2]. Triple Dairy uses this method of sharing knowledge, skills, and innovation transfer and wants to know how to use it most effectively. The research goal is to determine how blended learning is most effective in the global dairy sector and what added value this method has. This research aims to deliver a report, article, and consultation that clarify how blended learning can contribute to the development of dairy professionals in the global dairy sector.

The world population is increasing from 7 to almost 10 billion people in the coming decades [3]. Food production and allocation are not keeping pace with the growth of the population. There is a need for food security and safety. The UN set Sustainable Development Goals in 2015 to create a better world by 2030. In 2030 there should be no

Poverty (SDG 1), zero hunger (SDG 2), quality education (SDG 4), and responsible production (SDG 12) (Global Goals, sd). Milk and dairy products are a good source of nutrition, especially in countries where nutrition is scarce; they also present livelihood opportunities. The development of the global dairy sector contributes to improving food security and food safety and can reduce poverty [4]. Agricultural production should increase by 50 to 60 % in the future [3]. Increase it, and there is a need for more knowledge, innovation, and practical skills. Knowledge is crucial to making improvements. Knowledge transfer in the dairy sector worldwide is currently practiced inefficiently. The effect is limited and not future-proof [5]. There is increasing pressure on dairy farmers to produce more sustainably; there is a need for knowledge to continue producing and delivering in the future. Blended learning is the solution, as believes Triple Dairy. This study aims to determine how blended learning is most effective in the global dairy sector and what added value this method has.

2 Literature Review

Blended learning combines digital or online learning and classroom or face-to-face learning. With the same budget, one can reach a much larger target group than currently because knowledge can be offered independently from time and place. There are a few conditions for blended learning: 1). There must be a good internet connection. 2). Access to proper equipment such as smartphones or laptops [5].
Blended learning is more sustainable because it has a significant effect, it is approachable, lifelong learning is started, and the link is made from learning to taking action. Blended learning can also be more environmentally friendly because fewer travels are made. On the one hand, work identity must be the connecting framework for the knowledge, skills, and attitudes involved in the training process. On the other hand, the guiding principle for learning and training processes design. The confrontation between externally provided images and one's views and wishes and their reflection form the core of such identity learning [6].

Kolb's learning style is a cyclical process. Kolb has four learning phases; he states that all four learning phases must be completed. It is less critical which learning phase is started with, but he indicated a preference for using the same sequence. Kolb argues that learning becomes easier by going through all four learning phases [7]. He discovered that people develop a preferred style, namely the learning phase, in which they are naturally strong. One can define *lifelong learning* as a dynamic concept concerned with finding different ways of responding to change. Lifelong learning is seen asan important concept for improving the employability and adaptability of people. The innovative dimension of lifelong learning is in a new approach to the processand context of learning. Learning is seen as a 'continuous process,' encompassing purposeful learning activities over a lifetime [8]. The Dutch dairy sector exports not only dairy products but also knowledge, technology, and dairy farmers. An example is Friesland Campina's Dairy Development Program, in which knowledge and people are used to help local dairy farming in Asia and Africa. Experienced people help dairy farmers increase productivity and connect to the international quality system [9].

3 Methodology

This research examines whether the blended learning method can contribute to the improvement of dairy professionals and in what way. The primary research methods for this study are literature review, surveys, and interviews to get insight into blended learning and how blended learning can contribute to the development of dairy professionals in the global dairy sector.

3.1 Research design

This research is qualitative. Non-numerical data will be collected and analyzed to understand concepts, opinions, and experiences with blended learning and learning ofprofessionals in the global dairy sector. The research is limited to the research questions, and it should not go beyond it. These limits are prevention the study from going off-topic and staying within the boundaries of the subject of blended learning in the global dairy sector.

3.2 Data collection

A performed desk research gained important information about didactics, learning styles, teaching styles, and blended learning. But also about the developments in de global dairy sector. Main desk research sources used Google, Greeni, ResearchGate, Books. It seems an excellent way to hold surveys to collect data. Seven participants in the pilot training Master class 2030 by Triple Dairy have responded to the survey. Nine trainerstrained to train for the EARNED project answered to determine the learning and teaching styles of the dairy professionals. Interviews and conversations were held with professionals in the dairy sector and professionals in agro education. Nine stakeholders have answered the interviews with open questions. By participating in the development of the blended learning modules, such as making blueprints, validating modules with Triple Dairy. This information also contributes to the answers to the research questions.

3.3 Data processing

For the answers to the survey, Microsoft Office Suite Excel 2016 is used to process and analyze them. Processing and analysis have taken place after collecting the information. Microsoft Office Suite Word 2016 was used to write the report. A SWOT analysis describes the strengths, weaknesses, opportunities, and threats of the contribution of the training methodology blended learning among dairy professionals in the global dairy sector.

4 Result and Discussion

4.1 Blended learning

The definition of *blended learning* outlined in the literature is as follows:" a combination of digital or online learning and classroom or face to face learning" In the interviews with stakeholders in the transfer of knowledge in the dairy sector, a definition of *blended learning* method was asked. The descriptions vary, from concise "a mix of online and offline learning" to a more comprehensive definition of "a mix of didactic methods." Figure 4 shows the definitions of *blended learning* given by the stakeholders. *Blended learning* makes it possible to provide high-quality education efficiently. Using a platform, knowledge innovations and *practical skills* can be distributed faster in a blendedmanner. *Blended learning* makes use of *digital resources*.

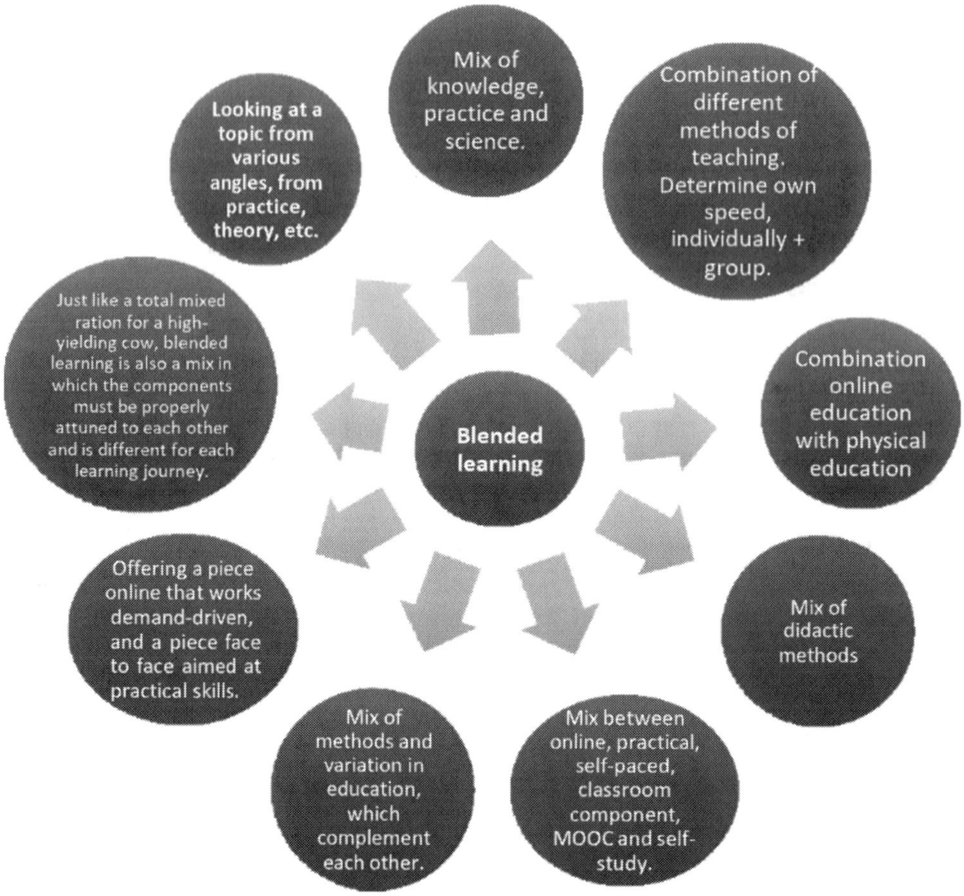

Fig. 1. Definitions of blended learning by involved stakeholders

The definitions from the literature and those given by the stakeholders are all formulated differently, but they come down to the same. There is a mixture of online resources and physical meetings to transfer knowledge, innovation, and *practical skills. Blended learning* includes *digital recourses*. In *developing countries* such as East Africa, with the EARNED project, digital resources are challenging when it is first necessary to inventory whether WIFI and computers are available.

The equation made by one of the parties during the interview is very appropriate for the dairy sector and covers the load well. The equation is as follows: "*blended learning* is like a total mixed ratio for a cow." *Blended learning* is also a mix that must be tailored to the needs and will be different and adapted in every situation.

4.2 Method of knowledge transfer in the international dairy sector

The knowledge, innovation, and *practical skills* sharing in the dairy sector occur differently. The literature describes the traditional way of sharing knowledge: a Dutch expert goes to an area, gives his training, and leaves again. In the interviews with those *involved, other methods also came up. Figure 5 describes those methods.*

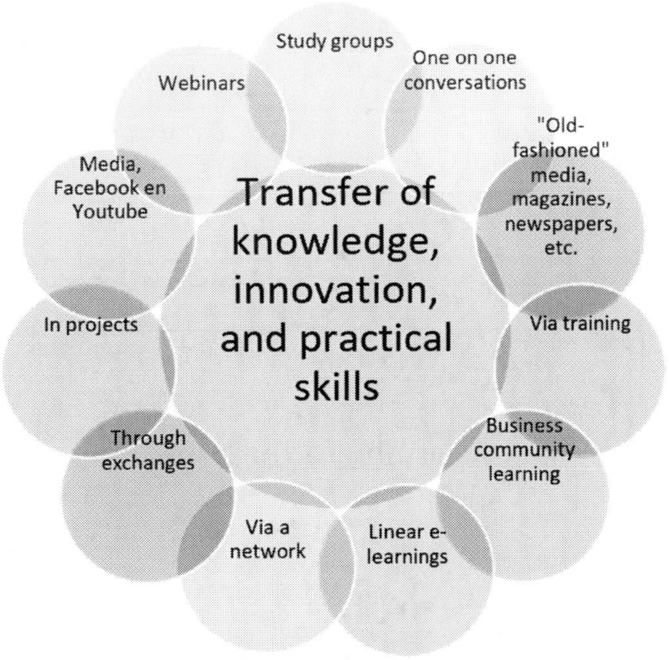

Fig. 2. Methods transfer of knowledge innovation and practical skills

Based on the experiences in the international dairy sector, the transfer of knowledge takes place in a very traditional way, just like informal education. In this traditional method, the focus is on knowledge, and practical skills and innovation are not or rarely addressed. There is an international lack of *practical skills transfer*, while practical experience is essential. The current method of knowledge transfer in the dairy sector is inefficient and focuses onknowledge, *practical skills,* and innovation are lagging.

4.3 Learning and teaching styles and needs of dairy professionals

The personalities with which the participants feel connected vary; Kolb states that all people differ; this applies to the Dutch farmers and the trainers from East Africa. The

The affinity that the Dutch farmers and trainers from East Africa preferred are "cow and entrepreneurship," crop, manure, and soil were the other aspects to be chosen.

Below, the survey results with the participants of the Master class 2030 are shown. The learning preference is mainly verbally shown in figure 3. According to the participants, figure 4 shows the ideal learning method is a classroom lesson. Furthermore, the choice for learning formation is working together as a team according to t 3.

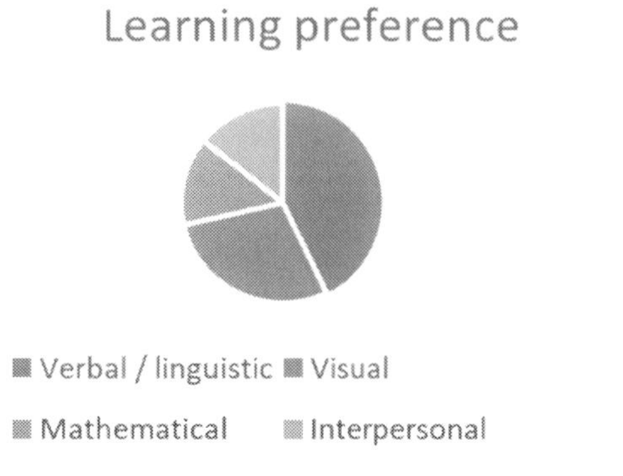

Fig. 3. Learning preferences of Dutch farmers

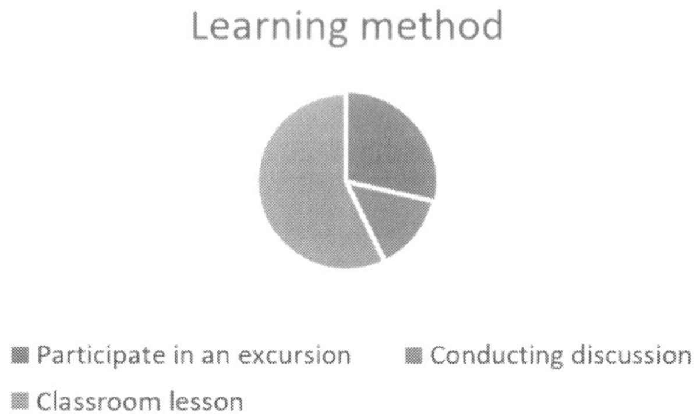

Fig. 4. Learning methods Dutch farmers

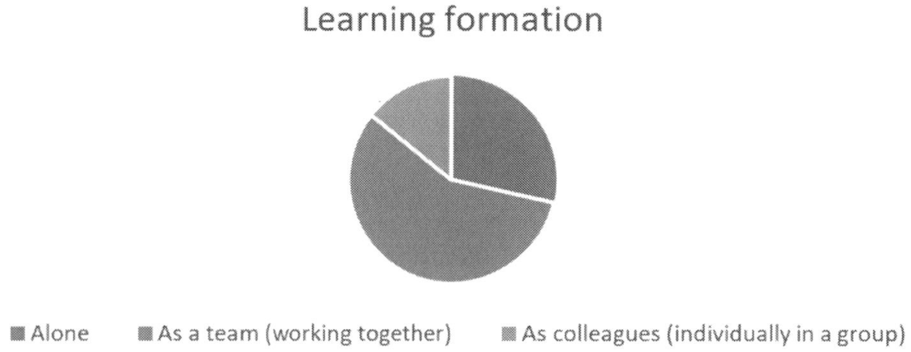

Fig. 5. Learning Dutch formation farmers

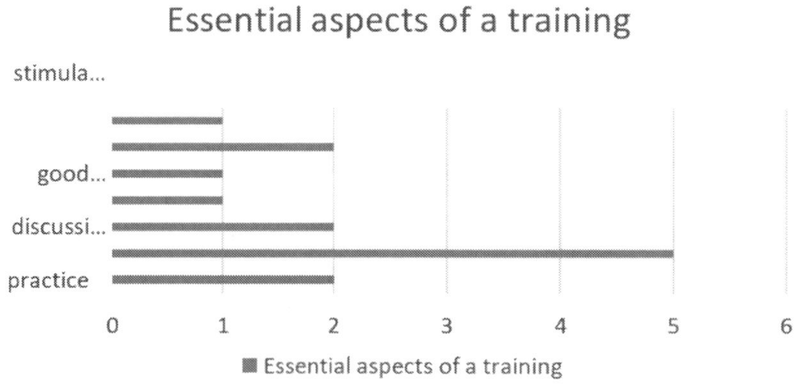

Fig. 6. Essential aspects of training by Dutch farmers

Figure 6 shows the essential aspects that should be present in training according to the Dutch farmers. It is also indicated how often these are mentioned. Interaction is an outlier, but practical discussion and own input have also been mentioned several times.

The Dutch farmers indicate that interaction practices are important aspects of training. The Dutch farmers are quite progressive. Because they are pioneers, they are less concerned with practice but more with entrepreneurship. Trainerslook from a different perspective; they focus on didactics. Furthermore, they are aware because they also receive training on a blended way of training in the Earned project.

There is little affinity with manure, soil, and crop, but these points are necessary to make the dairy sector more sustainable and improve. Personalities are different, as outlined in the literature. Interaction is an important point that came back from the Dutch farmers. The dairy professionals need to connect with practice and practical skills in learning, as shown by the interviews and surveys. As Confucius said, practice is so enormously important.

4.4 Opportunities for the development of the global dairy sector concerning blended learning

Based on pressure from the government and because farmers themselves need to be future-proof. Many farmers must produce more sustainably, and knowledge is needed to change to a more sustainable way of farming. In order to implement change and improvement, vision, skills, incentives, resources, and an action plan must be present, as mentioned in figure 3. Only if these components are present will a successful change take place. In figure 7, several sentences are shown that the persons interviewed gave. Some are positive, but others are skeptical or negative about blended learning in the dairy sector.

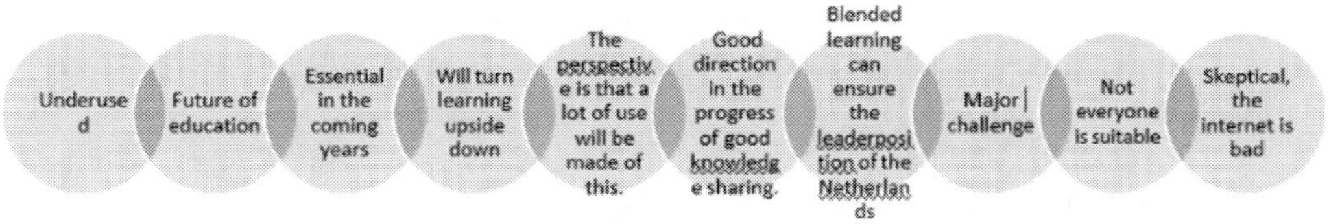

Fig. 7. Statements about integrating blended learning in the dairy sector

The Netherlands is a forerunner in dairy knowledge, but something must be done to stay one of the best. *Blended learning* on a platform such as Brainstud builds provides a stable basis to stay a frontrunner in dairy knowledge.

Blended learning can stabilize the position of the Netherlands and contribute to the improvement of the dairy sector worldwide.

Blended learning can help this success by providing vision, skills, incentives, and action plans. Sometimes the government or the secondary sector will have to facilitate resources or other lacking parts. Using blended learning enables more practice; using a mix of teaching methods can also positively influence the implementation of practical skills. Blended learning can also contribute to the commercialization of the dairy sector in developing countries.

5 Conclusions

Blended learning is a catch-all term in blended learning, the components of online learning and physical meetings in which different didactic methods are used and tailored to the situation. Blended learning uses digital resources, a challenge in the global dairy sector. The Covid pandemic makes digital resources essential, but blended learning can be put to good use in these times. Using blended learning enables more practice; for example, the theory can be discussed in advance so that there is more time for practice during the training. Using a mix of teaching methodscan also positively influence the implementation of practical skills.

Blended learning can contribute to the improvement of the sector if didactic methods are used properly. The right mix of the right methods is relevant and attractive to dairy professionals at the right time. Having local people train and setting up a platform on which the information can easily be ensures a large outreach.

Reference

1. U. Köse, "A blended learning model supported with Web 2 . 0 technologies," Procedia Soc. Behav. Sci., vol. 2, no. 2, pp. 2794–2802 (2010)
2. Leraar24, sd hoe effectief is blended learning. [Online] Available at: https://www.leraar24.nl /2621382/hoe-effectief-is-blended-learning/[Accessed 30 November 2020].
3. FAO, 2017. The future of food and agriculture, Trends and challenges. [Online] Available at: http://www.fao.org/3/a-i6583e.pdf [Accessed: 24 June 2020].
4. FAO, 2013. Milk and dairy products in human nutrition. [Online] Available at: http://www.fao.org/3/i3396e/i3396e.pdf [Accessed: 30 June 2020].
5. Panhuis, J. I. h.,. Een blended leeraanpak voor een efficiëntere internationale transfer van nederlandse agro kennis, sl: sn (2017)
6. Klarus, J. H. e. R, Handelen van opleiders in competentiegericht leren en opleiden. [Online] Available at: https://edepot.wur.nl/117555 [Access: 18 November 2020] (2007).
7. Galan, K. d. Ontwerpen met Kolb. In: Trainen een praktijkgids. Amsterdam: Pearson, p. 219 (2015).
8. Fragoulis, H. Innovations to address the challenges of lifelong learning in transition countries. Utrecht: sn (2002).

 https://doi.org/10.1051/e3sconf/202234800030

9. Nederlandse Zuivel Organisatie, de economische kracht van de Nederandse zuivelsector. [Online] Available at: https://www.consultancy.nl/media/Roland%20 Berger%20-%20Marktanalyse%20Nederlandse%20Zuivelsector-7014.pdf [Geopend 29 October 2020] (2013).

Correlation between teat length and lactation periods on the level of subclinical mastitis occurrence in Sappy Valley Farm

Tetty Barunawati Siagian[1*], and *Surya Hapsara* Amidjaya[1 2]

[1] Veterinary Paramedic of College of Vocational Studies, IPB University, Jl Kampus Kumbang Bo.14, Bogor, Jawa Barat, Indonesia
[2] Sappy Valley Farm, Distric Bogor, Jl Desa Pamijahan, Sentul, Jawa Barat, Indonesia

Abstract. This study aims to determine the correlation between dairy cattle teat length at several lactation periods and subclinical mastitis profile. This study used 43 dairy cattle at several lactation periods. The lactation periods of the cows studied were lactation periods 1, 2, and 3. Subclinical mastitis testing used IPB-1 Mastitis Test, and teat length was measured using a measuring tape (cm). The data is taken and then analyzed using the average and described. The results showed that the average teat length of 5.11 cm did not experience subclinical mastitis, while the average teat length was 5.48 cm, 6.09 cm, and 7.28 cm, respectively, had subclinical mastitis. The highest incidence of subclinical mastitis is influenced by teat length and lactation period. The length of the teat and the lactation period, the higher the severity of subclinical mastitis. These findings are essential for a practical clinical diagnostic approach in dairy cattle with subclinical mastitis.

1. Introduction

Cow's milk is one of the animal products consumed by Indonesian people. Indonesian people's consumption of cow's milk products is still very low compared to other developing countries. Consumption of cow's milk for the people of Indonesia only reaches 8 liters/capita/year. This value also includes processed products containing milk. Domestic fresh cow's milk production contributes about 25% of the national milk demand. Most of the national milk needs are still met from imported milk, either as raw materials or as processed products [5].

Efforts to meet milk needs in Indonesia are carried out on dairy farms. Sappy Valley Farm is one of the dairy farms in Bogor Regency. Sappy Valley Farm is a dairy farm with a population of 111 dairy cows. The dairy cattle population owned by the farm are calves, heifers, lactating cows, dry cows, and bulls. Sappy Valley Farm can produce approximately 500 liters of milk per day. The milk production is considered not good enough compared to the ideal milk production. Dairy cow's milk production is influenced by cage conditions, lactation period, weather, and disease [2].

Many factors cause milk production to be less than optimal, one caused by disease. The disease that dairy cows often experience is inflammation of the udder or mastitis. Currently, mastitis is still a significant problem in smallholder farms in Indonesia [21]. Mastitis can cause considerable losses in dairy farms due to poor milk quality, decreased milk production, increased costs of drugs and veterinary services, high numbers of cattle abandoned prematurely, and sometimes death from the disease [15]. Mastitis is inflammation of the internal tissues of the udder. Mastitis is divided into 2, namely clinical mastitis and subclinical mastitis [36]. Clinical mastitis causes clinical symptoms in the form of changes in quality and decreased milk production [43,27].

The incidence of subclinical mastitis in Indonesia reaches 80% and causes a decrease in milk production by 20% [6]. Subclinical mastitis is caused by bacteria [16]. Subclinical mastitis in cattle farming is like an iceberg phenomenon. Mastitis in a farm is influenced by animal and environmental factors. Predisposing factors also affect the prevalence of mastitis in dairy cows. Predisposing factors for subclinical mastitis in dairy cows are influenced by the condition and shape of the udder and the animal's age. The shape of the udder, one of which is length. Subclinical mastitis does not cause clinical symptoms. This occurrence in subclinical mastitis is like an iceberg phenomenon. This means that the incidence is high but not marked by physical changes in the udder, making it difficult to detect it [34]. Research by [22] states a relationship between nipple diameter and the level of mastitis infection. Other studies have also shown a relationship between nipple length and the rate of mastitis infection [4]. Research on the relationship between nipple length and lactation period has been reported by

Septiani [28], but the data are still minimal. Based on this, it is necessary to review the relationship between the length of the nipple and the incidence of mastitis in various lactation periods. This study aims to determine the correlation of the length of the cow's nipples at various periods of lactation with the incidence of subclinical mastitis.

2. Materials and Methods

The study was conducted from January 15 to February 14, 2019, at the Sappy Valley Farm cattle farm, Bogor Regency. The tools used in this study were a paddle, measuring tape (cm), cloth, and stationery. The materials used were IPB-1 Mastitis Test reagent and quaternary milk samples.

Data is obtained by observation, examination of subclinical mastitis and interviews with farmers. The study was conducted on 43 lactating cows owned by Sappy Valley Farm. Determination of the sample is based on cows in normal lactation conditions. Measurements were carried out in each udder quarter after the cow was bathed. The nipple length was measured using a measuring tape. These measurements are expressed in centimeters (cm). Subclinical mastitis testing using IPB-1 Mastitis Test reagent [32]. The method of examining subclinical mastitis with IPB-1 reagent is a Paddle filled with 2 ml of cow's milk. IPB-1 Mastitis Test reagent was added to the sample in a ratio of 1:1. Samples and reagents were homogenized horizontally for 15-30 seconds. The reaction that occurs is observed. According to [32], if the reaction is adverse, the milk and reagents remain homogeneous, a positive reaction forms a thin mucus, a positive reaction two forms thicker mucus, and a positive reaction three mucus is very thick like a gelatinous mass. The test principle of the IPB1 Mastitis Test is that the reagent will bind to DNA from the somatic cell nucleus to form a thick mass like gelatin. The thicker the mass, the greater the amount of somatic cell nuclear DNA that binds to the IPB-1 Mastitis Test reagent. Data analysis was carried out quantitatively and qualitatively. Quantitative data is calculated using percentages. The data are qualitatively described descriptively and supported by literature studies.

3. Results and Discussion

The subclinical mastitis test at Sappy Valley Farm showed that 29 of the 43 dairy cow population during lactation were positive for subclinical mastitis. The prevalence of subclinical mastitis at Sappy Valley Farm from January 15 to February 14, 2019, was 67% (Figure 1).

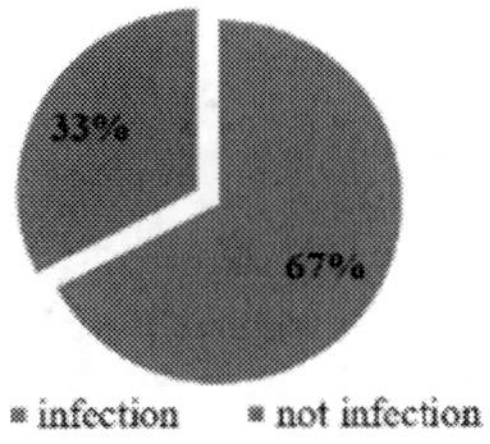

Fig. 1. Presentation of Subclinical Mastitis in Lactating Cattle at Sappy Valley Farm

The prevalence of subclinical mastitis in Sappy Vally Farm is lower than that of subclinical mastitis in cattle farms in Indonesia. The prevalence of subclinical mastitis in Indonesia until 2008 reached 85% [25]. The prevalence of subclinical mastitis in the Bogor KUNAK area is 81.4% [1]. According to [7], the prevalence of subclinical mastitis in other parts of Indonesia, namely East Java, reaches 80%-86%.

The prevalence of subclinical mastitis in Sappy Valley Farm occurred in the first, second, and third lactation. The predisposing factors for subclinical mastitis in dairy cattle were the condition and shape of the udder and the animal's age. The udder hangs very low, the nipple hole is too large, and the presence of sores (blisters) makes it easier for microorganisms to enter [35]. The animal's age also determines whether or not an animal is easily infected with subclinical mastitis [20]. The older the cow, the more sensitive it is to mastitis. This is because the mechanism of closing the nipple hole is decreasing, and healing is slower [11].

There is a relationship between subclinical mastitis and udder conditions [28]. The udder condition of cows susceptible to subclinical mastitis is seen from the udder anatomy, milk production capacity, and nipple length [8]. Subclinical mastitis can be influenced by nipple length, ligament attachment, feed, weather, lactation period, and genetic quality of cattle [14]. The study results on the length of the cow's nipples at Sappy Valley Farm showed that the length of the cow's nipples ranged from 4-8 cm. The study results of the relationship between nipple length and the severity of subclinical mastitis at Sappy Valley Farm can be seen in Table 1.

Table. 1. Correlation of Nipple Length with Subclinical Mastitis Rate at Sappy Valley Farm

No	Degree of Subclinical Mastitis	Minimum nipple length (cm)	Maximum nipple length (cm)	Average nipple length (cm)
1	-	4	6	5.11
2	+	5	7	5.48
3	++	5	7	6.09
4	+++	7	8	7.28

Based on table 2 above, shows that there is a relationship between the length of the cow's nipple and the degree of subclinical mastitis at Sappy Valley Farm. Table 2 shows that nipples with an average length of 7.28 cm are more susceptible to subclinical mastitis than nipples with an average length of 5.11 cm. Based on this, nipple length can be used as a predisposing factor for subclinical mastitis. Nipples with a mean length of 7.28 cm had a severity of subclinical mastitis at a positive level of 3 (+++). Nipples with an average length of 6.09 cm had a severity of subclinical mastitis at a positive level of 2 (++). Nipples with a mean length of 5.48 cm had a severity of subclinical mastitis at positive 1 (+). The condition of the nipple that did not experience subclinical mastitis was indicated by the nipple having an average length of 5.11 cm. Nipple length can be used as an indicator of the diagnosis of subclinical mastitis in dairy cattle. Based on the above study results, it was shown that dairy cows with a nipple length of 5.48-7.28 cm were likely to have subclinical mastitis.

The average nipple length of 6.5-7.5 cm in dairy cattle in Lembang had subclinical mastitis [28]. BIF (2011) [4], on Fresian Holstein dairy cows in Europe, stated that the ideal normal size for the front nipple is 6 cm in length with a diameter of 2.9 cm, while the rear nipple is 5 cm in length with a diameter of 2.6 cm. Cases of subclinical mastitis occur in the nipple with a length of 7–9 cm. Long nipples will facilitate the entry of microorganisms into the udder and develop in the udder tissue. This can cause an inflammatory reaction in the internal tissues of the udder. Nipples with a concise size will cause impaired milking and incomplete emptying of the udder, and the alveoli are not capable of selective power, thus triggering inflammation [19]. Short nipples can also complicate the milking process so that the alveoli do not secrete milk completely and cause subclinical mastitis [35].

The lactating dairy cows at Sappy Vally Farm consist of cows from the first to the third lactation period. The higher the lactation period, the older the cows. The effect of the lactation period on the level of subclinical mastitis can be seen in Figure 2. The percentage of the severity of mastitis levels positive one and positive two seems to decrease from lactation period one to the third lactation period. In contrast to the severity of mastitis level positive three, the percentage increased from the second lactation to the third lactation period.

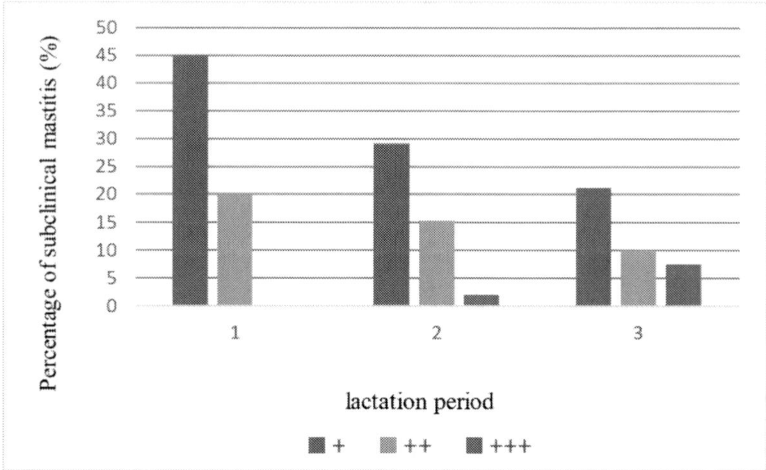

Fig.2. Percentage of incidence of subclinical mastitis

Figure 2 shows that the longer the lactation period, the more cows are at risk of being infected with subclinical mastitis. Following [31] statement, the incidence of subclinical mastitis will increase until approximately eight years of age cattle and continue to increase from the first lactation period until it reaches a peak, namely in the third and fourth lactation periods decrease according to the next lactation period. The older the cow and the higher the milk production, the more relaxed the nipple sphincter will be. Microorganisms will easily infect nipples with loose sphincters. The higher the milk production of a female cow, the longer it takes for the sphincter to close completely [35]. Figure 2 shows that during the first lactation period, the percentage of the severity of subclinical mastitis was the greatest compared to the second and third lactation. This was because the first lactating dairy cows had a high level of aggressiveness, so teat dipping did not go well.

4. Conclusion

Based on the study results, it was shown that dairy cows with a nipple length of 5.48-7.28 cm were likely to have subclinical mastitis. The severity of subclinical mastitis infection is influenced by nipple length and lactation period. The longer the nipple and the longer the lactation period, the higher the severity of subclinical mastitis.

5. Acknowledgments

Thanks to Sappy Valley Farm for providing the opportunity to do research

References

1. Alhansah SD. Prevalensi Mastitis Subklinis serta Pengetahuan, Sikap, dan Praktik Peternak terhadap Pengendalian Mastitis Subklinis di KUNAK Bogor [Skripsi]. (IPB, Bogor, 2016)
2. Archer SC, Green MJ, dan Huxley JN. Association Between Milk Yield and Serial Locomotion Score Assessments in UK Dairy Cows. *J Dairy Sci*. 93: 4045–4053 (2010)
3. Bhutto AL, Murray RD, Woldehiwet Z. The effect of dry cow therapy and internal teat-sealant on intra-mammary infections during subsequent lactation. *Res Vet Sci*. 90:316–320 (2011)
4. [BIF] Beef Improvement Federation. *BIF Guidelines for Unifrom Beef Improvement Program*. Ed ke-8. (Georgia Univ., Athena, 2011)
5. Daryanto A. *Peningkatan Daya Saing Industri Peternakan*. (PT Permata Wacana Lestari, Jakarta, 2007)
6. [DITJENNAK] Direktorat Jenderal Peternakan dan Kesehatan Hewan. *Statistik Peternakan* (Direktorat Jenderal Peternakan, Jakarta, 2006)
7. Effendi H. *Angka Prevalensi Bovine Mastitis dari Beberapa Peternakan Sapi Perah di Jawa Timur*. (Universitas Airlangga, Surabaya, 2008)
8. Farmer WS, Chrestman G. *Dairy Cattle Judging*. (Mississipi State University, Mississipi, 2006)
9. Haerah D. Deteksi *Staphylococcus aureus* Penyebab Mastitis Subklinis Pada Sapi Perah Di Kecamatan Cendana Kabupaten Enrekang [Skripsi]. (Universitas Hassanudin, Makasar, 2015)
10. Halasa T, Nielen M, Werven TV, Hogeveen. A simulation model to calculate cost and benefits of dry period interventions in dairy cattle. *J Liv Sci*. 129: 80-87 (2010)
11. Hidayat AP, Effendi AA, Fuad Y, Patyadi K, Taguchi, Sugiwaka T. *Buku Petunjuk Teknologi Sapi Perah di Indonesia untuk Peternak: Kesehatan Pemerahan*. (Sonysugema Pressindo, Bandung, 2002)
12. Hillerton JE, Berry EA. Treating mastitis in the cow is a tradition or an archaism. *J Appl Microbiol*. 98:1250-1255 (2005)
13. Hurley WL, Morin DE. *Mastitis lesson a. lactation biology*. [Internet]. [diunduh 2019 Januari 23]. Tersedia pada:http://http://ansci.illinois.edu/static/ansc438/ (2000)
14. Ikawati A. Analisis kandungan protein dan lemak susu hasil pemerahan pagi dan sore pada peternakan sapi perah di Wonocolo Surabaya [Skripsi]. (Universitas Airlangga, Surabaya, 2011)
15. Kumar R, Yadav BR, dan Singh RS. Genetic determinants of antibiotic resistance in *Staphylococcus aureus* isolates from milk of mastitic crossbred cattle. *J. Curr Microbiol*. 60:379–386 (2010)
16. Karimuribo ED, Fitzpatrick JL, Bell CE, Swai ES, Kambarage DM, Ogden NH, Bryant MJ & French NP. Clinical and subclinical mastitis in smallholder dairy farms in Tanzania: Risk, intervention and knowledge transfer. *Preventive veterinary medicine* (74) : 83-98 (2006)
17. Karimuribo ED, Fitzpatrick JL, Swai ES, Bell C, Bryant MJ, Ogden NH, Kambarage DM, French NP. Prevalence of subclinical mastitis and associated risk factors in smallholder dairy cows in Tanzania. Vet Rec. 163:16-21 (2008)
18. [LIPTAN] Lembar Informasi Pertanian. *Sanitasi kandang sapi perah* [Internet]. [20 Febuari 2019] Tersedia dari : http://www.deptan.go.id (2000)
19. Lukman DW, Sudarwanto S, Sanjaya AW, Purnawarman T, Latif H, Soejoedono RR, Pisestyani H. Higiene Pangan. Bagian Kesehatan Masyarakat Veteriner Departemen Ilmu Penyakit Hewan dan Kesmavet. (Institut Pertanian Bogor, Bogor, 2009)
20. Nurhayati I S. Kajian Pengendalian Mastitis Subklinis Melalui Pemberian Antibiotik pada Saat Kering Kandang di KPSBU Lembang Jawa Barat [Tesis]. (Institut Pertanian Bogor, Bogor, 2014)
21. Nurhayati I, Martindah E. Pengendalian mastitis subklinis melalui pemberian antibiotik saat periode kering pada sapi perah. *Wartazoa*. 25(2): 65-74 (2015)
22. Prasetyo BW, Sarwiyono P. Hubungan antara Diameter Lubang Putting terhadap tingkat kejadian mastitis. J. Ternak Tropika. 14(1):15-20 (2013)
23. Prihadi S. *Dasar Ilmu Ternak Perah*. (Universitas Gajah Mada, Yogyakarta, 1997)
24. Rahayu ID. The sensitivity of Staphylococcus aureus as mastitis pathogen bacteriae into teat dipping antiseptic in dairy cows. *J Protein*. 14:31-36 (2007)

25. Rahayu. Kerugian ekonomi mastitis subklinis pada sapi perah.[Internet]. [diakses pada18 Februari 2019]. Tersedia dari: http://www.umm.ac.id/fapet/ekonomi -mastitis.hlml (2009)
26. Raymundo LJ, Couch CS dan Harvell CD. *Coral Disease Handbook: Guidelines for Assesment, Monitoring and Management*. (The University of Queensland, Brisbane, 2008)
27. Samad MA. *Animal husbandry and veterinary science*. Vol. II. (Bangladesh Agricultural University, Mymensingh, 2008)
28. Septiani YN. 2013. Panjang Puting dan Periode Laktasi sebagai Faktor Predisposisi Mastitis Subklinis pada Sapi Perah di Koperasi Peternak Sapi Bandung Utara (KPSBU) Lembang Kabupaten Bandung[Skripsi]. (Institut Pertanian Bogor, Bogor, 2013)
29. Sharif A, Muhammad U, Ghulam M. Mastitis control in dairy production. *J Agric Soc Sci*. 5(3):102-105.
30. Siregar A Z. 2010. Pengaruh celup puting sari buah mengkudu (*Morinda Citrifolia L*) terhadap kasus mastitis subklinis pada sapi perah berdasarkan pemeriksaan total *plate count* [Internet] . [diakses pada 18 Febuari 2019] Tersedia dari : http://www.fkh.unair.ac.id/artikel1 /2010 /ARTIKEL%20IL MIAH%20A.pdf (2009)
31. Siregar S. *Jenis Teknik Pemeliharaan dan Analisa Usaha Sapi Perah*. (Penebar Swadaya, Jakarta, 1995)
32. Sudarwanto M. Pereaksi IPB-1 sebagai pereaksi alternatif untuk mendeteksi mastitis subklinis. Bogor (Institut Pertanian Bogor, Bogor, 1998)
33. Sudarwanto M, Latif, Noordin M. *The relationship of the Somatic cell counting to sub-clinical mastitis to improve milk quality*. 1st International AAVS Scientific Conference, Jakarta, Indonesia (2006).
34. Sudarwanto M, Sudarnika E. Hubungan antara pH susu dengan jumlah sel somatik sebagai parameter mastitis subklinis. *Med Pet*. 31:107-113(2008)
35. Subronto. *Ilmu Penyakit Ternak 1*. (UGM Pr., Yogyakarta, 2003)
36. Sudono A, Rosdiana RF, Setiawan BS. *Beternak Sapi Perah Secara Intensif*. (Agromedia Pustaka, Jakarta, 2003)
37. Sugiri YD dan Anri A. Prevalensi Patogen Penyebab Mastitis Subklinis (*Staphylococcus aureus* dan *Streptococcus agalactiae*) dan Patogen Penyebab Mastitis Subklinis Lainnya pada Peternakan Skala Kecil dan menengah di Beberapa Sentra Peternakan Sapi Perah di Pulau Jawa. (Balai Pengujian dan Penyidikan Penyakit Hewan dan Kesmavet (BP3HK), Bandung, 2010)
38. Sunarko C. 2009. *Petunjuk Pemeliharaan Bibit Sapi Perah*. (BBPTU Sapi Perah Baturraden, Banyumas, 2009)
39. Surjowardojo P. Tingkat kejadian mastitis dengan whiteside test dan produksi susu sapi perah friesien holstein. *Jurnal Ternak Tropika* 12: 46-55 (2011)
40. Supar. Mastitis subklinis pada sapi perah di Indonesia: Masalah dan pendekatannya. *Wartazoa*. 6 (2): 48-52 (1997)
41. Supar dan Ariyanti T. Kajian Pengendalian Mastitis Subklinis pada Sapi Perah. Prosiding: Prospek Industri Sapi Perah Menuju Perdagangan Bebas 2020. (Pusat Penelitian dan Pengembangan Peternakan, Jakarta, 2008)
42. Sutarti E, Budiharta S, Sumiarta B. Prevalensi dan faktor-faktor penyebab mastitis pada sapi perah rakyat di Kabupaten Semarang, Provinsi Jawa Tengah. *J Sain Vet*. 21:43-49 (2003)
43. Tyler DH, Ensminger ME. Dairy Cattle Science. Ed ke-4. (New Person Practice Hall, United states, 1993)
44. Timmereck TC. *Epidemiologi, Suatu Pengantar. Edisi Terjemahan*. (Penerbit Buku Kedokteran EGC, Jakarta, 2001).
45. Walder DN. *Dry cow therapy for mastitis control*. (Oklahoma State University, Oklahoma, 2007)
46. Winarso D. Hubungan Kualitas Susu dengan Keragaman Genetik dan Prevalensi Matitis Subklinis di Daerah Jalur Susu Malang sampai Pasuruan. J.Sains Vet. 26(2): 58-65 (2008)

Load Balancing Simulation Android Application as an Online Learning Media

Walidatush Sholihah[1*], *Ahmad* Abdul Malik[1], and *Inna* Novianty[1], *Nur* Aziezah[2]

[1]Computer Engineering, College of Vocational Studies, IPB University, Indonesia
[2]Informatics Management, College of Vocational Studies, IPB University, Indonesia

Abstract. Learning activities during the covid pandemic are carried out online. Learning materials are delivered synchronously or asynchronously. Asynchronous techniques allow students to learn independently, and materials can be studied repeatedly. One way that can be used to convey material is through simulation. The students will understand the learning material better using simulation. The simulation made in this paper is a load balancing material simulation. This material is studied by students in computer network studies at both high school and college. Load balancing is a technique for dividing the load on the server. The algorithms discussed in this simulation are round-robin and least connected. In this simulation, the user is invited to interact with the application and see the difference between the two algorithms. The method used to create the application consists of five stages. The stages are problem analysis, application design, development, and testing. This simulation application is made using GameMakerStudio 14.9 and can be installed on Android devices. The application was tested on 20 users with vocational education backgrounds and students. More than 50% of users say this application is easy to use and the material is easy to understand.

1 Introduction

The COVID-19 pandemic has led to the digitization process in some daily activities, including teaching and learning activities. Learning activities are carried out from home using the online method. This resulted in the need for improvement in online learning. Online learning is carried out synchronously and asynchronously. Synchronous activities can be interpreted as activities in real-time.
In contrast, asynchronous activities can be carried out between teachers and students [1]. Synchronous activities can be carried out with various virtual meeting applications. The problem with this system is that the internet connection used by each user is quite high. So another alternative to online learning is asynchronous. Asynchronous activities can use learning management system applications. In addition, it can also use simulation applications.

Cloud computing is a service that can be accessed anywhere, providing convenience and on-demand network access for configured computing resources [2]. In cloud computing, one

* Corresponding author: walidah@apps.ipb.ac.id

of the most important pieces of information to understand is how to share server loads or load balancing. Load balancing manages incoming requests and ensures no server is overloaded or overloaded [3]–[5]. Overloaded or overloaded servers can cause system failure [6], [7]. A survey conducted by Snow Software in 2020 showed that of 250 IT organization leaders worldwide, more than 80% had experienced an increase in the use of cloud computing [8]. An increase or trend towards something will also increase the need for information about it.

Learning or understanding load balancing should be easy to obtain thorough information circulating in cyberspace. However, most learning only contains text that is quite difficult for beginners to understand. The simulation was chosen as a learning method because simulation is considered easier to convey information. In the simulation of receiving information, the percentage of success is quite high [9]. People who receive information using simulation media will more easily remember information than those who only receive information from vocal delivery [10]. Simulation is also very often found in education in technology [11], [12]. Many kinds of research on the Round Robin algorithm have been carried out. The Round Robin algorithm can improve performance by reducing waiting time [13].

In a learning process, apart from the method, the media should also be considered because learning media can affect learning effectiveness. Android is the operating system with the largest number of active users [14], and during the pandemic, the worldwide rate of use increased by 70% [15]. Android also allows users to access android devices anywhere without being limited by space and time [16], [17] being why smartphones with the Android operating system are chosen as learning media.

The problem in this study is how to make an android application to simulate the load balancing process using the round-robin algorithm and the least connection algorithm. Then the purpose of this research is to create a balancing simulation android application using the round-robin algorithm and the least connection algorithm. The benefits of this research are to facilitate the delivery of information about load balancing, round-robin algorithm, and least connection algorithm.

2 Methodology

The methodology used in this study consists of (1) Problem Analysis, (2) Application Design, (3) Building Application, (4) Testing. The research method can be seen in Figure 1.

Fig. 1. Research Method

In the analysis of the problem, it was found that the problem of learning in this network requires a medium where students can learn the material repeatedly. In addition, students and teachers do not need to be simultaneous. Learning media or applications must be easily accessible and easy to use. From this problem, a solution has been found. One of the learning media that can be made is an android-based application. The Android application is easy to get through the play store. This application can be opened anytime, anywhere, and repeatedly. After the application is installed on the mobile device, there is no need for an internet connection to open it.

At the application design stage, a flow is created to show the processes that occur in the application. This android simulation android application was created using GameMakerStudio. The programming language used is Delph. In the testing process, this application was tested on 20 users. Users are asked to use the application and write their response to this application.

3 Results

This simulation application discusses two load balancing algorithms. There are Round Robin and Least Connection algorithms. These algorithms have different request distribution patterns. The difference in the request distribution pattern is used as the main material in the simulation application.

The difference in the request distribution pattern between the round-robin algorithm and the least connection algorithm lies in the determinant of the request distribution. In the round-robin algorithm, requests are distributed based on the order of the server at the time of installation of load balancing. This server order is then used as a distribution determinant for an incoming request. In the least connection algorithm, the determinant of the distribution of requests is the number of active requests on each server. This determination to use the number of active requests makes the least connection algorithm more flexible than the round-robin algorithm.

The round-robin algorithm is an algorithm that works by distributing every incoming request to the servers sequentially [18]. The least connection algorithm is an algorithm that works by distributing requests to servers with fewer active connections [18]. An illustration of how the round-robin algorithm works is shown in Fig 2.

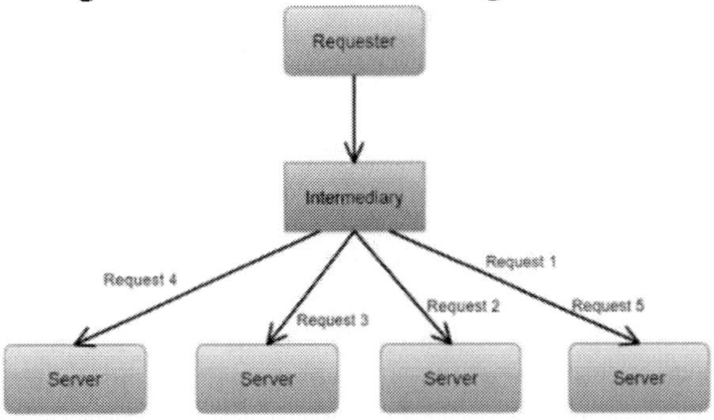

Fig. 2. Round Robin Algorithm (Source: codinginthetrenches.com/2016/07/18/dnsround-robin-failures/)

In Fig 2. the requester is a client that sends a request to the server. An intermediary is a software or hardware load balancer. Requests from clients will go to the load balancer. First, the first incoming request will be distributed, and then the next request. It can be seen that the incoming requests will be distributed sequentially to the existing server. Request 1 goes to the rightmost server sequentially until request 4 goes to the leftmost server. After that, request 5 is redistributed to the far-right server. This request distribution pattern is the distribution pattern of the round-robin algorithm.

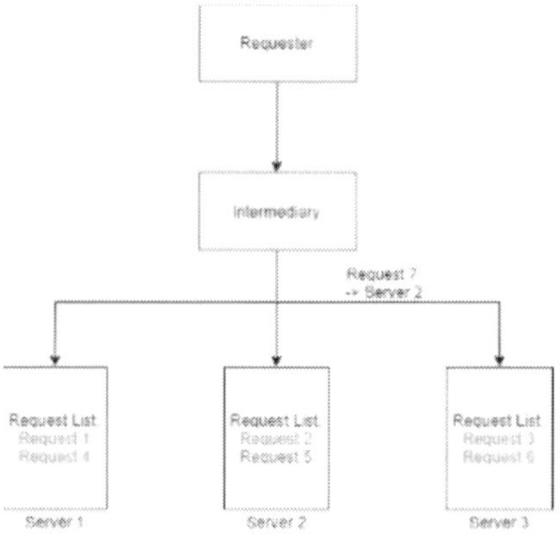

Fig. 3. Least Connection Algorithm

Fig 3. shows the distribution of requests from the least connection algorithm. A request List is a list of requests that exist on the server. It can be seen on server 1 that requests 1 and 4 are green, indicating that both requests are still active. On server 2, request 2 is active while request 5 is red, which indicates the status is inactive. On server 3, requests 3 and 6 are active. Due to the rules of the least connection algorithm, the next request will be distributed to the server with the fewest active connections (server 2).

This android application shows the concept of the round-robin algorithm and the least connection in load balancing. The application flow can be seen in Fig 4. application flow starts on Start. Users will be faced with a menu and choose from three topics presented: load balancing, round-robin, and least connection. After selecting the material, the user can choose the learning material, whether to study the introduction of the material or a simulation. In the simulation section, users will face different servers depending on the selected simulation material. There are two servers in the load balancing simulation material, four servers in the round-robin simulation material, and three servers in the least connection simulation material. The user acts as a load balancer in the simulation section and distributes incoming requests according to the specified rules. The request distribution rules depend on the selected simulation material.

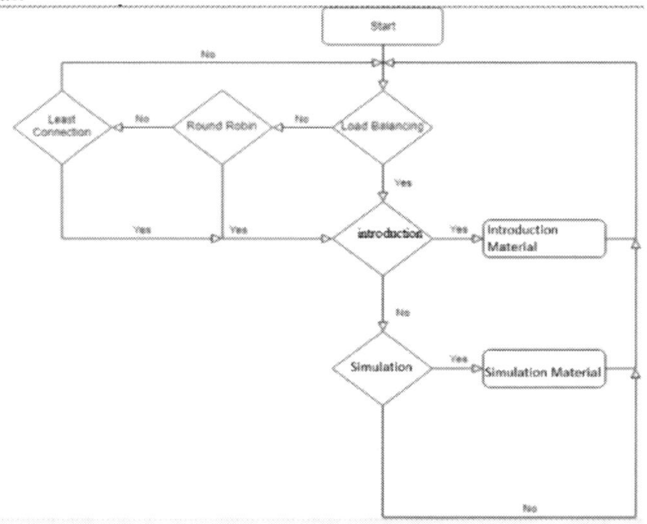

Fig. 4. Application flowchart

The design process was created using Adobe Photoshop software and the Sprite Editor at GameMakerStudio. UI design is made following the application flow where there are two

https://doi.org/10.1051/e3sconf/202234800032

menus, namely the main menu and the material menu. The main menu gives the user the option to select the material or read the guide directly. Figure 5 shows the UI design process for the material menu, but in Figure 5 the design is still in the form of a template which will later be filled through the Sprite Editor in GameMakerStudio 14.9. In the context menu, the user can choose between the introduction of load balancing, round-robin algorithm, and least connection algorithm.

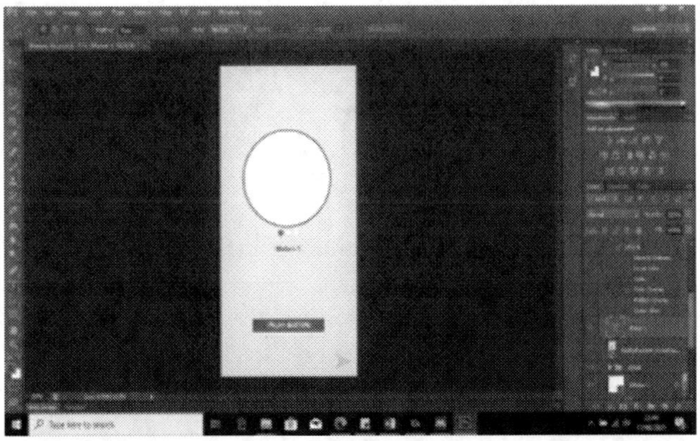

Fig. 5. Design process

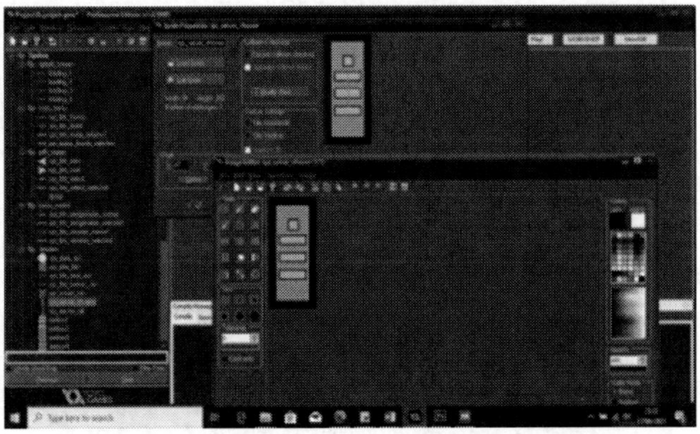

Fig. 6. Creating objects using the sprite editor

Figure 6 shows creating objects using the Sprite Editor in GameMakerStudio. On the left of Figure 9 there is a tree that lists objects that have been created. In the sprite properties tab, there are properties of objects that contain information and some object settings such as object height and width information in pixels, object center point settings, etc. In the sprite editor tab is the section where the object is created. The object will be formed in this section and given the appropriate color. The objects created will later be displayed in the application.

Application development is done using GameMakerStudio 14.9 software. GameMakerStudio is software that offers an easy software creation process with Drag-and-Drop (DnD) features combined with simple programming using the Delph programming language. The UI and previously designed objects will also be used in this process.

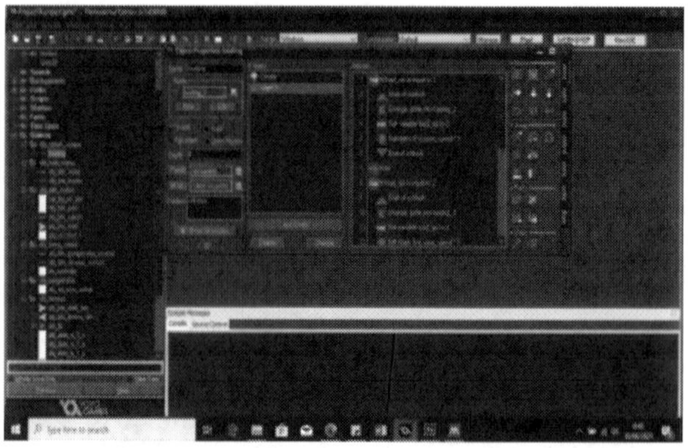

Fig. 7. Creating an object in GameMakerStudio (part 1)

In Figure 7, an object is created in GameMakerStudio. The object here is different from the previously designed object. Pre-designed objects have the designation sprite. The object uses the previously designed sprites as the display of the object. In this section, the object can be given commands such as if the user presses the screen, it will carry out one or more instructions that have been compiled. Figure 7 shows the creation of an object named loading displayed on the application splash screen. In Figure 8, there are objects named obj_button_start and obj_button_howto. Obj_button_start serves as a start button that will take the user to the context menu, and obj_button_howto serves as a guide button that will take the user to the wizard section. Both objects will be displayed on the main menu.

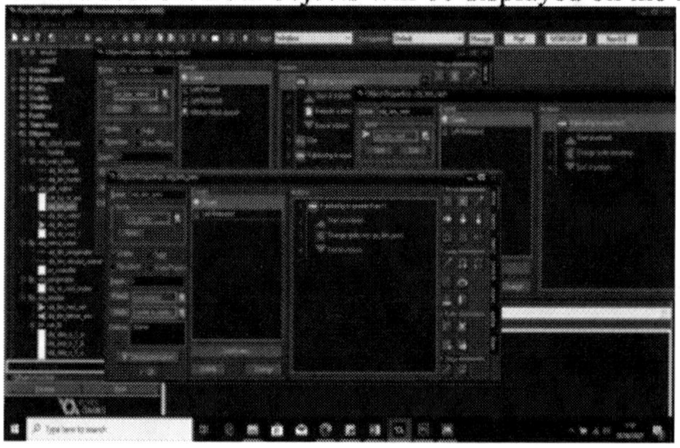

Fig. 8. Creating an object in GameMakerStudio (part 2)

A room is a page where there are objects that have been arranged in such away. Objects in the room will be displayed on the android screen. The object that has been designed will be used as a sprite. Then the sprite will be used as the object display in GameMakerStudio. The object that has been created is then arranged in the simulation room. The arrangement of objects in the simulation room varies depending on what material the user chooses.

Figure 9 shows the objects that have been arranged in the simulation room with load balancing material. There are three computer-shaped objects on the right. The three objects represent the client. There is an object in the form of a black striped circle with the main color blue symbolizing the internet, a square object with a line in the middle symbolizing the load balancer, a square object with several small squares symbolizing the server. The green line represents the connection path between objects. A box-shaped object with a sideways triangle inside is the play button to start the simulation.

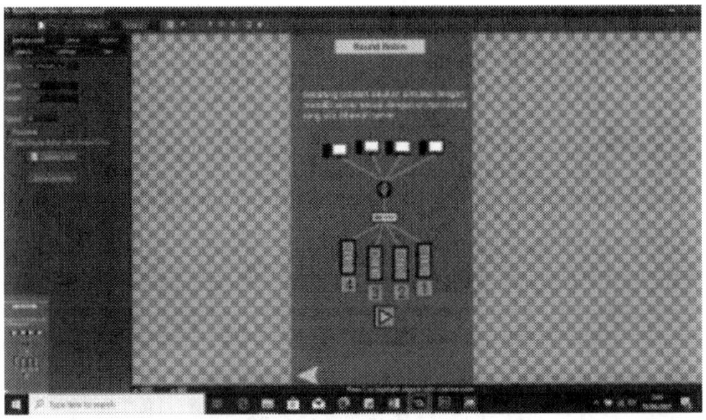

Fig. 9. The process of compiling objects

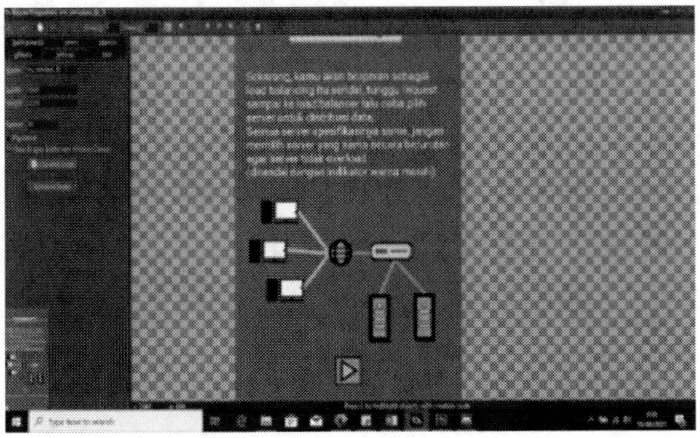

Fig. 10. Arranged objects for round-robin algorithm

Figure 10 shows the objects that have been arranged in the simulation room with round-robin algorithm material. There are four clients, and four servers, then there is a new object in the form of a box with numbers below it. This object serves as a marker. Users are required to select servers sequentially according to the rules of the round-robin algorithm.

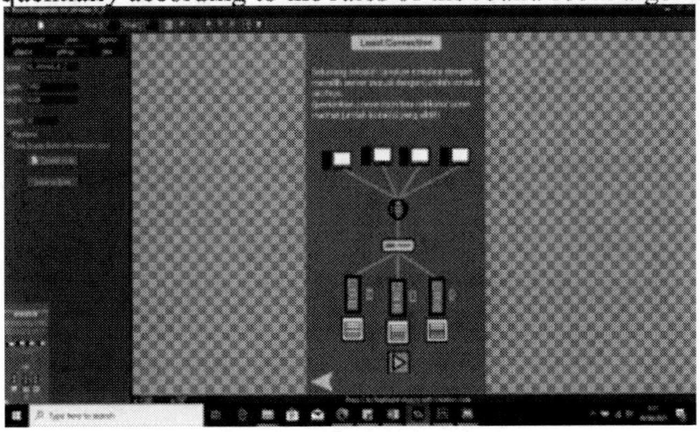

Fig. 11. Arrange objects for the least connection algorithm

Figure 11 shows the objects that have been arranged in the simulation room with the least connection algorithm material. There are four clients and three servers, then there are three small square objects with numbers in them which are markers of the server order, and three square objects with several small squares in them, which indicate the number of active connections on the server above the object. Users must select a server with the least number of active connections. If there are two or more servers with the same number of active connections, the user is required to select a server according to the serial number on the object beside the server.

Testing is done by using this application on the cellphone. During the Covid-19 pandemic, the application was installed on the cellphone, and the test was carried out at the. The features and functions of the application are tested one by one. Two lines of objects on the page are displayed on the android screen (Fig. 12). The first line consists of one client and one server. This line demonstrates a simple request flow from the client to the server. A small yellow circle represents the request. The demonstration went well without any problems. The second line consists of three clients and one server. This line demonstrates the flow of requests with more clients to the server. The number of requests will overload the server because it runs out of resources needed to serve incoming requests. As a result, the demonstration went well without any problems

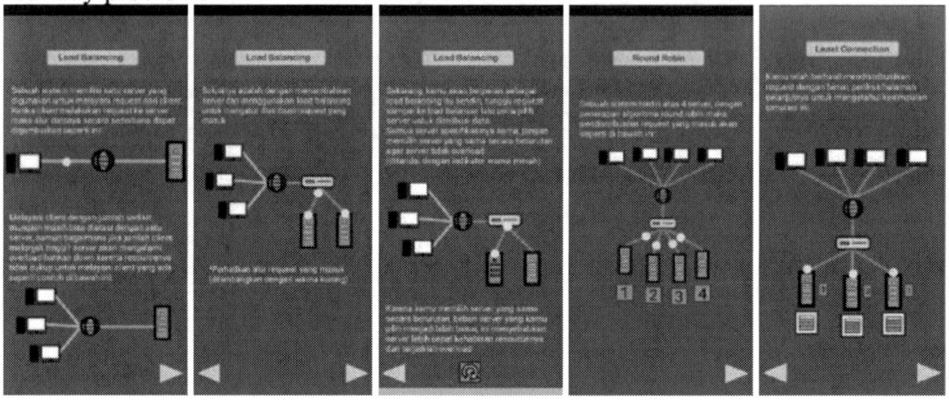

Fig. 12. Load balancing simulation display

This load balancing simulation application was tested on 20 users. The application testers have an IT background (organization/group/institution), especially in cloud computing. Testers were asked to use the app for one day. Then the testers are asked for feedback about the simulated android application that has been tried. Testers are asked to fill out a test form. The results of the feedback can be seen in Table 1.

Table 1. User feedback.

Material	User Feedback		
	Easy to understand	Not too difficult to understand	Difficult to Understand
Introduction to load balancing	13	7	0
Introduction to the round-robin	13	7	0
Introduction to least connection	14	5	1
load balancing simulation	14	6	0
round-robin simulation	15	5	0
least connection simulation	13	6	1

From Table 1, more than 50% of users stated that it was easy to use the application. Simulation of load balancing material can also be understood. Thus it can be concluded that this application can be used as a learning tool to understand the concept of load balancing.

4 Conclusion

The load balancing simulation application can be used as a learning module to understand the concept of load balancing. Users can download this application from the Google play store. More than 50% of users say this application is easy to use and the material is easy to understand.

5 References

1. W. Sholihah and A. Mardiyono, *Mengelola Kelas Online Dengan Moodle 3.8.* Yogyakarta: Deepublish (2020)
2. D. P. Agustino, Y. Priyoatmojo, and N. W. W. Safitri, "Implementasi Honeypot Sebagai Pendeteksi Serangan dan Melindungi Layanan Cloud Computing," in *Konferensi Nasional Sistem & Informatika 2017*, pp. 196–201 (2017)
3. G. Joshi and S. . Verma, "A Review on Load Balancing Approach in Cloud Computing," *Int. J. Comput. Appl.*, vol. 119, no. 20 (2015)
4. S. Kumar, "Cloud Computing based Load Balancing Architecture: A Study," vol. 8, no. 2, pp. 112–116 (2017)
5. D. Leman, "Load Balancing 2 Jalur Internet Menggunakan Mikrotik Round Robin," *Riau J. Comput. Sci.*, vol. 5, no. 2, pp. 137–143 (2019)
6. S. D. Riskiono and D. Pasha, "Analisis Metode Load Balancing Dalam Meningkatkan Kinerja Website E-Learning," *J. Teknoinfo*, vol. 14, no. 1, p. 22, Jan. (2020)
7. S. K. Mishra, B. Sahoo, and P. P. Parida, "Load balancing in cloud computing: A big picture," *Journal of King Saud University - Computer and Information Sciences*, vol. 32, no. 2. King Saud bin Abdulaziz University, pp. 149–158, Feb. (2020)
8. C. Hugh, "Study: How the `New Normal` is Changing Cloud Usage and Strategy," in *Snow Software*, p.1 (2020)
9. D. Wurara, S. R. U. A. Sompie, S. D. E. Paturusi, and H. V. Kainde, "Rancang Bangun Aplikasi Game Pembelajaran Dan Simulasi Sistem Bilangan Digital Berbasis Android," *J. Tek. Inform.*, vol. 15, no. 1, pp. 13–22, Mar. (2020).
10. A. Muafa, E. Ekohariadi, and A. Santoso, "Pengaruh Penggunaan Media Pembelajaran Simulasi Komputer Terhadap Hasil Belajar Siswa Sekolah Menengah Kejuruan Di Gresik," *J. Pendidik. Vokasi UNESA*, vol. 2, no. 01, p. 246525 (2014)
11. C. D'Angelo, D. Rutstein, C. Harris, R. Bernard, E. Borokhovski, and G. Haertel, "Simulations for STEM learning: Systematic review and meta-analysis," (2014)
12. O. Chernikova, N. Heitzmann, M. Stadler, D. Holzberger, T. Seidel, and F. Fischer, "Simulation-Based Learning in Higher Education: A Meta-Analysis," *Rev. Educ. Res.*, vol. 90, no. 4, pp. 499–541, Aug. (2020)
13. S. Ul Sabha, "A Novel And Efficient Round Robin Algorithm With Intelligent Time Slice And Shortest Remaining Time First," *Mater. Today Proc.*, vol. 5, no. 5, pp. 12009–12015, Jan. (2018)
14. L. Tung, "Bigger than Windows, bigger than iOS: Google now has 2.5 billion active Android devices," (2019)
15. A. Watson, "Coronavirus impact: global device usage increase by country 2020 | Statista," in *18 Juni 2020*, p.1 Jun. (2020)
16. D. Tolawo, A. Lumenta, and S. Karouw, "Perancangan Mobile Learning Algoritma dan Pemprograman," vol. 4, no. 2, pp. 1–5 (2014)
17. R. F. Rahmat, L. Mursyida, F. Rizal, K. Krismadinata, and Y. Yunus, "Pengembangan media pembelajaran berbasis mobile learning pada mata pelajaran simulasi digital," *J. Inov. Teknol. Pendidik.*, vol. 6, no. 2, pp. 116–126, Oct. (2019)
18. H. Triangga, I. Faisal, and I. Lubis, "Analisis Perbandingan Algoritma Static Round-

Robin dengan Least-Connection Terhadap Efisiensi Load Balancing pada Load Balancer Haproxy," *InfoTekJar (Jurnal Nas. Inform. dan Teknol. Jaringan)*, vol. 4, no. 1, pp. 70–75, Sep. (2019)

Artificial Coral Reef Growth Media Model Engineering from Ceramic

Ni Putu Muliawati[1], *I Made Dwi* Setiadi[*1], *Guino* Verma[1], *Maya Larasati* Donna Wardani[1], *Geby* Otivriyanti[1]

[1]Balai Teknologi Industri Kreatif Keramik (BTIKK), Badan Pengkajian dan Penerapan Teknologi (BPPT), Denpasar, Indonesia

Abstract. The experiment on the artificial coral reef was conducted at BTIKK using ceramic body mass code BL1. This engineering was carried out in two steps, the first step was making the design of the growth media, and the second step was testing the mass of the ceramic body and making the growth media model. The hexagonal design was used beginning from the design concept that applies geometric shapes with many sides, making it easier to arrange the formation under the sea. Aesthetically, the hexagonal shape can represent a design concept. It resulted from integration in the design of coral reef planting media with natural marine ecosystems and can produce new experiences and impressions to attract people's attention to pay more attention to the preservation of coral reefs. The assessment of fired shrinkage, water absorption, and apparent porosity on body mass for making the growth medium model was tested on 15 test pieces. The average value of fired shrinkage on 900°C temperature fired test pieces is 1.3%, the average water absorption is 19.7%, and the apparent porosity is 32.9%. Their value could be used to make artificial coral reef growth media.

1 Introduction

Indonesia is an archipelago country that consists entirely of islands. Territorial boundaries are drawn from the archipelago baselines covering land areas, interior waters, archipelago waters, and the territorial sea, including the airspace above it and the seabed and the land below the natural resources contained therein [1]. Indirectly, this condition means Indonesia has abundant natural resources potential, especially the sea. Due to its place between Indian Oceans and the Pacific Oceans, also flanked by Asia and Australian continents, the potential for natural resources exploitation by external and domestic parties is huge. As the fourth most populous country in the world [2], the population dynamics will impact marine biodiversity, especially shallow marine. Shallow marine ecosystems consist of several communities; one of them is coral reefs, which need to be preserved. Besides population activities, natural disasters such as earthquakes and tsunamis also affect to marine ecosystem especially coral reefs [3]. However, the natural factors could not be controlled, so the impact due to human dynamics is more controllable.

* Corresponding author: dwisetiadi@gmail.com

Coral reef conservation is a program that aims to restore the life of damaged coral reefs by planting artificial reefs. Artificial coral reefs have substantial positive effects on surrounding aquatic ecosystems. It can be built out of the metal framework, steel, steel-reinforced concrete, or concrete, as well as recycled plastics, ceramics, and fiberglass [4]. Ceramics are generally made from several mixtures and then go through a combustion process, brittle and durable. Artisans can use ceramics as a medium for coral reef conservation, considering the Covid-19 pandemic has an impact on the economic sector, especially MSMEs (Micro, Small, and Medium Enterprises) ceramic artisans. This paper was created as an initial study to develop ceramics as artificial coral reefs. In addition, it aims to empower MSMEs to help the government to conserve coral reefs.

2 Material and Method

Artificial reefs are a module to grow colonies of coral, can use local clay or earthenware. It can also be fired up to a maximum temperature of 1150oC from stoneware clay without a touch of glaze. Studies on coral recruitment frequently employ unglazed ceramic settlement tiles [5]. The surface of the artificial coral reef is not slippery by a layer of glazes so that coral animal seeds can live and thrive on this reef medium. It can be a growth medium as long as the surface is porous. This study uses a combination of materials commonly used in ceramics. The clay body mass used is code BL1. It was made from a mixture of quartz, kaolin, feldspar, and other materials. The processing of this clay used a ball mill with a production capacity of 350 kg. This research was divided into two steps. The first step was making an artificial reefs design, the second step was testing the mass of the ceramic body and making a model.

2.1 Design

This study used the semiotics method to create the design. Semiotics could be applied to various types of research, such as mass communication, visual communication, writing, and others. It has great potential in analyzing and interpreting data like text, music, photos, videos, and others [6]. Some literature and image were compiled and studied. They were analyzed by semiotics, and the outcome is design signs. It will be used to design the artificial coral reefs model.

a. Sturdy

   Heavy, solid, thick designs composed collectively are required to protect the substrate from ocean currents. Composed collectively means the placement of the substrate must be related to others, so the position is more solid and rigid. The substrate thickness is necessary to prevent the rate of erosion caused by some objects carried by ocean currents. It will be looked optimal when the wall thickness is 2-5 cm, and the weight is 5-10 kg. The join system between the substrates should be tight but not rigid because it can damage the substrate structure. The join system using ropes can be applied because it is more flexible and can respond to the movement of ocean currents well.

b. Hollow

   Apart from being a medium for growing reefs, the substrate is expected to be a hiding place for small fish from ocean currents and predators. The design's success is an interaction of organic beings with artificial models. The design was made by providing a small room and some circular holes.

c. Dynamic
Dynamic means a design composition is not monotonous does not move, and does not stagnate. The concept of geometric shapes could be applied by considering their interactions with the surrounding environment. Four or six sides' designs could be applied because they are easier to make a join system.

2.2 Testing & Modeling

2.2.1 Testing

Ceramic clay code BL1 was analyzed by testing fired shrinkage, water absorption, and apparent porosity. The steps to make the test piece to fulfill the analysis is by making a test piece measuring 10 x 2 x 1.5 cm. The test pieces were made of 15 pieces using the following tools: 1) clay, 2) molds of test bricks, 3) string strings, 4) alphabet stamps. Then they dried for four days at room temperature and fired using a Nabertherm electric furnace. The firing process takes 6 hours from a furnace temperature of 29°C to a temperature of 900oC. Fired shrinkage was measured to determine how many clay percentages would be used. It will help estimate the size of the model according to the design. After the sample had dried (controlled by weighing, the difference in weight was less than 0.5 g for two days), the distance of the line marks was determined with a micrometer precisely to 0.1 mm (p cm). The dry shrinkage (DS) is calculated based on the formula [8]:

$$DS = \left(\frac{10 - p}{10} \right) \times 100\% \quad (1)$$

The measured samples by the distance between the marking are p cm. They were burned in a furnace to the final temperature to determine the fired shrinkage. After the fired process was completed, it was cooled in the furnace. The distance of the line marks was determined with a micrometer to 0.1 mm (p1 cm) [8]. Calculation of fired shrinkage (FS) formulated as below:

$$FS = \left(\frac{p - p1}{p} \right) \times 100\% \quad (2)$$

Moreover, the total shrinkage (TS) formulated as below [8]:

$$TS = \left(\frac{9{,}9 - p1}{9{,}9} \right) \times 100\% \quad (3)$$

The steps for measuring water absorption are post-fired samples are dried in an oven at a 105–110◦C temperature, so the weight remains constant. Then they were cooled in a desiccator and determined the dry weight (D grams) with an accuracy of 0.1 g. Then soaked in water, take up and wipe with a dried cloth. The sample has weighed with a balance with an accuracy of 0.1 g (W gram) immediately [8], then the water infiltration was determined by the formula:

$$WA = \left(\frac{W - D}{W} \right) \times 100\% \quad (4)$$

Determination of the apparent porosity value was by weighing the fired weight after cooling (B) with an accuracy of 0.1 grams, the saturated weight measured when sinking at the bottom of a container with 1.000 ml of water (J), and the floating weight in water is symbolized (M), determined by the formula [8] :

$$PP = \left(\frac{J - B}{J - M} \right) \times 100\% \quad (5)$$

2.2.2 Modeling

Models or substrates were made by the hand-building technique using an electrical wheel. Clay was wedged to minimize air bubbles and make it easier to build. It could be added to some water slowly by hand. Then it dried at room temperature for 3-4 days and then fired in a gas furnace to 900°C temperature for 6 hours.

Fig.1. Wedging process (left), hand-building technique (center), and firing process (right).

3 Result & Discussion

3.1 Design Engineering

3.1.1 Using Ceramic

According to [7], showed that artificial coral reefs made of ceramics were non-toxic and pH neutral. Ceramics are made from several mixtures and then are brittle and durable through a fired process. It could be built in various designs, methods, and local clay materials.

3.1.2 Hexagonal Shape

Inspired by [7], the hexagonal design was used beginning from the design concept that applies geometric shapes with many sides, making it easier to arrange the formation under the sea. Aesthetically, the hexagonal shape can represent a design concept resulting from integration in the design of coral reef planting media with natural marine ecosystems and can produce new experiences and impressions to attract people's attention to pay more attention to the preservation of coral reefs.

3.1.3 Have a Coral Reef Seedling Development Space

This growth space is an area that can facilitate the need for growth space for coral reef seedlings. In addition, it has to be able to protect coral reef seedlings from the ocean currents during their growth period. The protection is applied by designing walls around the coral reef growth space.

3.1.4 Design Formation & Join System

Model or substrate placement under the sea has to be collective and mutually bound to each other. The purpose is to strengthen the position of the substrate and create new coral reef communities. The formation can be very varied and not monotonous with a hexagonal base shape. The hexagonal shape allows the composition of substrate composition to look like a honeycomb in the ocean. The arrangement between the substrates is to bring together the sides of each substrate. A rope could use as a joint system between substrates because it is more flexible and has longer durability in the sea. The recommended type of rope is a manila rope, plastic rope, or cable ties.

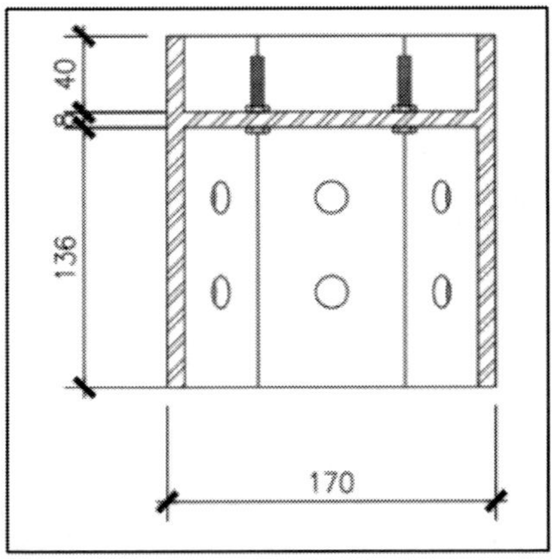

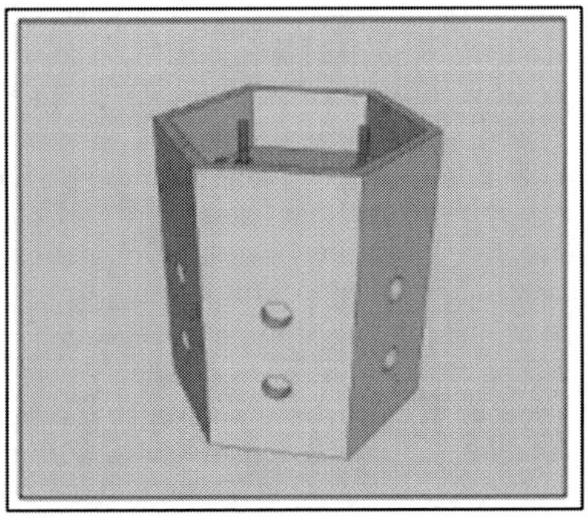

Fig. 2. Section view design (left) scale 1:5 and perspective view design (right).

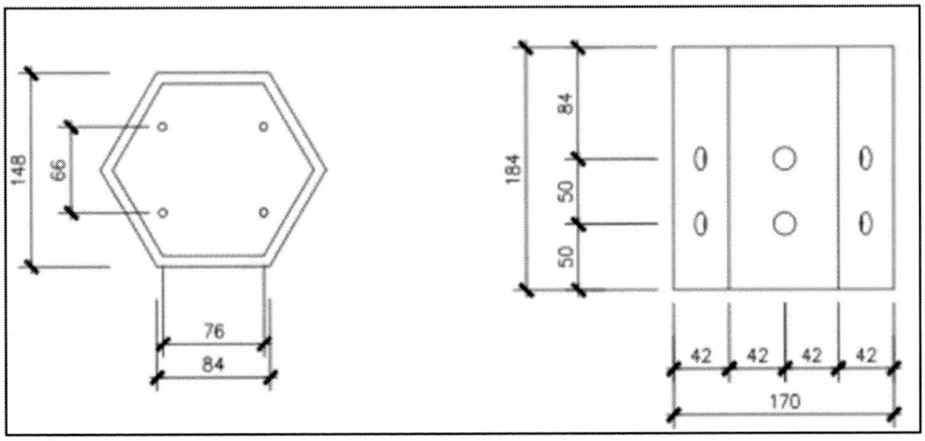

Fig. 3. Upper view design (left) and front view design (right) scale 1:5.

Fig. 4. Formation and join system design.

3.2 Ceramic Physical Properties

Ceramics' physical properties values were obtained by measuring the length and weight of the test pieces. The result was calculated using the equation. Based on the calculation results, the average value of dry shrinkage is 5.3%, fired shrinkage is 1.3%, and total shrinkage is 6.5%. Meanwhile, the average water absorption value is 19.7%, and the apparent porosity is 32.9%.

The shrinkage value would be compared with how much water absorption is after burning in ceramics. If the sintering temperature is higher, the shrinkage value will follow. It is due to the mass transport (diffusion) of atoms between particles that causes the formation of grains and the elimination of pores. The diffusion mechanism is the main cause of shrinkage during sintering, a volume with grain diffusion being the dominant mechanism in the sintering process of ceramic materials [9]. Shrinkage values could be used to estimate the use of materials when making a model or substrate. Water absorption shows how much fluid can enter through the ceramic pores. These pores can serve as a place for the deposition of $CaCO_3$ from the secretions of the coral reef itself and other marine organisms such as green algae or red algae. The persistence of coral reefs depends on $CaCO_3$ balance, deposition, and removal throughout the reef. In addition, other organisms such as calcareous green algae and red algae also contribute their hard parts to the coral reefs [10].

Fig. 5. Test pieces of ceramic, fired at 900°C temperature.

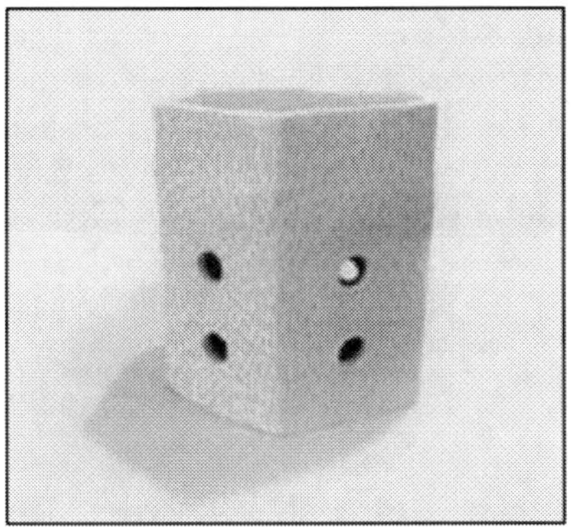 

Fig. 6. Front view of the model.

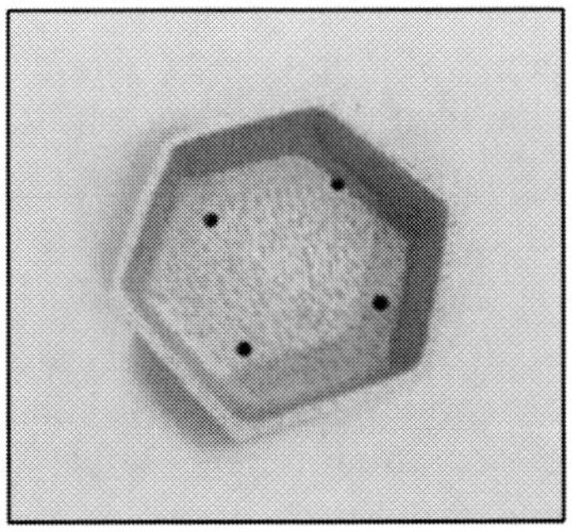

Fig. 7. Upper view of the model.

Table 1. Calculation of physical properties.

Sample Code	Lenght (cm)			Weight (gr)			Shrinkage (%)			Water Absorption (%)	Apparent Porosity (%)
	Wet	Dried	Firing	Firing	Saturated	Floating	Dried	Firing	Total		
1	9,9	9,3	9,2	38,2	45,0	22,6	6,1	1,1	7,1	17,8	30,4
2	9,9	9,4	9,2	37,5	44,8	22,3	5,1	2,1	7,1	19,5	32,4
3	9,9	9,4	9,3	37,7	45,2	22,8	5,1	1,1	6,1	19,9	33,5
4	9,9	9,3	9,2	35,9	42,1	21,1	6,1	1,1	7,1	17,3	29,5
5	9,9	9,4	9,2	37,3	44,1	22,0	5,1	2,1	7,1	18,2	30,8
6	9,9	9,3	9,2	37,3	45,0	22,6	6,1	1,1	7,1	20,6	34,4
7	9,9	9,4	9,3	37,5	45,0	22,8	5,1	1,1	6,1	20,0	33,8
8	9,9	9,4	9,3	37,0	44,2	22,2	5,1	1,1	6,1	19,5	32,7
9	9,9	9,4	9,3	36,6	43,9	22,0	5,1	1,1	6,1	19,9	33,3
10	9,9	9,4	9,2	36,5	43,9	21,6	5,1	2,1	7,1	20,3	33,2
11	9,9	9,4	9,2	36,4	43,8	21,8	5,1	2,1	7,1	20,3	33,6
12	9,9	9,4	9,3	37,7	45,7	22,7	5,1	1,1	6,1	21,2	34,8
13	9,9	9,4	9,3	37,3	44,9	21,9	5,1	1,1	6,1	20,4	33,0
14	9,9	9,4	9,3	36,9	44,7	22,3	5,1	1,1	6,1	21,1	34,8
15	9,9	9,4	9,3	36,9	44,3	22,1	5,1	1,1	6,1	20,1	33,3

4 Conclusions

The engineering of artificial coral reefs was made of ceramic. The steps to make it were making designs, measuring and testing the physical properties, and making the models. The hexagonal design was used to make making the formation easier and has aesthetic value. The value of the physical properties of ceramics can be used as a basis in this manufacture. Total shrinkage value can be used to estimate material requirements also the water absorption and apparent porosity as the parameters of CaCO3 circulation if this model is installed as an artificial coral reef on the seabed.

References

1. Presiden Republik Indonesia. UU No.32 Tahun 2014, Bab V, Pasal 1 dan Pasal 2, (2014)
2. U.S. Census Bureau, International Data Base (demographic data) and USA Trade Online (trade data), 1 July, (2021), (https://www.census.gov/popclock/world)
3. Hadi, T.A. et.al. The Status of Indonesian Coral Reefs. Research Center for Oceanography, p.1, Indonesian Institute of Sciences (2020)
4. Recfishwest. Artificial Reefs in Australia: A Guide to Developing Aquatic Habitat Enhancement Structures, P.2, Suite 3, 45 Northside Drive, Hillarys, Western Australia, (2014)
5. S.C.A. Ferse. Artificial reef structures and coral transplantation: fish community response and effect on coral recruitment in North Sulawesi/Indonesia. Dissertation to obtain the doctoral degree of natural sciences, p.17, Center for Tropical Marine Ecology, University of Bremen (2008)
6. B. Mudjiyanto, E. Nur. Semiotics In Research Method of Communication. Journal of Communication Research Informatics and Mass Media - PEKOMMAS, **Vol. 16**, No. 1 (2013)
7. Kalam A, Mieno T, Casareto BE. Development of Artificial Reefs Using Environmentally Safe Ceramic Material. J Ecosys Ecograph **8**: 253. (2018)
8. Effendi, M.D, et al. Identifikasi Mineral Lokal Batuan Gayo dan Pemanfaatannya sebagai Massa Raga Keramik Stoneware, *JST* (Journal of Sains & Technology), **Vol.1** No.7 (2018)
9. H. Magawe. Karakterisasi Hasil Sinter Reaction Bonded Alumina (RBAO) Dengan Penambahan Tepung Jagung. Thesis, Departement of Mechanical Engineering, University of Kristen Petra, Surabaya, Indonesia (2007)
10. Edwards, A.J. & Gomez, E.D. Konsep dan panduan restorasi terumbu: membuat pilihan bijak di antara ketidakpastian. Translate from Reef Restoration Concepts and Guidelines: making sensible management choices in the face of uncertainty. By: Yusri, S., Estradivari, N. S. Wijoyo, & Idris, iv + 38 pages Yayasan TERANGI, Jakarta (2008)

Performances of Debu and Kelabu Sentul Hens in the Different Rearing System at *Poultry Breeding Development Center* Jatiwangi Majalengka

Hanifah Fauziyyah Ihsani[1]*, *Rukmiasih*[2] and *Maya* Fitriati[3]

[1]College of Vocational Studies, IPB University (Bogor Agricultural University), Kumbang st 14, 16128, Bogor, Indonesia
[2]Faculty of Animal Science, IPB University (Bogor Agricultural University), Raya Dramaga st, 16680, Bogor, Indonesia
[3]BBPT Unggas Jatiwangi, Loji st km 35, Jatiwangi, Majalengka, Indonesia

Abstract. The pattern of raising domestic chickens in Indonesia is still maintained traditionally, used as a side business and savings. Besides that, Domestic chickens still have high broodiness, slow growth, and reproduction rates, so changes are needed in the rearing system. Observations and data collection were carried out at Poultry Breeding Development Center Jatiwangi. Debu and Kelabu Sentul hens have reached the age of 29 weeks in the deep litter system, and colony cages were observed for three months by observing egg production, FCR, and Depletion. Debu Sentul hens reared in the deep litter systems had lower egg production, higher FCR, and Depletion (28.34%, 9.34, 21.05%, respectively) than those reared in colony cages (36.08%, 7.04, 4.76%, respectively). Kelabu Sentul hens reared in the deep litter systems had lower egg production, higher FCR, and Depletion (24.64%, 12.01, 15.51%, respectively) than those reared in colony cages (41.64%, 6.06, 10.86%, respectively). In conclusion, the performance of Kelabu and Debu Sentul hens reared in colony cages was better than in deep litter system cages. The performance of Kelabu Sentul hens reared in colony cages was better than Debu Sentul hens.

1 Introduction

Domestic chicken is a type of Indonesian poultry that comes in various clumps, including Kedu chicken, Sentul chicken, Pelung chicken, Cemani chicken, Kokok Balenggek chicken, and others. Domestic chicken has a distinct flavor with chewy meat and low-fat content. Domestic chicken also boasts high selling value, making them popular among Indonesians. Furthermore, domestic chickens are highly adaptable. Based on [1], Sentul chicken shows the appearance and morphology as a dual-purpose type of local chicken.

The pattern of raising domestic chickens in Indonesia is still maintained traditionally, used as a side business and savings. It has an impact on the small number of populations that farmers keep. In addition, Domestic chickens still have high broodiness, slow growth,

* Corresponding author: hanifahfihsani@gmail.com

and reproduction rates. According to [2], almost all of Indonesia's native chicken breeds have low production and reproduction rates. They have a moderate growth rate and a low feed efficiency. So changes are needed in the rearing system.

Poultry Breeding Development Center Jatiwangi is one of the centers dedicated to preserving and developing West Java's native genetics, especially the Sentul chicken. Maintenance management, housing, feeding and drinking, reproductive management, health management, waste management, and production performance must be considered when producing Sentul chicks to create high-quality Sentul chickens [3].

This study aims to examine the performance of Sentul Debu and Kelabu chickens rearing in different cages.

2 Methods

2.1 Location

Observations and data collection were carried out at Poultry Breeding Development Center Jatiwangi. It is located in Majalengka Regency, West Java Province.

The Poultry Breeding Development Center Jatiwangi has a land area of 16.5 Ha owned by the West Java Provincial Government and has a right to use certificate no. 324 of 1989. The land is used for office purposes (17,802 m^2), Sentul chicken raising (24,750 m^2), and duck raising (25,974 m^2). Rice fields or fields and cliffs are examples of unused land. Poultry Breeding Development Center Jatiwangi has a hall, a service room, an official residence, a barracks, a hatchery, a prayer room, and a warehouse.

2.2 Data collection

The information gathered was based on both primary and secondary sources. Primary data was obtained through direct observation, while secondary data was obtained through interviews and agency data. The data collected is related to the report, specifically housing data, feeding and drinking, lighting, livestock reproduction, egg collection, livestock selection, livestock transfer, livestock health, maintenance performance, livestock waste, and production distribution.

2.3 Observed variable

Debu and Kelabu Sentul hens were reared in the deep litter system and colony cages. Debu and Kelabu Sentul hens that have reached the age of 29 weeks were observed for three months of rearing by observing Hen Day Production, FCR, and Depletion. Hen Day Production was measured using egg production and chicken population. FCR was calculated based on feed consumption and egg production weight. Depletion was the percentage of the population of chickens that died or were culled. According to [4], the formula for calculating HDP, FCR, and Depletion is as follows:

$$HDP = \frac{Total\, egg\, production}{Total\, population} \times 100\% \tag{1}$$

$$FCR = \frac{Feed\ consumption}{Total\ egg\ production\ \times average\ egg\ weight} \tag{2}$$

$$Depletion = \frac{Total\ population\ were\ died\ and\ culled}{Total\ population} \times 100\% \tag{3}$$

3 Result and discussion

3.1 Rearing Management

Poultry Breeding Development Center Jatiwangi aspires to be a reliable center and facilitator in the management and provision of seeds and the transfer of advanced local chicken and duck management technology in Indonesia, leading to an independent, dynamic, and prosperous West Java community. One of the goals of the Poultry Breeding Development Center Jatiwangi is to preserve domestic chicken and duck lines as germplasm in West Java. Sentul chicken is one of the leading domestic chicken lines in West Java. Existing Sentul chickens include Debu Sentul chickens and Kelabu Sentul chickens, raised in stages similar to local/breed chickens in general, namely the starter, grower, and layer period.

3.1.1 Housing

The housing system used at the Poultry Breeding Development Center Jatiwangi for rearing Debu and Kelabu Sentul Hens are open houses with a deep litter system, colony cages, and battery cages (fig. 1). The cage connects west and east. The shape of the roof used varies depending on the cage form. Colony cages have a monitor roof, whereas the deep litter systems and battery cages have a gable roof. The floor of the cage used in each cage is a litter system with husks at the cage's base.

The Poultry Breeding Development Center's cage density ranges from 3-6 birds/m^2, referring to [5] as many as six birds/m^2. The colony cage at the Poultry Breeding Development Center meets the minimum area requirement for chickens. Table 1 displays the cage's area and capacity. Based on Table 1 and the cage's reference density, the population of chickens kept for each available cage still needs to be increased.

Table 1. The area and capacity of the layer phase maintenance cage

Building	Type	Size (m^2)	Total cages (pcs)	Cages capacity (birds)	Housing capacity (birds)	Population (birds)
A1	colony cages	50	18	6	108	106
A2	deep litter system	20	-	-	120	66
A3	battery cages	40	126	1	126	81
B1	deep litter system	20	-	-	120	86
B2	colony cages	329	176	6	1056	578
B3	deep litter system	20	-	-	120	71
C1	colony cages	189	96	6	576	368
C2	colony cages	406	140	6	840	104
C3	deep litter system	20	-	-	120	102
D1	colony cages	189	96	6	576	358
D2	colony cages	392	212	6	1272	232
D3	deep litter system	20	-	-	120	119
E1	colony cages	189	21	6	126	60
E2	deep litter system	32	-	-	192	68
F1	deep litter system	20	-	-	120	63
F2	deep litter system	20	-	-	120	72

(a) (b) (c)

Fig. 1 (a) Deep litter systems (b) Colony cages (c) Battery cages

3.1.2 Feeding management

Debu and Kelabu hens reared at the Poultry Breeding Development Center Jatiwangi are fed daily using a restricted feeding system. The feed was given on a limited basis and provided commercial seed chicken feed. Every day, 100 grams of feed were given to each head. A hanging feeder with 3-4 pieces per flock is used in the deep litter system cage. In battery cages and colony cages, PVC feed bins are used.

Water was provided on an ad libitum basis for the treatment of Debu and Kelabu hens at the Poultry Breeding Development Center Jatiwangi. The water supply is groundwater that was previously stored in a water tower. There are numerous drinking water locations. The deep litter system cage has a gallon capacity of 7 liters that is changed twice a day and an automatic drinking water container that is cleaned once a week. Nipple drinkers are used in battery cages and colony cages.

The feed provided is obtained through an auction. There was a change in the feed given at the time of the research because the quality of the feed was being prepared for the next feed auction during the trial period. Efforts are made to prevent stress caused by feed changes, and adaptation is carried out by gradually mixing old feed with new feed. Table 2 compares the nutritional content of the old (feed A) and new (feed B) feeds.

Table 2. The content of nutrients needed and the nutritional content given to hens

Number	content of nutrients	Unit	Requirements	Feed A	Feed B
1	Water level (max)	%	14	13	13
2	Crude protein (min)	%	16	15-17	18
3	Crude fat (min)	%	3	4	3
4	Crude fiber (max)	%	8	6	7
5	Total ash (max)	%	14	13.5	14
6	Calcium (Ca)	%	2.75 - 4.25	3.5	3.25 - 4.25
7	Phospor(P) total	%	0.6 -1	0.7	0.6
8	Aflatoxin (max)	µg/kg	50	20	50
9	Metabolism Energy (ME) (min)	kcal/kg	2500	2700	2700

Source: BSN (2013)

The nutrition of the feed provided was following the standards established by [6].

3.1.3 *Lighting*

The layer phase of Debu and Kelabu Sentul Chicken at Poultry Development Center Jatiwangi receives natural lighting for 12 hours per day. The chickens in the colony cages were exposed to additional light for 5-7 hours during the night. From 9 p.m., additional lighting is provided. This treatment is given to the chickens to be active at night. In the cage, there are about 6-10 lights. The lamp is an 18-watt fluorescent lamp and a 28-watt bulb with 50 lx intensity.

3.1.4 *Management of reproduction*

Debu and Kelabu Sentul chickens in the layer phase at the Poultry Breeding Development Center Jatiwangi had a male to female ratio of 1:5 in deep litter systems and colony cages. Each cage type has a different mating system for Debu and Kelabu Sentul chickens in the layer phase at Poultry Breeding Development Center Jatiwangi. Chickens used natural mating in deep litter systems and colony cages. Chickens in battery cages are mated using artificial insemination. Artificial insemination was performed on Sentul chickens in one line and between lines. Artificial insemination of a single line is used to obtain replacement stock with pure blood from Debu Sentul or Kelabu Sentul chickens. Artificial insemination between strains is used to produce crossbred chickens from the selection of both Debu and Kelabu lines and to facilitate the evaluation of each seedling (seed value index). Artificial insemination occurs twice a week, on Tuesdays and Fridays. The semen used in Artificial Insemination is liquid semen collected and diluted before the procedure. The sperm was diluted in a 1:2 ratio with 0.9% NaCl.

3.1.5 *Collection and handling of eggs*

Eggs from Debu and Kelabu hens are collected once a day at the Poultry Breeding Development Center Jatiwangi. The collected eggs are assigned a flock number and recorded on the daily egg recording and egg delivery sheet. Eggs are collected and placed in the hatching room before being selected. The egg is chosen based on its weight and physical condition. Consumption eggs are eggs that do not meet the requirements for hatching. The eggs must be between 45-55 grams in weight, not cracked, have a smooth surface, and be of normal shape to hatch (ovoid).

3.1.6 Selection

Chicken selection occurs at the start of maintenance, specifically when the chicken will transfer to the cage. The poultry entering the pen is chosen based on qualitative and quantitative criteria. Bodyweight and age are two quantitative properties that are taken into account. The bodyweight of a hen ranges between 1300 and 1600 grams, while a rooster ranges between 2000 and 2600 grams. The chickens chosen were between the ages of 19 and 24 weeks. Table 3 and Figure 1 show the following qualities were considered when selecting males: coat color, gender, eye color, health, and comb shape.

Table 3. Qualitative properties of Debu and Kelabu Chickens

Criteria	Debu Sentul Chicken	Kelabu Sentul Chicken
Fur color	white dominant	grey dominant
Shank color	cream	cream/grey
Beak color	cream	grey
Eye color	orange/bright yellow	
Comb shape (rooster)	*single/pea comb*	

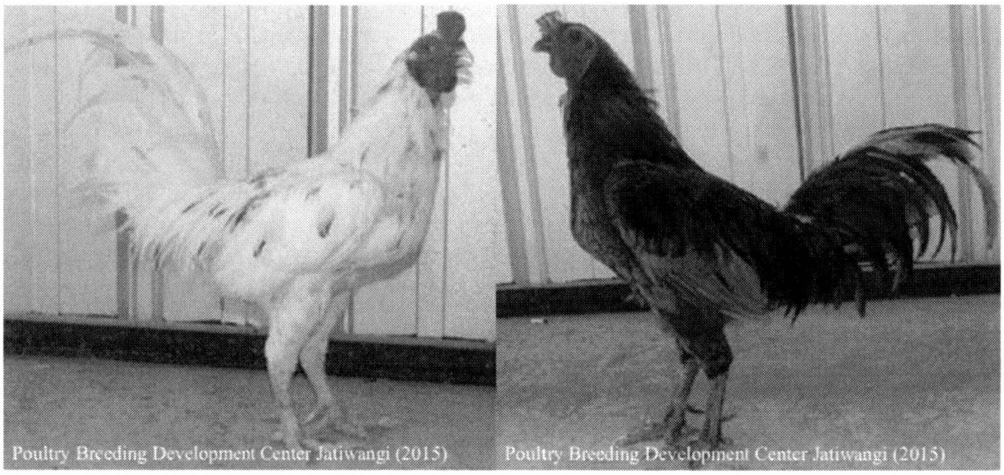

(a) (b)

Fig. 2 (a) Debu Sentul chicken (b) Kelabu Sentul chicken

3.1.7 Transfer

Debu and Kelabu Sentul Chickens are reared with a continuous system. Transfer activities were carried out in rearing Debu and Kelabu chickens at the Poultry Breeding Development Center Jatiwangi to fill empty cages with replacement stock. Livestock transfer was carried out during the transfer of the rearing phase from the grower to the layer. Livestock transfer can also be performed during the transfer of rejected chickens. Livestock transfers are performed 2-4 times per week. When the Hen Day Production reaches 5%, the rearing cage is transferred from the grower period to the layer period. Chickens aged more than 2-2.5 years, with a productivity of 10%, disabled or sick chickens, and having unhealthy physical characteristics are the criteria for transfer chickens to be culled.

3.1.8 Health management

Health management implementation at the Poultry Breeding Development Center Jatiwangi follows the same schedule every four months. Routine activities are carried out to take blood samples, vaccination, and administering deworming drugs. Blood samples were collected twice every four months before and after vaccination to test AI and ND antibody titers. During the layer period, Debu and Kelabu Sentul received the ND-AI vaccine. Vaccination is administered every four months. The vaccine used is an inactive vaccine injected as much as 0.5 cc into the chest muscle. The blood sampling results performed a month before and after vaccination are used to determine vaccine effectiveness. Deworming is administered two weeks before and two weeks after vaccination. Deworming is administered through feed at 2 grams per kilogram of feed.

Biosecurity were applied such as disinfection, sanitation, personal hygiene, and traffic control. Disinfection is accomplished by providing a dipping pool at both the institution's and the cage's entrances. Sanitation is accomplished by changing the litter and washing the equipment. The replacement of husks is carried out when the husks in the cage are wet and emit an unpleasant odor. The litter is usually changed every 2-4 weeks. The equipment to be washed is a drinking place with a washing frequency of twice a day for a gallon drinking water container and an automatic drinking water container every day. Personal hygiene of workers is prior before entering the cage. In addition, workers in a cage must wear masks and boots. To prevent the disease from getting out and spreading from the cage, the workers clean themselves by taking a shower. Traffic control is accomplished by limiting entry permits and regulating the cage's entry flow. Another activity is cutting grass around the cage to ensure team members' safety and minimize pests and predators.

3.1.9 Performance

Debu and Kelabu Sentul hens that have reached the age of 29 weeks were observed for three months of rearing by observing Hen Day Production, FCR, and Depletion. Hen Day Production was measured using egg production and chicken population. FCR was calculated based on feed consumption and egg production weight. Depletion was the percentage of the initial population of chickens that died or were rejected. Table 4 shows the performance of Debu and Kelabu Sentul Hens in the deep litter system.

Table 4. Performance of Debu and Kelabu Sentul Hens in deep litter system cages.

Type	Ages (Week)	Initial Total Population (birds)	Final Total Population (birds)	Total production (eggs)	HDP (%)	FCR	Depletion (%)
Debu Sentul hens	29-32	76	71	604	29,18	8,57	6,58
	33-36	71	62	414	20,07	12,46	12,68
	37-40	62	60	648	35,78	6,99	3,23
Kelabu Sentul hens	29-32	58	53	549	34,53	7,24	8,62
	33-36	53	51	211	13,10	19,09	3,77
	37-40	51	49	379	25,76	9,70	2,00

Table 4 shows the performance of Debu and Kelabu Hens reared in the deep litter system. Egg production (Hen Day Production) of Debu and Kelabu Sentul Hens that are the age of 33-36 weeks decreased from the previous age (first month of the study) due to feeding changes in the second month of rearing. The egg production of Kelabu Sentul Hens increased again in the third month of the study (age 37-40 weeks), but not as much as the egg production of Debu Sentul hens. Debu Sentul Hens, aged 37-40 weeks, produced the most eggs, followed by Kelabu Sentul Hens, aged 29-32 weeks. According to [3], egg production is still lower than 40% that of local chickens reared in an intensive system.

According to [3], the FCR produced by Debu and Kelabu Sentul Hens in deep litter system cages is higher than the FCR produced by local chickens reared with an intensive system of 4.9 - 6.4. The high FCR is caused by a change in feed, which causes the chickens to adapt to the new feed, resulting in a decrease in egg production. Furthermore, a high FCR rate can occur when a lot of feed is scattered around the area due to chickens scrambling to eat, and the nature of local chickens is still aggressive.

The depletion rate of Debu and Kelabu Hens in deep litter system cages was quite high. The high Depletion rate was caused by the large number of chickens that died after becoming infected by worms. According to [7], the primary prevention of worm infection is regular cage sanitation by cleaning chicken manure. The transmission mechanism is through worm eggs that come out with feces and are then eaten by other chickens. It is highly recommended that the cage be disinfected regularly.

Table 5 shows the performance of Debu and Kelabu Hens reared in colony cages.

Table 5. Performance of Debu and Kelabu Hens reared in colony cages

Type	Ages (Week)	Initial Total Population (birds)	Final Total Population (birds)	Total production (eggs)	HDP (%)	FCR	Depletion (%)
Debu Sentul hens	29-32	42	42	408	34,69	7,21	0,00
	33-36	42	40	402	31,14	8,03	4,76
	37-40	40	40	509	42,42	5,89	0,00
Kelabu Sentul hens	29-32	46	45	562	43,84	5,70	2,17
	33-36	45	43	606	45,19	5,53	4,44
	37-40	43	41	460	35,91	6,96	4,65

Table 5 shows that the third month's production of Kelabu Sentul hens was lower than the previous two months. It is due to a damaged cage, which causes uneven feeding. In general, egg production of Kelabu Sentul hens was higher than that of Debu Sentul hens when reared in a colony cage. Egg production increased by 41.61% when fed the same nutrient content. [8]

The FCR produced by Debu and Kelabu hens in colony cages was higher than the study's [8] result of 5.17. The high FCR is due to the lack of weighing the remaining feed, which means that the actual amount of feed consumed is unknown.

The depletion rate of Kelabu Sentul hens in colony cages was higher than Debu Sentul hens. The high rate of Depletion was due to death caused by cannibalism. Cannibalism is triggered by feather pecking behavior, which is common in colony rearing systems, according to [9]. This behavior can be caused by both internal (genetic) and external (environmental) factors (environment and feed). Some efforts that can be made to prevent the appearance of feather pecking behavior in a chicken farm include genetic improvement efforts, beak cutting, a better feed management approach, and a better cage and environmental management approach.

Table 6 shows the performance of Debu Sentul hens reared in deep litter systems and colony cages.

Table 6. Performance of Debu Sentul hens reared in deep litter systems and colony cages

Type	Ages (Week)	Initial Total Population (birds)	Final Total Population (birds)	Total production (eggs)	HDP (%)	FCR	Depletion (%)
Deep litter systems	29-32	76	71	604	29,18	8,57	6,58
	33-36	71	62	414	20,07	12,46	12,68
	37-40	62	60	648	35,78	6,99	3,23
Koloni cages	29-32	42	42	408	34,69	7,21	0,00
	33-36	42	40	402	31,14	8,03	4,76
	37-40	40	40	509	42,42	5,89	0,00

Table 6 shows that Debu Sentul hens' egg production in deep litter systems was lower than that of those reared in colony cages. According to [10], maintenance on battery cages is more effective and efficient than other maintenance systems. According to [11], Birds in the cages produced significantly higher rates and laid heavier eggs than those on deep litter systems. Based on [12], although productivity in alternative systems is frequently lower than in conventional, intensive layers, eggs from alternative systems have been shown in numerous studies to have better nutritional properties. Debu Sentul hens reared in colony cages had lower FCR and depletion rates than Debu Sentul hens reared in deep litter systems, implying that Debu Sentul hens reared in colony cages performed better than Debu Sentul hens reared in deep litter systems.

Table 7 shows the performance of Kelabu Sentul hens reared in deep litter systems and colony cages.

Table7. Performance of Kelabu Sentul hens reared in deep litter systems and colony cages.

Type	Ages (Week)	Initial Total Population (birds)	Final Total Population (birds)	Total production (eggs)	HDP (%)	FCR	Depletion (%)
Deep litter systems	29-32	58	53	549	34,53	7,24	8,62
	33-36	53	51	211	13,10	19,09	3,77
	37-40	51	49	379	25,76	9,7	2,00
Colony cages	29-32	46	45	562	43,84	5,7	2,17
	33-36	45	43	606	45,19	5,53	4,44
	37-40	43	41	460	35.91	6,96	4,65

Table 7 shows that Kelabu Sentul hens' egg production in deep litter systems was lower than that of those reared in colony cages. According to [10], maintenance on battery cages is more effective and efficient than other maintenance systems. According to [11], Birds in the cages produced significantly higher rates and laid heavier eggs than those on deep litter systems. Based on [12], even though productivity in alternative systems is frequently lower than in conventional, intensive layers, eggs from alternative systems have been shown to have better nutritional properties. Kelabu Sentul hens reared in colony cages had lower FCR and depletion rates than Kelabu Sentul hens in deep litter systems. Implying that Kelabu Sentul hens reared in colony cages performed better than Kelabu Sentul hens reared in deep litter systems.

3.2 Waste management

The waste generated by raising Debu and Kelabu chickens at the Poultry Breeding Development Center Jatiwangi is solid waste. Carrion and litter are the main types of solid waste produced. Every day, solid waste in the form of carrion is collected from each cage and burned in an incinerator, and if capacity is exceeded, the carrion is buried. At the time of cleaning the cage, solid waste in the form of litter is collected in sacks, and the litter sacks are then used as fertilizer for agriculture around the hall.

3.3 Production and Distribution

The Poultry Breeding Development Center has a production distribution program, which includes a grant of 50,000 Sentul chicken seeds in the form of DOC for farmer groups. This program's expected outcome is the ability to print brood hens in the area. The livestock group that received the DOC grant received 1,000 DOC, starter period feed, and a package of medicines and disinfectants.

As a result of production, the Poultry Breeding Development Center Jatiwangi provides chicken eggs, DOC, and rejected chickens. The price of chicken eggs sold at BPPT Unggas Jatiwangi is arranged in [13], with whole chicken eggs costing IDR 1000 and cracked chicken eggs costing IDR 500. Rejected chickens are priced at IDR 25000 for young rejected chickens and IDR 20.000 for adult rejected chickens.

4 Conclusion

They were rearing Debu and Kelabu hens at Poultry Breeding Development Center Jatiwangi intensively in open house cages. The population in the cage is still below the cage capacity. Debu Sentul hens reared in deep litter systems produced fewer eggs than those in colony cages (28.34% vs 36.08%). Kelabu Sentul hens reared in deep litter systems produced fewer eggs than those in colony cages (24.64% vs 41.64%). Kelabu Sentul hens reared in colony cages produced more eggs than Debu Sentul hens (41.64% vs 26.08%). Debu Sentul hens reared in deep litter systems had higher FCR and depletion rates (9.34 and 21.05%, respectively) than those in colony cages (7.04 and 4.76 %, respectively). Kelabu Sentul hens reared in deep litter systems had higher FCR and depletion rates (12.01 and 15.51%, respectively) than those reared in colony cages (6.06 and 10.86 %, respectively). The performance of Kelabu and Debu Sentul hens reared in colony cages was better than in deep litter system cages. The performance of Kelabu Sentul hens reared in colony cages was better than Debu Sentul hens.

References

1. Nataamijaya, A. G., *Karakteristik penampilan pola warna bulu, kulit, sisik kaki, dan paruh ayam Pelung di Garut dan ayam Sentul di Ciamis* (2005)
2. Henuk, Y. L., & Bakti, D, *Husbandry systems of keeping native chickens in tropical areas of Indonesia*, Tropical Animal Science and Production (TASP, 2016)
3. C. Hidayat, S. Asmarasari, *Native Chicken Production in Indonesia: A Review*, Jurnal Peternakan Indonesia (Indonesian Journal of Animal Science), **17,** 1 (2015)
4. Bell DD, Weaver WD, *Commercial Chicken Meat and Egg Production. Springer Science+Business Media. Inc.* (2002)
5. Kementan, Keputusan Menteri Pertanian Nomor 31/Permentan/OT.140/2/2014 tentang Pedoman Budidaya Ayam Pedaging dan Petelur yang Baik (2014)
6. BSN, Pakan ayam buras : bagian 3- Layer SNI 7783.3:2013 (2013)
7. S. Iskandar, *Panduan Teknis Usahatani Ayam Kampung* (2017)

8. T Widjastuti, R. Wiradimadja and S. Darana, *Energy and protein requirement based on the efficiency of protein utilization of Sentul chickens and manifested in the production performance*. Sci. Pap.–Ani. Sci. Ser: Seria Zootehnie, **63**, 24-28 (2015)
9. C. Hidayat, *Upaya Menekan Perilaku Mematuk Bulu yang Memicu Perilaku Kanibalisme pada Peternakan Ayam*, Jurnal Peternakan Indonesia (Indonesian Journal of Animal Science), **21**, 163-175 (2019)
10. BPTP Sumatera Utara, *Teknologi Budidaya Ayam Buras dan Itik* (2015)
11. J.A. Oluyemi, Yomi O. Roberts, *The Cage Versus the Deep Litter System for the Management of Layers in the Humid Tropics*, Poultry Science, **54**, 1982-1989 (1975)
12. RAKONJAC, S., BOGOSAVLJEVIĆ-BOŠKOVIĆ, S., PAVLOVSKI, Z., ŠKRBIĆ, Z., DOSKOVIĆ, V., PETROVIĆ, M., & PETRIČEVIĆ, V., *Laying hen rearing systems: A review of major production results and egg quality traits.* World's Poultry Science Journal, ***70***(1), 93-104 (2014)
13. Peraturan Daerah Provinsi Jawa Barat, Perda Jabar no. 06 tahun 2018 tentang Perubahan Kedua atas Peraturan Daerah Provinsi Jawa Barat nomor 14 tahun 2011 tentang Retribusi Daerah (2018)

The Impact of Brand Awareness, Brand Association, and Perceived Quality towards Brand Loyalty (A case study of New Product)

Silvia Dewi Sagita Andik[1*] *Annisa fitri* Rachma[1]

[1]College of Vocational Studies, IPB University

Abstract. This study examines the effects of brand awareness, brand association, and perceived quality towards brand loyalty of new development products. A total of 30 respondents were given an online questionnaire related to food brand equity (A case study new development product: Yobrow). Linear regression and multiple regression tests were used to test the data and infer the results. Data were analyzed using SPSS software. The results show that perceived quality leads to brand loyalty, brand awareness, brand association and simultaneously impacts brand loyalty. In the context of perceived quality, the consumer will evaluate the product's perceived quality. Perceive quality can give subjective judgment to a consumer's mind. The good quality of the product will increase brand loyalty product.

1 Introduction

Brand loyalty has an essential role in strategic marketing [10]. Brand loyalty also has strategic benefits to the marketer, such as increasing the market share, expanding the new consumers, supporting the brand extension, strengthening the brand to the competitive threats [4]. The marketer needs to focus on brand loyalty, in particular, to build a new brand. Several variables impact brand loyalty, such as brand awareness, association, and perceived quality. [2] reveal that, to increase customer loyalty, the marketer should build customer confidence. It is not important to enhance customer confidence to ensure consumers are aware of that brand. In addition, the brand association has a role in increasing brand loyalty [12], perceived quality as a determinant factor in brand loyalty.

A brand is important for consumer markets. Consumers are more aware and trust the brand that links to loyalty. For a new brand, it is important to build brand loyalty in terms of the food brand. In addition, Food product issues related to food safety, hygiene, and safety product delivery. The high risk of the food product is influenced by several factors, thus, this study focuses on maintaining brand loyalty for the new brand. Yobrow is one of the new food brands that sell healthy brownies. Most previous research examines brand loyalty for a popular brand, but only a few studies examine brand loyalty for a new brand. Meanwhile, many new brands need to be maintained due to new brands in the first step of the life cycle product. This study examines the impact of brand awareness, brand association, and perceived quality on brand loyalty, focusing on new brands.

2 Methods

2.1 Data Collection and Procedure

The study was held from June - to July 2021. The total sample in this research is 30 respondents. The majority of respondents were in Jakarta city. Jakarta city is the capital city of Indonesia. This location is determined based on several criteria, i.g (1) Jakarta is a metropolitan city with a large population (2) Jakarta city is central to food trade. Sampling respondents were selected based on these specific criteria: (1) the customer of Yobrow brand (2) the consumer who has purchased healthy chocolate products from Yobrow. Determination of total sample regarding the theory from Roscoe (1975). Sample size from 30 to 500 samples. The technique used for sampling in this study was non-probability sampling. Convenience sampling was chosen because of the respondents' ease of access and willingness to be interviewed.

2.3 Measure and Analysis

The variables in this study included independent variables, i.e., brand awareness, brand association, perceived quality, and brand loyalty as the dependent variable. All of the variables are dimensions of brand equity. A Likert scale of 1 - 5 was used as the measurement scale in the questionnaire, and the data analysis applied descriptive and Regression analysis using SPSS software. Descriptive analysis was used to calculate respondent characteristics, while regression analysis examined the causality among variables.

2.4 Hypotheses Development

Brand loyalty is the level of consumer interest in a product brand. Consumers who have strong loyalty will repurchase products. While many brands appear in the market that offers superior products, consumers prefer to choose with a good experience. Brand loyalty provides added value to the company [3]. Some factors increase brand loyalty. The factors belong to the dimension of brand equity. Dimensions of brand equity consist of brand awareness, brand association, perceived quality, and brand loyalty (brand loyalty) [11].

Brand awareness influences brand loyalty. High levels of brand awareness will lead to higher leads on brand loyalty [12]. [1] also reveal that brand awareness has a significant and positive relationship with brand loyalty. The definition of brand awareness is the consumer's ability to recognize the specific brand in their minds [9]. In addition, the brand association also has a role in increasing brand loyalty [12]. In line with research conducted by [5], there was a significant relationship between brand association and brand loyalty in private clubs. The definition of brand association is related to brand attributes, the consumer target, the consumer need, which is the foundation of brand loyalty and consumers' purchasing decisions [14]. Perceived quality relates to consumer brand engagement, is has an essential role in a consumer's perception of the reliability and dependability of a product or service, and is attached to customers' preference, consumer satisfaction, and purchase intention [13]. Perceived quality also determines brand loyalty. The nature of brand loyalty starts at the first stage of product purchasing, which is often made regarding that particular product's perceived quality. Based on the literature review, this study develops three hypotheses, therefore,

H1: Brand awareness has impacts on brand loyalty
H2: Brand association has impacts on brand loyalty
H3: Perceived quality has impacts on brand loyalty

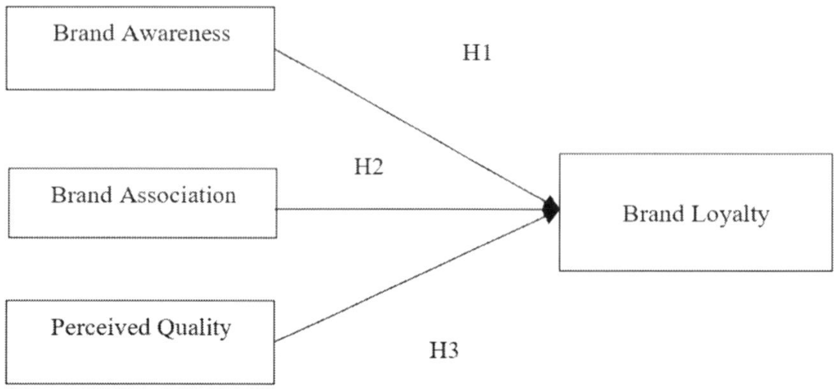

Fig. 1. Hypotheses framework

3 Results and discussion

3. 1 Respondent Characteristics

The characteristics of the respondents consist of gender, age, education, income, and family members. Table 1 below presents the demographic characteristics of the 30 respondents.

Table 1. Respondent characteristics

Characteristics	Total	Percentage
Gender		
Female	10	33
Male	20	67
Total	**30**	**100**
Age		
< 20 years old	0	0
21 – 25 years old	7	23
26 – 30 years old	7	23
31 – 35 years old	5	17
> 35 years old	11	37
Total	**30**	**100**
Level of Education		
High School	8	27
Diploma/Bachelor's Degree	20	67
Undergraduate	2	6
Total	**30**	**100**
Income (IDR)		
< 500.001	2	7
500.001 - 1.000.000	2	7
1.000.001 - 5.000.000	11	36
5.000.001 - 10.000.000	6	20
> 10.000.000	9	30
Total	**30**	**100**

Based on Table 1, most respondents were male (67%) and female (33%). The majority are adults aged above 35 (37%). Most of their education is graduated with a diploma or a bachelor's (67%). Based on revenue level, most respondents are on a monthly income of Rp. 1,000,001 – Rp. 5,000,000 (36%). This finding shows that the majority of respondents have adequate income. Yobrow's customers are dominated by adult people who have adequate income. Most adult people are more aware of their healthy life.

3.2 Path Analysis

Path analysis is used to measure a causality relationship among predetermined variables based on a theoretical. To examine whether there is an impact of intervening was used in path coefficient comparison. The path coefficients were calculated to obtain the regression model of relationship hypothesis variables.

Table 2. Model R Square

R	R Square	Adjusted R Square	Std. Error of the Estimate
.699	.489	.430	.5758

a. Predictor: (Constant), Perceived Quality, Brand Awareness, Brand Association

Regarding Table 2, the value of R square is 0.489. It means that brand awareness, association, and perceived quality simultaneously influence brand loyalty by 48,9%. If the value of R square is close to the 1 (range is 0 – 1), it means that the effect is bigger.

3.3 The Impact of brand awareness, brand association, and perceived quality on brand loyalty

There are dependent and independents variable in this study where the independent variables are brand awareness, brand association, and perceived quality. The dependent variable is brand loyalty. The regression analysis is used to examine the impact of independent variables on the dependent variable. The results showed in Table 3.

Table 3. The Path Analysis

Model	Coefficients unstandardized		Standardized Coefficients	T	Sig.	Hypotheses
	B	Std. Error	Beta			
(Constant)	-1.034	1.114		-.928	0.362	
Brand Awareness	0.440	0.289	0.257	1.521	0.140	H1 Not Accepted
Brand Association	0.117	0.295	0.093	0.395	0.696	H2 Not Accepted
Perceived Quality	0.606	0.294	0.460	2.063	**0.040***	**H3 Accepted**

Based on Table 3, the results show that score = 0.460, which means the correlation between the variables on brand loyalty and perceived quality has a positive significance level with a significant count smaller than the standard value of $0.04 < 0.05$. Thus, the third hypothesis is accepted. This finding stated that perceived quality has an impact on brand loyalty. This finding shows that the good quality of the product increases the customer's brand loyalty. In the context of perceived quality, the consumer will evaluate the product's perceived quality. Perceived quality is a vital factor in determining brand loyalty. In particular, it is important to develop higher quality food for healthy food. Food products with good quality become a critical point for gaining a competitive advantage in the food and beverage industry [6]. In the particular new product, it is important to ensure that product has good quality to attract consumer trust. Yobrow brand is a new start-up business in the food industry. Thus, it is important to ensure the perceived quality from their customer. There are four indicators in the perceived quality aspect, and perceived quality refers to the consumer's assessment of the product's superiority. This study also indicates that Yobrow brand has added value, which is the product provides benefits for consumer health. However, Yobrow brand is a new product that still needs more promotion to build consumer trust.

Regarding the results, brands awareness and brand association do not impact brand loyalty. The score of brand awareness = 0.257 and brand association = 0.093, a significant level is bigger than the standard value ($p\text{-value} > 0.05$). this finding does not support the previous research that stated brand awareness and brand association have a big role in increasing consumer loyalty to a particular brand. In addition, Yobrow brand is a new brand, so most consumers did not trust the Yobrow brand. One of the strategies to increase the attention towards healthy food is food promotion [7]. Promotion increases consumer awareness towards the Yobrow brand. As we know, there are many kinds of promotions, such as through social media (Instagram, Whatsapp, Tiktok, Youtube, Facebook, etc.). On the other side, conventional promotion also impacts increasing brand awareness. [8] also stated that conventional media has a positive impact and effectiveness in attracting consumers. To determine the kinds of promotions, the marketer should adjust with the market target.

4 Conclusion

This study concludes that perceived quality is essential in increasing brand loyalty, primarily for the new brand. Consumers will trust a brand that has good quality. Thus, the marketer needs to focus on the quality of the product. In particular food products, marketers offer a good taste of brownies "yobrow". In addition, it is important to give the added value of the product. The marketer is also concerned with promotion to increase consumer awareness.

There are several limitations to this study. First, this study used a small sample size, and second, this study used non-probability sampling. For future study, it should be to enlarge the sample size to represent the population. The technique method is better-used probability sampling to obtain the results more accurate. In addition, the future study can apply the model causality, such as SEM (structural equation model).

References

1. Alkhawaldeh, Abdelbaset and Al-Salaymeh, Marwan and Alshare, Fathi and Eneizan, B. Mohammad. European Journal of Business and Management, **9**, 36, (2017)
2. A. Marliawati, D. Cahyaningdyah. *Manag Analys J.* **9,** 2 (2020)
3. AF. Hanafi. Pengaruh brand equity mobil Toyota Yaris terhadap kepuasan konsumen (studi pada pengguna mobil Toyota Yaris di Yogyakarta) [skripsi]. Yogyakarta: Universitas Muhammadiyah Yogyakarta (2016)
4. E. Atilgan, S.Aksoy, S.M.Akinci. Marketing Intelligence and Planning, **23**, 3, pp. 237-248 (2005)
5. E.M.Azadi. Biomed Pharm. J, **8**, 2, pp. 1-8 (2015)
6. F.A. Konuk. J. Retailing and Cons. Servic. 50 (2019)
7. F. Folkvords, Hermans, R.C.J. Food Addiction. **7** (2020)
8. I. N. Wirata , Satiadji A.R, Citrawati L.P,Arjaya K, Sutarini I.A., Asmarani I.R. Proceedings of the 1st Padjajaran Communication Conference Series 2019 (2019)

9. J. Sasmita, N. M. Suki. International Journal of Retail & Distribution Management, **43**, 3, 276-292. (2015)
10. K. Memon, S. Khan, S. IKSP J of Bus and Economics, **2**, 1, 18-27 (2021)
11. N. Khan, S. Hamed, R. Rahmani, Y.H. Hong, B.C Tan. Int. J. Bus. Manag. **10**, 1, 172–181 (2015)
12. R. Chinamona, E.F.Mazariri. *J of Bus and Retail Management Research (JBRMR)*, **12**, 1 (2019)
13. S.R. Nikhashemi , N. Valaei, A.K Tarofder. Glob. Bus. Rev. **18**, 3, 108–127 (2017)
14. T.W Len, M. Cindy, M.M. Lynn. J of Mark Manag, **3**, 1, 137-155 (2007)

Financial Feasibility Analysis of Product Modification Katuk and Spinach Brownies Tartlet as an Alternative Breastfeeding Mother's Snack

RA.Hangesti Emi Widyasari[1*], *Mela* Nurdialy[2], and *Jihan* Fadhilah[3]

[1] Food Service and Nutrition Industry Management, College of Vocational Studies – IPB University, Kampus IPB Cilibende, Bogor 16151, Indonesia
[2]Accounting, College of Vocational Studies – IPB University, Kampus IPB Cilibende, Bogor 16151, Indonesia
[3]Student of Food Service and Nutrition Industry Management, College of Vocational Studies-IPB University, Bogor 16151, Indonesia

Abstract. Katuk and spinach brownies tartlet products from katuk leaves as a source of galactagogue and spinach, rich in iron, can help breastfeeding mothers in exclusive breastfeeding for six months. This study aims to modify the katuk and spinach brownies tartlet products and analyze these products' financial feasibility. This research makes brownies tartlet. This tartlet uses katuk leaves and spinach. This study indicates that the energy content in one product is 288 cal, carbohydrate 30.2 g, Protein 5.2 g, fat 17.9 g, and iron 0.8 mg. Financial feasibility analysis results showed that BEP of production of 37.820 pieces, BEP of sales of IDR 189.102.195; Return on Sales of 23.4%; Return on Investment (ROI) of 189.5%; NPV of IDR 91.831.077 and PBP during 0.55 year, Discounted Payback Period (DPBP) of 0.60 years, Profitability Index of 2.38 and Accounting Rates of Returns (ARR) of 52.6%. Based on scenario analysis, katuk and spinach brownies tartlet is feasible for business if minimum sales of 72% of products. Based on the financial feasibility analysis comparison data above, katuk and spinach brownies tartlets are more profitable than pineapple carrot juice business; chocolate spread with coconut ingredients business, and banana chips

1 Background

A breastfeeding mother is one of the groups in the life cycle that need attention. Breastfeeding is a process of giving breast milk which is the best source of nutrients with a bioactive composition that can improve the health status of mother and child. According to [1], 80% of a child's brain development starts in the womb until age 3, known as the golden period. Therefore, at this time, it is necessary to give exclusive breast milk for six months that can protect the infant against invasive infections [2]. One of the global nutrition targets of Sustainable Development Goals is to promote the rate of mothers performing exclusive breastfeeding within six months up to at least 50% in 2025 [3]. WHO and UNICEF emphasize keeping exclusive breast milk feeding in the first six months of life [4].

Based on the research results conducted in the United States, babies who get exclusive breast milk six months lower have ISPA (respiratory tract infection) by 72% and have diabetes by 30%. In addition, breast milk can reduce the risk of SIDS (sudden infant death syndrome) by 36% [5]. According to data monitoring nutritional status in Indonesia in 2017,

* Corresponding author: hangesti@apps.ipb.ac.id

the scope of exclusive breastfeeding for the first six months by a mother to her baby is still low. 65% of babies are not exclusively breastfed for six months at birth. This figure is quite far from the target of exclusive breast milk coverage in 2019 set by WHO or the Ministry of Health, 80% [6]. One of the most common factors associated with the failure of exclusive breastfeeding practices is that the breast milk factor has not come out in the first week after birth [7].

The use of galactagogue is one way to increase the secretion and production of breast milk. Developing local food crops that function as galactagogue can be one strategy to overcome the failure of exclusive breastfeeding due to the low secretion and production of breast milk.

Katuk and spinach brownies tartlet is a modified product development of brownies tartlet. In general serves as a sweet snack (dessert) that is currently popular as an alternative snack for breastfeeding mothers—fortifying local foodstuffs in this product by adding katuk leaves and spinach. Katuk leaves *(Sauropus androgynous*) have been known in traditional medicine in South and Southeast Asia as a breast milk-enhancing drug because it contains galactagogue compounds to help stimulate, maintain or increase breast milk production. In addition, Spinach *(Amaranthus Viridis)* is a superfood that contains many nutrients, so it to refer to as the king of vegetables.

This research aims to modify katuk and spinach brownies tartlet products as an alternative snack for breastfeeding mothers. While the main objective is to calculate the energy and nutritional content of the product and analyze the business and production cost of katuk and spinach tartlet brownies.

2 Literature Review

2.1 Tartlet

Tart derived from French can be translated as pie or tart or any baked dish filled with something sweet or savory such as custard or pastry cream and then decorated on the top (topping). A tart is a form of rigid pastry, thick filling that does not overflow [8]. The main ingredients of tart from wheat flour and eggs. Tartlet refers to a mini tart that has a thickness of less than 1 inch (2.5 cm) made according to individual portions. In general, this food is a dessert. There are many types of tarts, with popular varieties including Tart Treacle, Tart Meringue, Tart Tatin, and Tart Bakewell.

2.2 Brownies

Brownies are one of the most popular cake-based wheat flour, a chocolate cake with attractive colors, a delicious aroma, and a less-fluffy texture [9]. The recipes of brownies first appeared In the book Home Cookery called Service Club Book in 1904. Brownies are initially chocolate cakes that fail because the texture does not expand and is slightly wet on the inside, unlike cakes in general that are soft and porous. But in reality, brownies turned out to be well received by consumers.

2.3 Katuk leaf *(Sauropus androgynus)*

One of the traditional medicinal plants in Indonesia that facilitates breastfeeding is katuk. *Sauropus androgynus* is known as cekur manis (Malaysia), Pak-Wanban (Thailand), rou ngot (Vietnam), or a multivitamin plant (India). Katuk leaf contains galactagogue that can stimulate the mother's breast milk gland by increasing the mammary's secretion cells [10]. A mother commonly consumes galactagogue after birth. A regular or delayed production occurs in this stage since progesterone decreases sharply after placenta removal as prolactin, cortisol, and insulin levels increase [11]. Breast milk production increases because katuk leaves contain alkaloids and sterol [12]. Based on the table of List Food Composition, every 100 g of katuk leaves have of 59 cal energy, protein 4.8 g, fat 1 g, carbohydrate 11 g, calcium 204 mg, phosphor 83 mg, iron 2.7 mg, vitamin A 103,705 SI, vitamin D 0.1 mg, vitamin C 239 mg, and water 81 g. Recent studies indicated that katuk, a traditional medicinal plant to facilitate breastfeeding, proved beneficial as an antioxidant. The leaves are an antioxidant because they contain vitamin C. Vitamin C in katuk leaves is relatively high, approximately 85.65% 92.43 to 92.18 mg / 100g [13]. They were consuming katuk leaf extract equivalent to 2 g of simplistic in capsule form based on research. A dose of 2x1 for 15 days can increase breast milk production in breastfeeding mothers [14].

2.4 Spinach *(Amaranthus Viridis)*

Spinach belongs to the Chenopodiaceae family (also known as goosefoot), a family of nutritional powerhouses including beets, chard, and quinoa. People have used spinach in various cultures throughout history, notably in the Mediterranean, Middle-Eastern, and South-East-Asian cuisines. It is cheap and easy to prepare [15]. Spinach is a type of leaf vegetable that is much in demand by the community. According to data obtained at ADS (Agribusiness Development Station), there is a trend of spinach demand that has increased every year since 2013. Based on World Healthiest's Food Ratting results, spinach is a green vegetable rich in various nutrients, especially iron (Fe), which is good to consume for breastfeeding mothers. The recommended iron adequacy rate for breastfeeding mothers per day is 18 mg [16]. Based on the table of List Food Composition that in every 100 g, raw spinach contains 36 cal energy, protein 3.5 g, fat 0.5 g, carbohydrate 6.5 g, calcium 267 mg, phosphor 67 mg, iron 3.9 mg, vitamin A 6090 SI, vitamin B1 0.08 mg, vitamin C 80 mg, and water 80.9 g.

2.5 Breastfeeding Mother

During breastfeeding, a mother needs additional energy to improve the quality and quantity of breast milk. The calculation of energy need of breastfeeding mother using the Harrist Benedict method results was obtained plus 500 kcal for the first six months and 550 kcal for the second 6 months. Based on [16], the average energy need of adult women is 2000 cal, so the energy need of a breastfeeding mother is about 2500-2550 cal per day. The macronutrients of breastfeeding mothers are carbohydrates 60-70% of the energy, protein 12-15%, fat 15-25%, and enough vitamins and minerals, especially Fe. The secretion of breast milk averages 800-850 ml per day and contains 60-65 kcal, protein 1-1.2 g, and fat 2.5-3.5 g every 100 ml. These substances are from the mother's body, and a daily food supply can replace them [17]. According to WHO and FAO, the recommended iron for breastfeeding mothers is 10-30 mg/day [18]. Exclusive breastfeeding for up to 6 months without iron supplementation may compromise the hematologic status of children [19]. Snacking for breastfeeding mothers can improve the quality and quantity of breast milk [20]. The snack needed by breastfeeding mothers takes 15%-20% of the energy required a day or about 375 cal-500 cal. If the mother lacks the mother's energy and nutritional need, apart from the quality of breast milk and maternal health is disturbed, it will affect the period for the mother to produce breast milk.

2.6 Financial Feasibility

According to Investopedia [21], financial analysis is a process to evaluate businesses, projects, budgets, and other finance-related transactions to determine their performance and suitability. Financial feasibility assessment using capital budgeting analyses changes with long-term effects that vary over time [22]. There are several techniques available to analyze to evaluate the feasibility of a business by considering the time value of money, risk, and rate of return on investment. Return on Investment (ROI) is one of the most popular performance measurement and evaluation metrics used in business analysis. The project's feasibility depends on analyzing the cash flows generated by the project and its costs. The analyst can use capital budgeting methods such as Payback Period, Net Present Values, Internal Rates of Returns, Accounting Rates of Returns, and Profitability Index to evaluate the financial feasibility [23]. The analytical techniques to determine and predict risk and influence are sensitivity analysis and scenario analysis [24]. Scenario analysis is a behavioral technique that uses multiple potential alternate results to assess the variability of the measured return by NPV when asking what-if questions [25], [26]. The commonly used scenario approach to estimate NPV is along with pessimistic (worst), most likely (base), and optimistic (best) cash inflow estimates [25], [26], [27].

3 Research Method

3.1 Time and Place Observation

The research was conducted on June 8, 2021, aiming to work from home during the pandemic in Subang West Java.

3.2 Type and How to Retrieval Data

Type and how to retrieve data in table 1 as below:

Table 1 Type and how to retrieve data

No	Data	Type of Data	Method of Retrieval Data
1.	Literature	Secondary	Via the internet
2.	Production process	Primary	Active participation by preparing the need for tools and ingredients
3.	Energy content and nutrients	Secondary	Via table of list food composition
4.	Production cost	Primary	Analysis directly

3.3 Need of Tools

The need for tools for users to make katuk and spinach brownies tartlet is in table 2 as below:

Table 2 Need of tools Source of primary data 2021

Name of Tool	Specification	Amount
Preparation Tool		
Big bowl	Porcelain	2
Spatula	Plastic	1
Small sieve	Stainless	1
Side plate	Porcelain	3
Tartlet mold	Stainless	15
Paintbrush	Wood	1
Scale	Plastic	1
Fork	Stainless	1
Knife	Stainless	1
Processing Tool		
Mixer	Stainless	1
Oven	Stainless	1
Flat baking sheet	Stainless	1
Balon whisk	Stainless	1
Presentation Tool		
Cake cup	Paper	15
Small mica	Plastic	15

3.4 Need of Ingredients

Need of Ingredients for used to make katuk and spinach brownies tartlet is in table 3 as below:

Table 3 Need of ingredients Source of primary data 2021

Ingredients	Unit Price (Rp)	Size	Total Price (Rp)
Spinach	3,000/tie	35 g	1,500
Katuk leaves	2,500/tie	25 g	625
Low protein wheat flour	8,000/kg	200 g	1,600
Medium protein wheat flour	10,000/kg	100 g	1,000
Yolk	2,000/ egg	2 eggs	4,000

Egg	2,000/ egg	1 egg	2,000
Sugar	13,000/kg	130 g	1,690
Margarine	5,500/sachet 250 g	150 g	3,300
Milk powder	2,500/sachet 27 g	27 g	2,500
White chocolate	12,500/stem 250 g	150 g	7,500
Cooking oil	13,000/kg	40 g	520
Almond	12,500/1/4ons	15 g	3,750
Total Price			29,985

3.5 The Flow of Production Process Plan

Based on figure 1, the flow of production process plan katuk and spinach brownies tartlet starting from the procurement of foodstuffs to serving. The total time needed in the production process flow is 2 hours over 25 minutes.

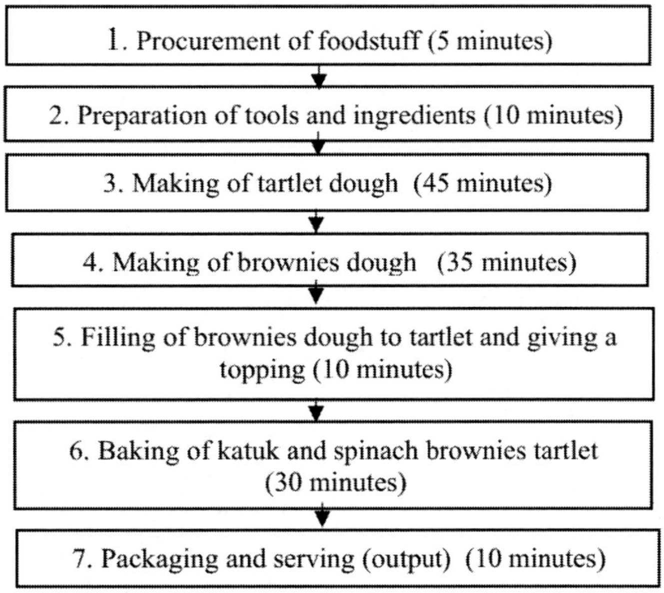

Fig. 1. The flow of the production process plan

3.6 Financial Analysis

This research was quantitative descriptive. The data was collected in the form of primary data and secondary data. Primary data was obtained by observation and interviews. At the same time, secondary data was obtained from various kinds of literature. The data analysis method used was an analysis of financial feasibility. Financial feasibility is an analytical tool to evaluate an investment's economic viability. Financial feasibility is a tool for examining the financial decision has expected return and risk [28]. This study will assess the financial feasibility of the Katuk and spinach brownies tartlet business based on the investment criteria of Break-Even Point (BEP), Return on Investment ROI), Net Present Value (NPV), Internal Rate of Return (IRR), Payback Period (PBP), Discounted Payback Period (DPBP), Profitability Index and Accounting Rates of Returns (ARR). The results of the financial feasibility analysis will provide information on whether this business is feasible or not to be undertaken. The data analysis divide into four steps.

The first step is to construct pro forma financial statements, consisting of each account assumption determined and transformed into pro forma financial statements. To calculate depreciation by using the straight-line method. The second step is to calculate Break-Even Point (BEP). The third step is to conduct the feasibility analysis using Return on Investment (ROI), NPV, IRR, Payback period, Discounted Payback Period (DPBP), Profitability Index, and Accounting Rates of Returns (ARR). The fourth step is to perform the risk assessment. Although the project can evaluate in step 3 based on its feasibility, we need to consider the risk to deal with the assumptions' uncertainty. The sensitivity analysis calculates risk assessment by changing the variable input by creating five scenarios.

4 Result and Discussion

4.1 Description of Katuk and Spinach Brownies Tartlet Product

Katuk and spinach brownies tartlet is a form of modified product development by changing the appearance, color, taste, texture, and aroma and adding energy and nutrients to brownies tartlet. This product is as a dessert. The processing technique is baking. The way of serving is before processing in the form of one piece with a natural weight of 30 g while the ripe weight is 27 g. This product is an alternative snack for breastfeeding mothers because it uses an additional 25 g of katuk leaves and 35 g of spinach. Based on the research [29], this is reinforced by 50 panelists of breastfeeding mothers who were given katuk biscuits as much as 4 g/day for 42 days showed that giving katuk biscuits helped increase breast milk production and supported the achievement of exclusive breastfeeding weight loss < 8%. The characteristic of katuk and spinach brownies tartlet is in table 4 as below:

Table 4 Characteristic of product Source of primary data 2021

Characteristic	**Katuk and Spinach Brownies Tartlet**
Appearance	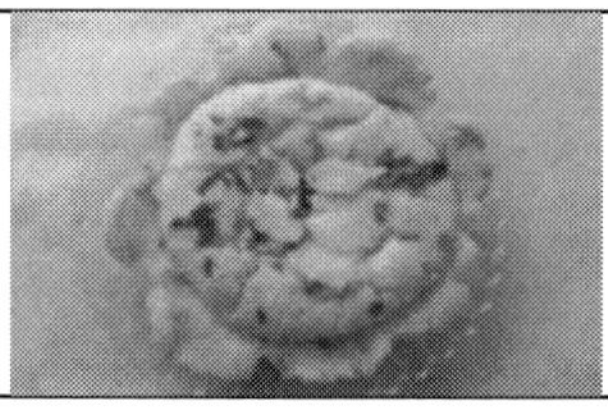
Colour	Yellow has a green spot
Taste	Sweet, savory
Aroma	The smell of cake and vegetable
Texture	Crunchy (tartlet), less-fluffy (brownies)

4.2 The Superior Value of Katuk and Spinach Brownies Tartlet

The Superior value of katuk and spinach brownies tartlet products are :

a. Developing snacks for breastfeeding mothers tare hat rare in society
b. The popular dessert that can improve mood and provide satiety when consuming
c. As an idea of utilizing local ingredients at economical prices that can create more interesting products
d. Using the katuk leaves serves as a galactagogue, and spinach is rich in iron good for breastfeeding mothers.
e. One of the products contains energy that can fulfill the needs of a daily snack for breastfeeding mothers.
f. It has a long shelf life, can last for two weeks if stored in the refrigerator at 5 ° C - 10 ° C and one week in an airtight jar at 25 ° C

4.3 Production Process

Standard operating procedures for katuk and spinach brownies tartlet production start from the procurement of foodstuffs to the presentation of dishes. The production process of katuk and spinach brownies tartlet is as follows :

a. Based on figure 2, Procuring of foodstuff takes about 5 minutes. This process includes the entire preparation and selection of ingredients needed and selected strictly in quality, size, freshness, and legal completeness, so the products produced to follow the quality standards determined. The purchase of ingredients once when going through the production process (conventional production system). Foodstuff consists of dry ingredients and wet ingredients.

Fig. 2. Procuring of foodstuff.

b. Based on figure 3, Preparing tools and ingredients takes about 10 minutes. This process consists of washing the ingredients and weighing the ingredients according to the structured recipe and the amount produced.

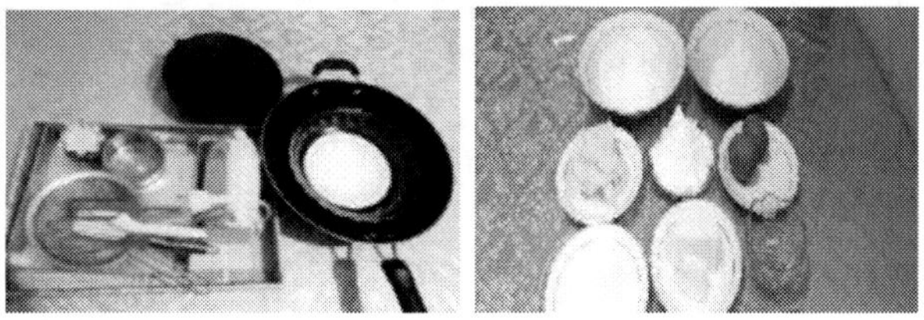

Fig. 3. Preparing of tools and ingredients

c. After the preparation process, the next is chopping katuk leaves and spinach. It takes about 5 minutes, so the following results are :

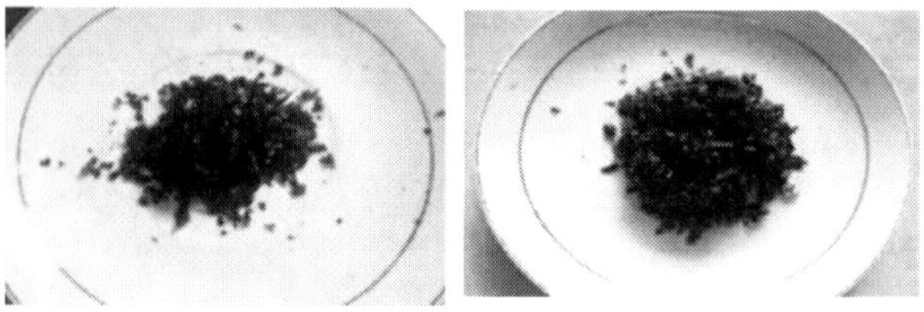

Fig. 4. The result of chopped leaves katuk and spinach

d. Based on figure 5, making tartlet takes about 45 minutes, including dough, printing dough, and baking dough

Fig. 5. Tartlet

e. Based on figure 6, making brownies takes about 35 minutes. First, beat the sugar and eggs until dissolved and cool. After that, mix the melted chocolate, oil, margarine, and additional wheat flour and milk powder.

Fig. 6. Brownies

f. Based on figure 7, the following process is filling brownies dough into a tartlet and giving a topping of chopped spinach and almonds that takes 5 minutes.

Fig. 7. Filling of brownies dough to tartlet and giving a topping

g. Based on figure 8, the process of baking *katuk and spinach brownies tartlet* in the oven with a temperature of 150°C for 30 minutes until ripe

Fig. 8. Process of baking

h. Based on figure 9, the process of packaging takes 5 minutes. Packaging of katuk and spinach tartlet brownies product use plastic mica size 6x and paper cake cup with a diameter of 4 cm with the following result.

Fig. 9. Serving (output)

4.4 The Energy and Nutrient Content

Based on the table, the method of calculating the nutritional value of the product can be seen from the table of the Food Ingredients List or now known as the Indonesian Food Ingredients Table. The energy and nutrient content of katuk and spinach tartlet brownies in 1 recipe (15 pieces) contains the energy of 4,324 cal, 77.9 g of protein, 268 g of fat, and 453.4 g of carbohydrate. In comparison, one part of the product contains the energy of 288 cal, 5.2 g of protein, 17.9 g of fat, 30.2 g of carbohydrate, and 0.8 mg of iron.

Table 5 The energy and nutrient content

Source of primary data 2021

No	Name of ingredient	Weight (g)	Energy (Cal)	P (g)	L (g)	KH (g)	Fe (mg)
1.	Spinach	35	13	1,2	0,2	2,3	1,4
2.	Katuk leaves	25	15	1,2	0,3	2,8	0,7
3.	Wheat flour	300	1,095	26,7	3,9	231,9	3,6
4.	Yolk	110	397	17,9	35,0	0,8	0,9
5.	Egg	55	89	7,0	6,3	0,4	1,5
6.	Sugar	130	473	-	-	122	0,1
7.	Margarine	150	1,080	0,9	121,5	0,6	-
8.	Powder milk	27	137	6,6	0,1	9,8	0,2
9.	White chocolate	150	572	13,5	52,5	80,4	3,0
10	Cooking oil	40	361	-	40,0	-	-
11.	Almond	15	92	2,9	8,2	2,4	0,5
Total one recipe (15 pieces)			4,324	77,9	268	453,4	11,9
Total one piece			288	5,2	17,9	30,2	0,8

4.5 Energy Contribution

Based on table 6, katuk and spinach brownies tartlet in one serving contribute energy by 11.5% to the needs of nursing mothers < six months and by 11.3% contribute energy to the needs of nursing mothers ≥ six months. Consumption a day is enough. One product already fulfills the energy need of a breastfeeding mother's snack (15%)

Table 6 Energy contribution

Source of primary data 2021

Category of Breastfeeding Mother	AKE (Cal)	Energy Contribution (%)
Breastfeeding mother < six months	2500	11,5
Breastfeeding mother ≥ six months	2550	11,3

4.6 The Result of Financial Analysis

The initial investment is the funds needed to start a business. The initial investment is IDR 66,520,750, clustered into machinery & equipment, legal cost, and Halal Certificate and Pre Operating Expense. These assumptions for financial feasibility analysis on interviews and relevant secondary data.

- The project lifetime is four years.
- The capacity of production per day is 360 pieces (= 3 x 7 pans x @ 20 tartlets = 28 recipes). Assumptions: there are 5% defective goods.
- Twenty-five workdays in a month.
- The price of tartlet is assumed to be fixed based on the market price at IDR 6,00 per piece for four years.
- Revenue growth/year is 5%.
- Inflation is assumed to be 4%
- Salvage Value is 10%
- The cost of capital is assumed to be 10%
- This business is subject to a final tax of 0.5% of turnover.
- The owner deposited 30% initial capital and 70% from a loan with an interest rate of 10%.
- Owner rents space for business

4.6.1 Project Proforma Financial Statements

Using the assumptions described previously, 75% of products sold and the projected proforma financial statements are :

Table 7 Income statement

Sources: author analysis

Income Statement				
	Year 1	Year 2	Year 3	Year 4
Projected Sales	538,650,000	565,582,500	593,861,625	623,554,706
Less : Cost of Goods Sold	299,736,750	430,821,622	446,624,243	449,872,325
Gross Profit	238,913,250	134,760,878	147,237,382	173,682,381
Operating Expenses				
Salaries (Indirect Labour)	30,000,000	31,050,000	32,136,750	33,261,536
Rent	30,000,000	31,050,000	32,136,750	33,261,536
Utilities	600,000	621,000	642,735	665,231
Telephone	1,320,000	1,366,200	1,414,017	1,463,508
Marketing & Promotion	24,000,000	24,840,000	25,709,400	26,609,229
Depreciation	1,305,000	1,305,000	1,305,000	18,000,000
Distribution	18,000,000	18,630,000	19,282,050	19,956,922
Office Supplies	300,000	310,500	321,368	332,615
Interest Rate	4,656,453	3,653,123	2,549,460	1,335,432
Total Operating Expenses	110,181,453	112,825,823	115,497,530	134,886,009
Earnings Before Interest and Taxes (EBIT)	128,731,798	21,935,055	31,739,852	38,796,372
Tax (0.5%)	2,693,250	2,827,913	2,969,308	3,117,774
Net Profit	126,038,548	19,107,143	28,770,544	35,678,599

Income statement data show this business has always generated a profit since the first year. Return on Sales is 23.4%. The following projected statement of financial position is :

Table 8 Balance sheet

Sources: author analysis

Balance Sheet				
	Year 1	Year 2	Year 3	Year 4
Assets				
Cash	126,038,548	19,107,143	28,770,544	35,678,599
Inventory	99,912,250	82,727,343	64,217,100	44,309,799
Total Current Assets	225,950,798	101,834,486	92,987,644	79,988,398
Equipment	15,800,000	12,470,000	9,140,000	5,810,000
Less: Accumulated Depreciation	3,330,000	3,330,000	3,330,000	3,330,000
Intangible Goods	10,000,000	7,500,000	5,625,000	3,750,000
Less: Accumulated Amortization	2,500,000	1,875,000	1,875,000	1,875,000
Total Fixed assets	19,970,000	14,765,000	9,560,000	4,355,000
Total Assets	245,920,798	116,599,486	102,547,644	84,343,398
Liabilities and Stockholder's Equity				
Liabilities	36,531,229	25,494,604	13,354,316	0
Stockholder's Equity	209,389,568	91,104,881	89,193,327	84,343,398
Total Liabilities and Stockholder's Equity	245,920,798	116,599,486	102,547,644	84,343,398

Cash at the end of the year is always positive. It means cash flow from income is more than cash out, and the amount continues to increase. Average Cash Flow is (IDR) 46,587,577. The projected cash flows are as follows :

Table 9 Cash flow

Sources: author analysis

Cash Flow					
	Year 0	Year 1	Year 2	Year 3	Year 4
Cash Flow from Operating Activities					
Cash In Flow					
Loan	46,564,525				
Capital	19,956,225				
Net Profit + Depreciation		131,868,548	24,937,143	34,600,544	41,508,599
Cash OutFlow					
Machinery & Equipment	14,800,000				
Legal cost and Halal Sertificate	10,000,000				
Other Civil Work					
Payment of Priciple Loan		10,033,296	11,036,625	12,140,288	13,354,316
Pre Operating Expense	41,720,750				
Total Cash Outflow	66,520,750	10,033,296	11,036,625	12,140,288	13,354,316
Net Cash In Flow - Cash Out Flow	-	121,835,252	13,900,517	22,460,256	28,154,282
Cash Flow Beginning		-	121,835,252	135,735,769	158,196,025
Cash Balance End	-	121,835,252	135,735,769	158,196,025	186,350,308
Annual Cash Flow	-	121,835,252	13,900,517	22,460,256	28,154,282
Average Cash Flow		46,587,577			

4.6.2 Break-Even Point (BEP)

The Break-even point is the point with sales revenue is equal to total cost and zero profit [30]. Table 10 shows the point of sale when it reaches the break-even point and the number of units produced each year. Break-even analysis determines the sales needed to break even [31]. This business will get a break-even point when sales go IDR 189,102,195 and 37,820 units in the first year. The profit margin in sales per unit was 58,27% in the first year.

Table 10 Break-even point (BEP)

Sources: author analysis

Break - Even Point				
	Year 1	Year 2	Year 3	Year 4
Sales (IDR)	718,200,000	754,110,000	791,815,500	831,406,275
Fixed Cost (IDR)	110,181,453	112,825,823	115,497,530	134,886,009
Variable Cost (IDR)	299,736,750	430,821,622	446,624,243	449,872,325
Contribution Margin Ratio	58.27%	42.87%	43.59%	45.89%
BEP - Sales (IDR)	189,102,195	263,180,142	264,933,519	293,932,097
BEP - Unit (units)	37,820	52,636	52,987	58,786

Financial analysts frequently divide costs into two types: variable costs and fixed costs. Variable costs change as the output changes, and they are zero when production is zero. Cost of direct labor and raw materials are examples of variable costs [31]. The cost must be issued in fixed amounts during a specific period. The size is not dependent on the size of operating volume or production volume during the period is the fixed cost [32].

4.6.3 The Feasibility Analysis

A feasibility study is an analysis that considers all of a project's relevant factors—including economic, technical, legal, and scheduling considerations—to ascertain the likelihood of completing the project successfully. Whether a project is feasible can depend on several factors, including the project's cost and return on investment, meaning whether the project generated enough revenue or sales from consumers.[33].

According to Investopedia [34], ROI is a performance measure used to evaluate the efficiency of an investment or to compare the efficiency of some different investments. Return on Investment (ROI) calculates the benefit (return) divided by the investment's cost. The capital budgeting method analyses and evaluates the feasibility of long-term investments to maximize the owner's wealth over a certain period [25], [35]. ROI analysis (when applied correctly) is a powerful tool for evaluating existing information systems and making informed decisions on software acquisitions and other projects [36].

According to Investopedia [37], the gross profit margin is metric analysts use to assess a company's financial health by calculating the amount of money leftover from product sales after subtracting the cost of goods sold (COGS).

Net Present Value (NPV) measures profitability on investment by calculating the present value of future cash flows minus the present value of the cost of the investment. The project is profitable if the net cash flow value is greater than zero [31].

The Internal Rate of Return (IRR) is the discount rate that makes a zero NPV of the investment. In other words, IRR is the expected compound annual rate of return that will earn on a project or investment [25], [26], [38].

The Payback period offers a rough measure of the investment's liquidity [28]. The company subjectively determines the maximum length of acceptable payback time based on project type, product circumstances, and project perceived risk [25].

According to Investopedia [39], the discounted payback period is a capital budgeting procedure used to determine the profitability of a project. A discounted payback period gives the number of years it takes to break even from undertaking the initial expenditure by discounting future cash flows and recognizing the time value of money. The metric is used to evaluate the feasibility and profitability of a given project.

According to Investopedia [40], the profitability index (PI) measures a project's or investment's attractiveness. The PI calculates by dividing the present value of future expected cash flows by the initial investment amount in the project. A PI greater than 1.0 is deemed a good investment, with higher values corresponding to more attractive projects. Under capital constraints and mutually exclusive projects should be selected, only those with the highest PIs.

According to Corporate Finance Institute [41], the accounting Rate of Return (ARR) is the average net income an asset is expected to generate divided by its average capital cost, expressed as an annual percentage. The ARR is a formula used to make capital budgeting decisions. It is used in situations where companies decide whether or not to invest in an asset (a project and an acquisition.) based on the future net earnings expected compared to the capital cost.

According to the feasibility analysis results from table 11, all the results met the criteria. The business of katuk and spinach brownies tartlet is feasible.

Table 11 Feasibility analysis

Sources: author analysis

Technique	Value	Acceptance Criteria	Result

Return on Investment ROI)	189.47%	Positive ROI (ROI > 0)	Accept
Gross Profit Margin	44.35%	Positive gross margin (GM > 0)	Accept
NPV (IDR)	91,831,077	Positive NPV (NPV > 0)	Accept
IRR	106.1%	IRR > interest rate	Accept
Payback Period	0.55 year = 6.6 months	Before the project's lifetime (4 years)	Accept
Discounted Payback Period (DBP)	0.60 year = 7.2 months	Before the project's lifetime (4 years)	Accept
Profitability Index	2.38	> 1 acceptable, < 1 not acceptable	Accept
Accounting rate of return (ARR)	52.6%	> 50% acceptable, < 50% not acceptable	Accept

Table 12 Comparison of financial feasibility analysis

Sources: author analysis

Technique	Katuk and spinach brownies tartlet's value	Pineapple carrot juice's value	Chocolate spread with coconut ingredient's value	Banana Chips
Return on Investment ROI)	189.47%			
Gross Profit Margin	44.35%			
IRR	106.1%	17.29%	**1526%**	8.25%
NPV (IDR)	91,831,077	**102,128,626**	20,389,635.9	31,024,617
Payback Period	6.6 months	2,7 months	1.93 years	**2 months**
Discounted Payback Period (DBP)	7.2 months			
Profitability Index	2.38			
Accounting rate of return (ARR)	52.6%			

Based on the financial feasibility analysis comparison data above, katuk and spinach brownies tartlets are more profitable than pineapple carrot juice business [42]; chocolate spread [43] with coconut ingredients business and banana chips [44].

4.6.4 The Scenario Analysis

Risk in capital budgeting is defined as uncertainty regarding the cash flows generated by the project or the degree of cash flow variability [25]. The feasibility analysis was carried out on the assumption of current conditions, which may change in the future. Thus, risk assessment is necessary. We use sensitivity analysis in financial modeling to analyze how the different values of independent variables affect a specific dependent variable under certain conditions [45]. Sensitivity analysis is uncertainty analysis by altering variables assumed from a base case to evaluate how they affect the measured results of the project [38]. If the NPV show to be highly sensitive to relatively small changes in the projected value, the forecast risk involved with the variable is significant [26]. Sensitivity analysis is carried out based on assumptions that have been described previously by making five scenarios. From scenario analysis that calculates the risk assessment, this business is feasible if it can sell at least 72% of its products. By conducting a risk assessment of several scenarios, if it can sell products at least 72%, all the results meet the criteria, and the business is feasible.

Table 133 Scenario analysis

Sources: author analysis

Sensitivity on Revenue					
% of Products Sold	**71%**	**72%**	**75%**	**80%**	**85%**
Return on Sales	19.1%	20.2%	23.4%	28.2%	32.4%
Return on Investment ROI)	146.5%	157.2%	189.5%	243.2%	296.9%
Gross Profit Margin	41.2%	42.0%	44.4%	47.8%	50.9%
Average Cash Flow (IDR)	15,787,036	23,487,171	46,587,577	85,088,253	123,588,930
IRR	*not applicable*	42.7%	106.1%	180.9%	246.0%
NPV (IDR)	(5,237,337)	19,029,767	91,831,077	213,166,595	334.502.112
Payback Period	0.71 years	0.66 years	0.55 years	0.42 years	0.34 years
Discounted Payback Period (DBP)	0.78 years	0.73 years	0.60 years	0.46 years	0.38 years
Profitability Index	0.92	1.29	2.38	4.20	6.03
Accounting rate of return (ARR)	40.7%	43.7%	52.6%	67.6%	82.5%

5 Conclusions and Suggestions

5.1 Conclusions

1. Based on the energy and nutrient content in 1 piece of katuk and spinach brownies, tartlet product contains the energy of 288 cal, 5.2 g protein, 17.9 g of fat, 30.2 g of carbohydrate, 0.8 mg of iron.
2. The analysis of business and cost of production of *katuk and spinach brownies tartlet* is feasible with a minimum of 72% of products sold.

5.2 Suggestions

1. Based on the energy and nutrient content result, katuk and spinach brownies tartlet has fulfilled the mother's daily snack to be an alternative snack for breastfeeding mothers.

2. Katuk and spinach brownies tartlet can be an alternative healthy food business.
3. Financial feasibility analysis results showed that BEP of production of 37.820 pieces, BEP of sales of IDR 189.102.195; Return on Sales of 23.4%; Return on Investment (ROI) of 189.5%; NPV of IDR 91.831.077 and PBP during 0.55 year, Discounted Payback Period (DPBP) of 0.60 years, Profitability Index of 2.38 and Accounting Rates of Returns (ARR) of 52.6%. Based on scenario analysis, katuk and spinach brownies tartlet is feasible for business if minimum sales of 72% of products.
4. Based on the financial feasibility analysis comparison data above, katuk and spinach brownies tartlets are more profitable than pineapple carrot juice business; chocolate spread with coconut ingredients business and banana chips.

References

1. The Indonesian Ministry of Health. Exclusive Breastfeeding Situation and Analysis. Jakarta (ID): Infodatin. (2014)
2. Agostoni C, Braegger C, Desci, Kolacek, Koletzko. Breast-feeding: A Commentary by the ESPGHAN Committee on Nutrition. Journal of Pediatric Gastroenterology and Nutrition 49:112–125. (2009)
3. W. Schultink. Why nutrition and breastfeeding are crucial to sustainable development. (2015). https://blogs.unicef.org/blog/author/wernerschultink/.(accessed August 27, 2021)
4. Hoseini L, Vakili R, Khakshour A, Saeidi M. Maternal Knowledge and Attitude toward Exclusive Breast Milk Feeding (BMF) in the First 6 Months of Infant in Mashhad. International Journal of Pediatrics, Vol.2, N.1, Serial No.3 (2014)
5. American Academy of Pediatrics. Breastfeeding and the Use of Human Milk. Journal of American Academy of Pediatrics.Vol.129 (3): 827-841 (2012)
6. Ministry of Health RI. Nutritional Status Monitoring Pocket Book 2017. Jakarta (ID): Healthy living Community Movement. (2017)
7. Syarief H, Damanik M, Sinaga T. Utilization of Bangun-bangun Leaf in the Development of Functional Supplementary Food Products for Breastfeeding Mothers. Indonesian Journal of Agricultural Sciences: SSN 0853 – 4217 Vol. 19 (1): 38 42. (2014)
8. Luxita S, Puneeta A. 2019. Organoleptic and Physicochemical Properties of Tarts Developed from Quinoa, Chickpea, and Oats Flour and Their Ranking by Topics Method. Journal food and nutrition of Amity University: No. (**2**) 2019, 457-468. (2019)
9. Wahyuni S, Holilah, Asranudin, Noviyanti. Estimation of shelf life of Wikau Maombo Brownies Cake Using Accelerated Shelf Life Testing (ASLT) Method with Arrhenius Mode. IOP Conf. Series: Earth and Environmental Science 122 (2018) 012082. (2018)
10. Ilmi B, Arini A, Novriyanti B. Formulation of Snack Bar from Torbangun Leaf as Food Source for Iron and Calcium. Indonesian Journal of Nutritional Science, Vol.01, No.1 (2021): 21–26. (2021)
11. Ngadiardi I, Agusta N, Srimati M, Fayasari A. Sauropus androgynus, Papaya Leaves, and Mung Beans as Mixed Galactagogue Drink for Urban Postpartum Mothers. Binawan university: Food Nutrition Journal 16 (2):101-10. (2019)
12. Rahmanisa, S, and Tara. The Effectiveness of Extraction Alkaloids and Sterols from Katuk leaves (Sauropus androgynus) on breast milk production. Majority Vol.**5** (1):117-121.(2016)

13. Hayati A, Arumingtyas L, Indriyani S, Hakim L. Local Knowledge of Katuk (Sauropus androgynus (L.) Merr) in East Java, Indonesia. University of Islam Malang: International Journal of Current Pharmaceutical Review and Research; 7(4); 210-215. (2016)
14. Rosdianah, Irma S. Giving Katuk Leaf Extract to Breast Milk Smoothness in Breastfeeding Mothers. JKM (Journal of Midwifery Malahayati), Vol 7(2): 45-50. (2021)
15. Al-gumbos N, Abu-Naser S. Spinach Expert System: Diseases and Symptoms. Al-Azhar University: International Journal of Academic Information Systems Research (IJAISR) ISSN: 2000-002X Vol., Pages: 16-22 9. (2019)
16. AKG. The Recommended Nutritional Adequacy Rate for The Indonesian People. Regulation of The Ministry of Health Republic Indonesia Number 28. (2019)
17. Zahro W, Pangestuti R, Widajanti R. Patterns of Breastfeeding and Nutritional Status of Breastfeeding Mothers in The Work Area of Kedung Mundu Public Health Center. Semarang (ID).Journal of Public Health e-Journal: Vol 4 (3). (2016)
18. Mangaroni, Cetin I, Verduce E, Canzone G, Glovannini. Maternal Diet and Nutrient Requirements in Pregnancy and Breastfeeding. An Italian Consensus Document. MDPI: Review. (2016)
19. Rosa, Marques, Jose, Taddei, Fabio, Lopez. Breastfeeding Exclusively and Iron Deficiency Anemia During the First 6 Months of Age. Rev Assoc Med Bras: 60 (1):18-22. (2014)
20. Noviyanti. S, Swamilaksita D, Angkasa D. The Potential of Fenugreek Seed Cookies with Variations of Red Bean Flour, Whole Wheat as a Companion Snack for Breastfeeding Mothers.Jakarta (ID): Journal of Esa Unggul University. (2015)
21. Financial analysis. In Investopedia. Retrieved August 23, 2021, from https://www.investopedia.com/. (2021)
22. Christensen, L. P. Raisin Production Manual. UCANR Publications. (2000)
23. Menya, S. O., &Gichinga, L. The Effects of Capital Budgeting Techniques on the Growth of Micro-Finance Enterprises in Mombasa, 4 (5), 6. International Journal of Science and Research (IJSR). Volume 4. (2015)
24. Karanovic, G., Barca, S., & Bogdan, S. Techniques For Managing Project Risk In Capital Budgeting Process, 1(2), 13. (2010)
25. Gitman, L. J., Zutter, C. J. Principles of managerial finance. The Pearson Series in finance (Fourteenth edition.). Boston: Pearson. (2015)
26. Ross, S. A., Westerfield, R. W., Jordan, B. D. Fundamentals of Corporate Finance, 922. Mc Graw Hill. (2003))
27. Brigham, E. F., & Houston, J. F. Fundamentals of financial management (12th ed.). Mason, OH: South-Western Cengage Learning. ((2009)
28. Fabozzi, F. J., & Peterson, P. P. Financial Management and Analysis (2nd edition.). (2003)
29. Pinem Br S, Ginting Br S, Simamora L. Effect of Consumption Biscuits and Oxytocin Massage on the Succes of Exclusive Breastfeeding for Breastfeeding Mothers at Simalingkar Health Center in 2019. Health Dynamics Journal of Midwifery a Nursing Vol 10 No.2 2019 (ISSN: 2086-3454 EISSN: 2549-4058) url: http://ojs.dinamikakesehatan.unism.ac.id DOI: https://doi.org/10.33859/dksm.v10i2 (2019)
30. D.R. Hansen, M.M. Mowen. Accounting Managerial 8th edition. Cengage Learning. (2006)

31. S. A. Ross, R. W. Westerfield, J. Jaffe, Corporate Finance, 9th edition. (2010)
32. Kastaman, R. Engineering Economics for Entrepreneurial Development.Pustaka. (2001)
33. Feasibilty Study. (2021). In Investopedia. Retrieved March 27, 2022, from https://www.investopedia.com/terms/f/feasibility-study.asp
34. Return on investment – ROI. (2011). In Investopedia. Retrieved August 23, 2021, from https://www.investopedia.com/. (2021)
35. Clark, J. J., Hindelang, T. J., & Pritchard, R. E. Capital Budgeting: Planning and Control of Capital Expenditures. (1989)
36. Alexei Botchkarev & Peter Andru, (2011). A Return on Investment as a Metric for Evaluating Information Systems: Taxonomy and Application. Available from: https://www.researchgate.net/publication/216356259_A_Return_on_Investment_as_a_Metric_for_Evaluating_Information_Systems_Taxonomy_and_Application [accessed August 23, 2021]
37. Gross Profit Margin. (2022) In Investopedia. Retrieved February 2, 2022, from https://www.investopedia.com/terms/g/gross_profit_margin.asp.
38. Van Horne, J. C., &Wachowicz, J. M. Fundamentals of financial management (13th ed.). Harlow, England ; New York: Financial Times/Prentice Hall. (2008)
39. Discounted Payback Period. (2022). In Investopedia. Retrieved February 2, 2022, from https://www.investopedia.com/terms/d/discounted-payback-period.asp.
40. Profitability Index. (2021) In Investopedia. Retrieved August 23, 2021, from https://www.investopedia.com/.
41. Accounting Rate of Return. (2022) In Corporate Finance Institute. Retrieved February 2, 2022, from https://corporatefinanceinstitute.com/resources/knowledge/accounting/arr-accounting-rate-of-return/.
42. M. Sari, A. Nurmaydha, D. U. M. Rohmah. Financial Feasibility Analysis of Pineapple Carrot juice Business. Journal of Physics: Conference Series. IOP Publishing. 1381 (2019) 012034. doi:10.1088/1742-6596/1381/1/012034. (2019)
43. Dyah Ayu Savitri, Herlina, Noer Novijanto. Financial Feasibility Analysis of Chocolate Spread with Coconut Ingredients as Agroindustrial Product. Journal La Bisecoman. Vol. 02, Issue 02 (014-024), (2021)
44. Ruswajia, Puguh Cahyono. Analysis of Business Feasibility of Banana Chips. Journal of Humanities and Social Studies. Volume 04, Number 02, September 2020, Page 197 – 200. e-ISSN: 2598-120X; p-ISSN: 2598-117X.
45. Sensitivity analysis. (2021). Retrieved August 23, 2021, from https://corporatefinanceinstitute.com/.

Regulatory Impact Assessment Analysis of Regulations on Pollution Load Capacity and Carrying Capacity of Lake Toba for Aquaculture Fisheries

Dahri Tanjung[1*], *Manuntun* Parulian Hutagaol[2], *Agit* Kriswantriyono[3], *Yuni* Puji Hastuti[4], *Kukuh* Nirmala[5], *Yulia* Wulandari[6]

[1]College of Vocational Studies, IPB University, Jl Kumbang No.14 Bogor 16151, West Java, Indonesia
[2]Department of Economic, FEM, IPB University, Jl. Agatis IPB Dramaga Campus, Bogor 16680, West Java, Indonesia
[3,6] CARE, IPB University, Jl. Pajajaran IPB Baranangsiang Campus, Bogor 16144 West Java, Indonesia
[4,5] Department of Aquaculture, Faculty of Fisheries and Marine Science, IPB University, Jl. Agatis, IPB Dramaga Campus, Bogor 16680, West Java, Indonesia

Abstract. The Government of Indonesia has determined Lake Toba be one of the super-priority tourist destinations. In order to realize the program, the government has recently been more aggressively structuring the Lake Toba area. One of the arrangements is to lift and reduce the number of floating net cages (FNC) so that fish production is gradually reduced from 70,000 tons to 10,000 tons per year in 2023. The FNC arrangement activity caused restlessness among FNC actors because they had to drastically reduce their production or even close their business, which would be detrimental to thousands of local FNC actors. This continued situation will be related to disrupting their main source of livelihood. This study aims to analyze the Governor of North Sumatra Decree. The method used is the survey method, collecting socio-economic data directly, focus group discussions/FGD with the parties involved. Then the regulation was analyzed using the RIA method. This analysis proposes three alternative options: business, implementing the Governor's Decree No. 188.44/2013/KPTS/2017, revising the Governor's Decree with a carrying capacity of 42,000 tons/year aquaculture. It is proposed to revise the Governor's Decree.

1. Introduction

The central government has determined Lake Toba be one of the four super-priority tourist destinations in addition to 4 other locations. In order to realize this program, the government has recently intensified various arrangements, such as lifting and reducing the number of floating net cages (FNC) so that production is reduced from 70.000 tons currently to 10.000 tons per year in 2023. The legal basis for this arrangement is The Governor of North Sumatra

* Corresponding author: ir.da@apps.ipb.ac.id

https://doi.org/10.1051/e3sconf/202234800037

Decree No.188.44/213/KPTS/2017 of 2017 concerning Pollution Load Carrying Capacity and Lake Toba Capacity for Aquaculture. The decree states that the capacity of Lake Toba is 10.000 tons and will be achieved in 2023 [1].

The various FNC structuring activities to reach 10.000 tons by 2023 have caused restlessness among FNC actors because they have to drastically reduce their production or even close their business, which will be very detrimental to thousands of local FNC actors. This continued situation will be related to disrupting their main source of livelihood. The Governor's Decree used is a regulatory product that has been in place since 2017. The basis of the decree is the result of research by the North Sumatra Regional Environmental Agency, which conducted research and found that the water quality of Lake Toba is polluted (mesotrophic). It carries the capacity for aquaculture ranges from between 9,900 to 81,000 tons. Furthermore, in the Governor's Decree, the capacity is set at 10.000 tons per year in 2023.

As part of the legal system, the Governor's Decree is also a regulatory instrument to provide a basic reference for implementing daily floating net cage aquaculture in Lake Toba. When compared to various general conditions, the fact is that many problematic regulations have been canceled by the Government (Ministry of Home Affairs), the majority of which are local regulations that hinder investment and create intolerance and discrimination. Therefore, it is very important to ensure a good regulatory formulation process [2].

As a public policy, regulation should be able to answer problems effectively, directly hit the core of the community's problems, be implemented effectively, fairly, give hope for improvements to the condition of the community, and from an economic perspective, can contribute to increasing regional competitiveness. The government and the community that use the waters of Lake Toba need to collaborate in the management of the waters of Lake Toba. The collaborative management model is, known as "co-management," has been widely applied to fish-rich waters in many countries such as the Americas [3].

The preparation of regulations is one of the better regulation program framework components. The existence of regulation provides legality to the local government to enforce provisions and rules to the community. It explains the community's rights to services, principles, and descriptions of the rules and regulations in service delivery that the government must comply with.

It is a common practice to review the implementation of regulations that have been passed. Through the review, it is hoped that the latest information on the status of the implementation of regulation can be known, identify obstacles to implementation in the field, recognize positive changes that have occurred, and ensure that the substance of the regulation is still under the situation on the ground or requires adjustments. One of the tools used is the Regulatory Impact Assessment (RIA) to conduct this review. In principle, RIA has been conceived as a key instrument for improving regulatory quality and good governance by ensuring more coherent and transparent policies. Making regulation more effective and efficient facilitates the interdepartmental process and often involves public consultation [4].

Regulations close to five years old are selected for review, namely the Governor of North Sumatra Decree No. 188.44/213/KPTS/2017 concerning Pollution Load Capacity and Carrying Capacity of Lake Toba for Aquaculture. The application of RIA also aims to see the effectiveness of the RIA method in supporting the preparation of effective regulations for the development of the fisheries sector and other aspects [5].

For this reason, this study aims to analyze the Governor of North Sumatra Decree using the RIA method. This study is an input for the government on whether to stick with the current regulations or make revisions.

2. Research Methods

This activity was carried out in the Lake Toba area, North Sumatra from January 2020 until October 2021. The methods used survey methods, collecting socio-economic data directly, and focused group discussions/FGDs with the parties involved. Pollution load carrying capacity and aquaculture capacity were carried out by analyzing the water quality of Lake Toba. Furthermore, the Governor's Decree regulation was analyzed using RIA Analysis. In this analysis, there were three alternative options: business as usual, implementing the Governor's Decree No. 188.44/2013/KPTS/2017, or revising the Governor's Decree with the latest aquaculture carrying capacity.

Regulatory Impact Assessment (RIA) is an approach to analyze an **existing** or **new regulation** supported by **empirical data and communicate** to decision-makers about the **impact**, both in terms of **costs and benefits** of a regulation. The RIA approach is widely used for regulatory reform to obtain quality regulations. World Bank researchers suggest that RIA uptake in the Global South increases in making reform sustainable more well-documented [6].

The RIA approach is a systematic process consisting of logical steps, which can also be positioned as logical thinking. The diagram below is a schematic of the RIA activity steps [7].

1. Formulation of the problem. The initial stage was the mapping of issues relevant to the regulation's objectives and substance and supported by data and facts on the ground.
2. Goal setting. In simple terms, the statement of purpose was a form of negation of the problem statement.
3. Formulation of Alternative Actions. This step was to identify what action options are available to address the problem and meet the objectives. In the RIA approach, the first choice of action is "do nothing" or "do nothing", which in the next stage would be considered as the initial condition (baseline) to be compared with various other options. At this stage, it is important to involve stakeholders from various backgrounds and interests to get the broadest picture of what options are available and can be taken (Figure 1).

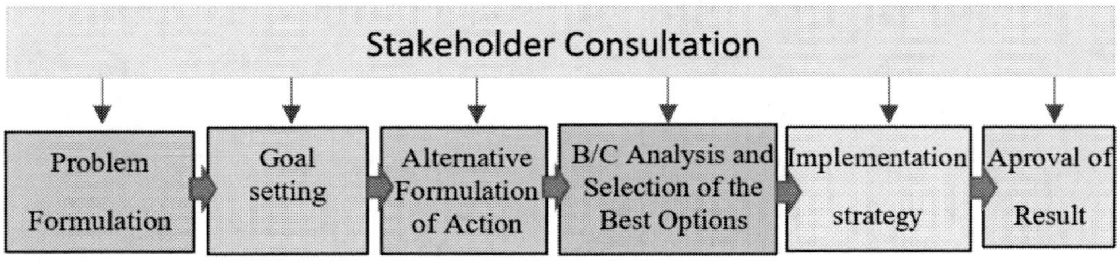

Figure 1. Schematic of RIA Activity Steps

4. Cost-Benefit Analysis. After formulating several action options, the next step was to assess the benefits to be obtained and the costs arising from each choice so that it could be decided which is the best choice. The benefit and cost analysis stage is an important part of policy decision-making regarding budget issues [8].
5. Formulation of an implementation strategy. The selection of the best course of action was a form of policy that cannot run itself once it has been set. Therefore, the next step was to formulate a strategy to implement the policy.
6. Result Approval. After all the steps above had been carried out, it was necessary to prepare a report or statement of results (RIA statement). This RIA statement would be submitted to the decision-maker, in this case the regional head who would decide whether to approve the best course of action chosen by the formulation team.

For the record, RIA places great emphasis on participatory processes; therefore, it is necessary to involve stakeholders through a public consultation process at every stage. The public or society is the party whose voice must be heard because they will ultimately receive the impact, both good and bad, of the existence of this policy.

The regulation chosen for review using the RIA methodology is the Governor's Decree No. 188.44/213/KPTS/2017 concerning Pollution Load Capacity and Carrying Capacity of Lake Toba for Aquaculture. This regulation was chosen because it is close to five years old and is a matter of debate.

There were at least three reasons to review an existing regulation, namely:

- The emergence of problems caused by the enactment of regulations.
- some changes or dynamics occur in the community so that regulatory changes/revisions are needed.
- Problems that the regulation is trying to solve still exist (in whole or in part).

3. Results and Discussion

3.1. Review Approach

After identifying the reasons for doing RIA, we can start the review by taking the first step of RIA, namely mapping the problems or changes in the community and the field related to the substance of the regulation. It is important to get evidence in supporting data and information that strengthens the problem or change. Mapping problems and collecting evidence can be carried out by internal parties (local governments) to oversee the implementation of these regulations. By RIA principles, mapping the problem or symptom and then formulating it into the main problem must involve a stakeholder consultation process. Namely, the parties affected by or have concerns about the regulation.

Before conducting phase two of RIA, setting goals, the next step is to assess or check the relevance of the problem map as mentioned above with the substance contained in the body or articles in the regulation being reviewed (Figure 2). There are two possible results of checking, as presented in the diagram below, as follows:

- Issues are irrelevant or not within the scope of the topic or regulatory context under review, so they can be ignored.
- Issues relevant to or related to the topic or regulatory context under review.

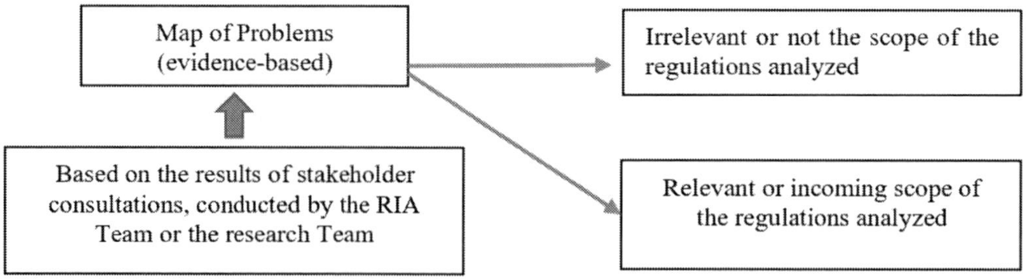

Figure 2. Problem Relevance Check Scheme

From the above steps obtained, a collection of problems relevant to the regulations. Next is to analyze the substance of the relevant problem categories more deeply, namely juxtaposing it with the statement of objectives and contents of the articles (body) in the regulations and conducting an assessment.

In this case, there are four possible conditions and what are the implications for their follow-up, as shown in the following diagram:

- The issue has been regulated in the regulations. After all, the need is strengthening for the implementation of the article related to the issue (Figure 3);

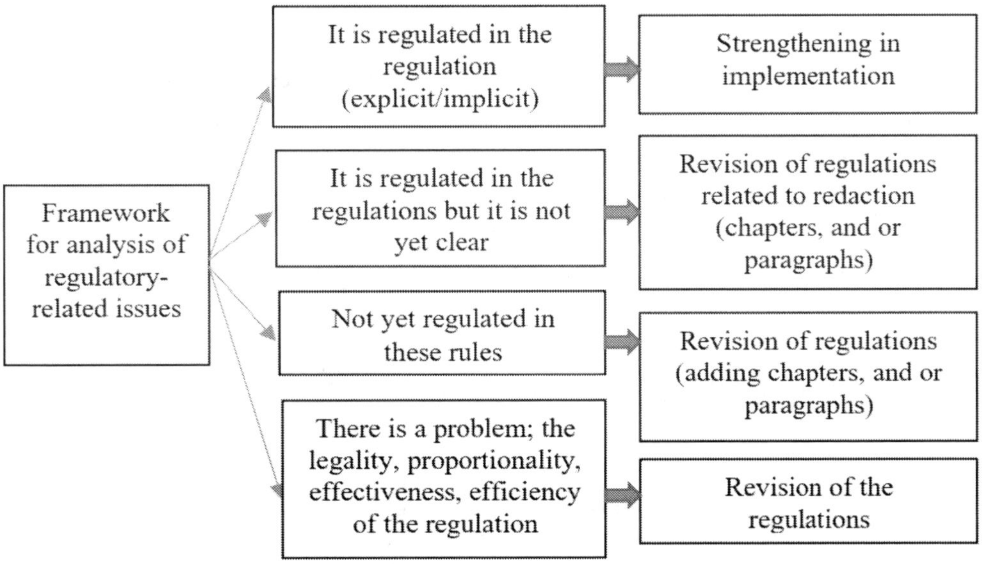

Figure 3. Problem Analysis Framework, Categories, and Implications

- The substance or issue of the problem has been regulated in the regulation but is not yet clear, so it requires editorial related revisions;
- The substance or issue of the problem has not been regulated in the regulations, so it requires additional explanations in the regulations (chapters, articles, and or paragraphs);
- There are legality, proportionality, effectiveness, and efficiency issues in the regulation that need to be revised.

Based on the analysis of the problem above, it is possible to revoke the regulation if conditions are found according to Law no. 12/2011 concerning the Formation of Laws and Regulations, as follows:

1. Systematics of laws and regulations change
2. The material of the legislation has changed by more than 50%
3. The essence changes

3.2. Review the Governor of North Sumatra Decree No. 188.44/213/KPTS/2017

The North Sumatra Provincial Government issued a Governor's Decree (SK Gub) No. 188.44/213/KPTS/2017 concerning Pollution Load Capacity and Carrying Capacity of Lake Toba for Aquaculture. This regulation, passed in August 2017, provides a legal basis for the planned development of marine cage fisheries management in Lake Toba. These regulations are expected to have major implications for the community. Among them is a target of no later than five years from enacting the regulation (2023). All parties implementing marine cage aquaculture in Lake Toba must have met the technical and administrative requirements stipulated in the regulations, namely the total production of Floating Net Cage (FNC) of 10,000 tons. In addition, this regulation is expected to accelerate the recovery of Lake Toba water back to the oligotrophic.

3.1.1. Review Process

The RIA process for reviewing this regulation was carried out by a research team and gathered input from several parties such as the Fisheries and Marine Service, the Environment Agency, Bappeda, and FNC actors. The regulatory review process is carried out according to the stages of RIA activities by following the problem analysis approach as above. The RIA process begins with a brief briefing to the Team on the RIA methodology. The RIA activity stage requires five meetings from February 2020 to August 2021.

3.1.2. Review Results

Formulation of the problem

From several meetings between the research team and with the community and private stakeholders, several data and facts in the field have been mapped, which can then be narrowed down into the formulation of the main problem as follows:

"Lake Toba is still polluted due to several things, namely domestic waste, hotel, restaurant, cafe waste, floating net cage waste, agricultural waste, livestock waste, and forestry waste."

After more than four years of implementing regulations, there are still problems that do not support the objectives of the regulation, so this becomes the basis for a regulatory review. Details of the problem mapping results and the supporting data can be seen in the Appendix of the RIA Report to the Governor's Decree of Lake Toba's Supporting Capacity. Any issues or problems that arise were checked for relevance to the substance of the regulation.

3.1.3. Goal Setting

With the statement above, a statement of objectives can be set and will determine the choice of action, which is in line with the objectives of the regulation as follows:

"Reducing the level of water pollution of Lake Toba to meet the specified quality standards."

3.1.4. Formulation of Alternative Actions

After going through a process of discussion and discussion of the issues and problems that were collected, including checking and collecting data in the field in 2021 and trying to compare them with factual conditions in the field, the research team concluded three alternative solutions to overcome the problem, as follows:

Option I: Do Nothing, which means not taking any action in response to the problems encountered and still applying the North Sumatra Governor's Decree of No. 188.44/213/KPTS/2017 concerning Pollution Load Capacity and Carrying Capacity of Lake Toba for Aquaculture as it is. In RIA, the "do nothing" option is a form of policy that decision-makers may choose.

Option II: Perform a series of actions that include:

1) Implementing the Governor's Decree No. 188 of 2017 concerning the Determination of the carrying capacity of Lake Toba;
2) Prepare the Governor's Decree on the Implementation Team of the Governor's Decree No. 188/2017 concerning the Determination of the carrying capacity of Lake Toba;

Option III: Perform a series of actions that include:

1) Revising the Governor's Decree No. 188 of 2017 concerning the Determination of the carrying capacity of Lake Toba by the results of the latest research conducted by the IPB University Team;
2) Prepare the Governor's Decree on the Implementation Team of the New Governor's Decree concerning the Determination of the carrying capacity of Lake Toba.

3.1.5. Parties affected by the policy

Governor's Decree No, 188.44/213/KPTS/2017 regarding the carrying capacity of Lake Toba for aquaculture has implications for the parties. The Determination of fish production is only 10,000 tons per year in Lake Toba until 2023. The parties affected by this Governor's Decree are as follows (Table 1):

Table 1. Parties affected by the North Sumatra Governor's Decree

The party	Forecast of the impact that will be received
A. Parties directly affected	
1. Company FNC actors	• Production decline • Profits decreased • Investment loss
2. Community FNC actors	• Production decline • Decreased / lost income • Loss of initial capital
3. Country and city	• Decrease in domestic fish production • Decrease in regional income from taxes/levies
4. Workers work	• Loss of fieldwork
5. Feed Seller	1. Decreased feed turnover
6. Fish seed seller	• Decrease in sales of fish seeds
B. The affected are not directly	
1. Business actors supporting FNC cultivation	• Production decline • The decline in turnover and profit venture
2. Tourism activities	
3. Market	
4. Sales of equipment cages	• Decrease in business turnover
5. Fish transportation Activities	• Decrease in fish, seed, and feed transport activities
6. Services stall eating, coffee, and cigarettes	• Decrease in sales volume

The parties directly affected by the implementation of The Governor's Decree are FNC owners, both FNC companies and the communities. Currently, two companies are involved in corporate FNC and 5,000 people for community FNC. Two FNC companies currently produce 45,000 tons per year, while community FNCs produce 3,000 tons per year.

The total production value of FNC in Lake Toba in 2020 is IDR 3.5 Billion (assuming the price of fish is IDR 25,000/kg). Production FNC companies all export and foreign exchange are generated from exports have reached IDR 1,474.6 Billion. Meanwhile, the production of community FNC is marketed at the local level to meet the needs of residents in 7 regencies around Lake Toba, plus the Medan city and its surrounding.

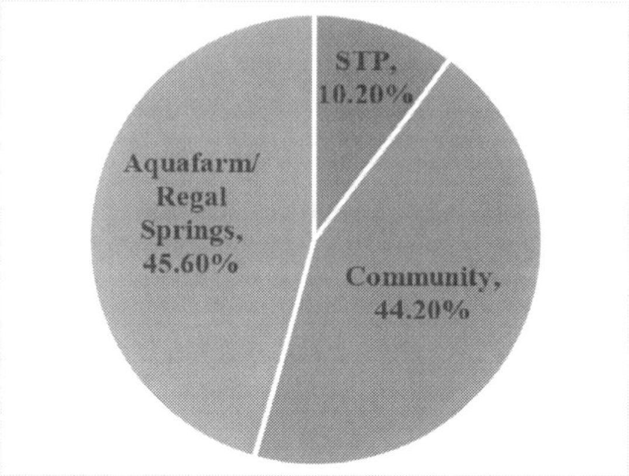

The selling price of live fish is IDR 25,000/kg** and the total fish production **is 74.584 tons per year.

Data GPMT Sumut 2020

Figure 4. Total Tilapia Production in the Lake Toba region

Judging from the amount of investment that has been allocated either by the company or by the community can be estimated as follows for company A where the production capacity is 34,010 tons per year (the number of FNC is as many as 418 holes). In comparison, for company B the production is 7,608 tons with FNC 184 holes.

For community FNC, the value of initial capital issued is calculated based on the capital issued for each floating net hole, IDR 7.5 million. So for all FNC owned by the community, the amount of investment that has been issued is Rp. 80 billion.

3.1.6. Analysis of Benefit-Cost and Selection of Options Best

A cost-benefit analysis was carried out to select the best option from the three options above, in detail in the Lake Toba RIA Report. The formulation team conducted a qualitative and quantitative cost-benefit analysis. The Formulating Team agreed to choose the third option as the best option (Table 2).

Table 2. Benefit-Cost Analysis

Scenario 1: Do Nothing (74,000 tons)					
No	***Group***	***Benefit***	***+/0/-***	***Cost***	***+/0/-***
1	*FNC Actors*	*Revenue (IDR)*	*2,478,443,030,000*	*- Operational Cost*	+
2	*Feed Industries*	*Revenue (IDR)*	*1,281,000,000,000*		+
3	*Nurseries*	*Revenue (IDR)*	*133,875,000,000*	*- Profit*	-
4	*Employments*		*29,300,400,000*		-
5	*Food Stalls*	*Revenue (IDR)*	*3,000,000,000*		-
6	*Material FNC*		*500,000,000*	*Water Pollution*	- - -
7	*Transportation*		*2,000,000,000*		
8	*Fisheries materials*	*plastic, gas*	*200,000,000*		
	Total		***3,928,318,430,000***		
Scenario 2: Governor's Decree (10,000 ton)					
No	***Group***	***Benefit***	***+/0/-***	***Cost***	***+/0/-***
1	*FNC Actors*	*Revenue (IDR)*	*406,327,693,560*	*- Operational Cost*	+
2	*Feed Industries*	*Revenue (IDR)*	*171,752,654,725*		+
3	*Nurseries*	*Revenue (IDR)*	*28,719,296,364*	*- Profit*	-
4	*Employments*		*3,928,510,136*		-
5	*Food Stalls*	*Revenue (IDR)*	*800,000,000*		-
6	*Material FNC*		*100,000,000*	*Water Pollution*	-

7	*Transportation*		*450,000,000*		
8	*Fisheries materials*	*plastic, gas*	*50,000,000*		
	Total		***612,128,154,785***		
Scenario 3: Analysis Result of The revision of Governors Decree (67,000 ton)					
No	***Group***	***Benefit***	***+/0/-***	***Cost***	***+/0/-***
1	*FNC Actors*	*Revenue (IDR)*	*2,264,861,527,270*	*- Operational Cost*	+
2	*Feed Industries*	*Revenue (IDR)*	*1,170,608,959,460*		+
3	*Nurseries*	*Revenue (IDR)*	*122,338,231,420*	*- Profit*	-
4	*Employments*		*26,775,418,230*		-
5	*Food Stalls*	*Revenue (IDR)*	*2,741,472,970*		-
6	*Material FNC*		*456,912,160*	*Water Pollution*	- -
7	*Transportation*		*1,827,648,650*		
8	*Fisheries materials*	*plastic, gas*	*182,764,860*		
	Total		***3,589,792,935,030***		

Source: primary data, 2021

Based on the analysis conducted, the best option is Alternative Action III, namely The Revision of the Governor's Decree in 2017. The arguments of this election are 1) in terms of benefits, Option III provides higher benefits (improvements in environmental quality, as well as FNC fisheries business opportunities due to government arrangements); 2) In terms of cost, there is a difference in cost implications, but the magnitude is not too significant; 3) better guarantee the sustainability of the Lake Toba management program because: strengthening the supervisory function; regulate the number of FNC; clarify the roles and responsibilities of the parties; and 4) in terms of time, option III requires additional time to revise Governor's Decree, but not too significant because the revised regulatory product is Governor's Decree which is relatively faster than revising the regulation.

If it is related to the magnitude of the benefits obtained between the FNC business and tourism, the FNC business is much larger. The main action of the solution choice is to revise The Governor's Decree because, in the analysis, it is found that several articles need to be adjusted and strengthened. The regulatory substances that need to be revised are, among others, presented in the previous chapter [9].

3.1.7. Implementation Strategy

Team formulator designing strategy and activities for implementation option that has been selected.

1. Revision of the Governor's Decree 188/2017. The government will make the regulation after delivering the result of the RIA to the Governor, who is expected to support the choice of the Formulating Team, who chose the third option. Revising the Governor's Decree is not expected to take a long time because the Legal Division is a member of the Formulating Team who is aware of this step and will speed up the revision process.
2. We are issuing the Governor's Decree on the Implementation Team to manage Lake Toba's carrying capacity and load capacity. This activity is seen not require a large amount of time and money.
3. Develop a work plan for the Implementation Team of The New Governor's Decree. The action requires careful and effective scrutiny activities. The following tables are the proposed strategy and activity plan for the implementation team.

Table 3. Strategy and Plan of Activity Team Implementation

No	Strategy and Plan of Activity	Executor
1	Develop a plan and implement the socialization of the new Governor's Decree to the community, especially to answer problems that arise in RIA, for example, the number of FNC, location, depth, and zoning.	Bappeda
2	Carry out activities to accelerate the target of achieving the implementation of the Governor's Decree until 2023 by, among other things, referring to the Lake Toba Master Plan 2025, including location/zoning of FNC, Fulfillment of requirements;	Bappeda, Government Fishery Office, PDAM (local water company)
3	MOU PDAM and PHRI on wastewater management by hotels and restaurants	Government Tourism office

3.1.8. Stakeholder Consultation

The RIA Team conducted two public consultation meetings to review the Decree of the Governor of the Management regarding Lake Toba's carrying capacity and load capacity, the first when mapping the problem and the second when submitting the RIA results report. Stakeholders invited to the consultation event were: Community leaders and Village Community Empowerment Institutions (LPMD) and others, District Apparatus, Management of the Indonesian Hotel and Restaurant Association representing the private sector, and Non-Governmental Organizations (KSM).

3.1.9. Submission of Results to Decision Makers

The activity was held in December of 2021 by an Audience with the Governor of North Sumatra to deliver results on the Governor's Decree of RIA of Management of Carrying Capacity and Capacity of Lake Toba. The RIA team and several related parties conveyed the results of this analysis, and the next decision depends on the Governor of North Sumatra.

3.1.10. RIA Follow-up

Immediately after the audience with the Governor, the Fisheries Service initiated a meeting to discuss the revision of the Governor's Decree, which the RIA Team attended. In addition to the main substance proposed from the RIA review, the revision of the Governor's Decree also improves several articles in the general provisions concerning definitions. Seeing that there were quite a lot of changes, the RIA team from the Legal Department proposed to make a new decree that replaced Governor Decree no. 188/2017 concerning Management of the Carrying Capacity and Load Capacity of Lake Toba.

4. Conclusion

The based analysis above, it can be drawn that the Decree of the Governor needs to be reviewed back. The basis of the proposal that is the result of the analysis is that if Governor's Decree is applied to determine the production FNC only 10,000 tons per year, it will negatively impact the great against the community's economy, especially that engaged directly with the cultivation of FNC. It is estimated that there will be an economic loss of IDR 3.5 trillion per year. Subsequently, after the analysis continued, the obtained results of the power capacity of Lake Toba on aquaculture fishery amounted to 67,000 tons per year. For this reason, it is proposed that the Governor's Decree be revised its carrying capacity from 10,000 tons per year to 67,000 tons.

The research was financially supported by the LPDP Institute, the Ministry of Finance of the Republic of Indonesia, and supported by the Department of Marine Affairs and Fisheries of North Sumatra.

References

1. D Tanjung. and P Hutagaol. Analysis of potential social conflicts in ecotourism development in the Lake Toba Region, North Sumatra. ISENREM Proceeding (2019)
2. IUWASH. Regulatory Impact Assessment (RIA) Dokumentasi Uji Coba Pendekatan RIA untuk Review Regulasi Air Bersih dan Sanitasi. USAID/Indonesia (2016)
3. C. Albornoz and J. Glückler. MDPI J. E **7**, 104 (2020)
4. K. Jacob, S. Weiland, J. Ferretti, D. Wascher, D. Chodorowska, Integrating The Environment In Regulatory Impact Assessments, OECD (2011)
5. T. Kurniawan, M. Azis Muslim and E. Sakapurnama, KJSS 39, 105 (2018)
6. World Bank Group. Global Indicators of Regulatory Governance: Worldwide Practices of Regulatory Impact Assessments (2018)
7. Kamkhaji, P. Ladegaard and P. Lundkvist, EJRR 10, 187 (2019)
8. C. R. Belfield, A.B. Bowden & V. Rodriguez, AJE 40, 335 (2019)
9. D. Tanjung, R. Syarief, and P Hutagaol. Estimated Recreational Value of Lake Toba using the CTM and CVM Method. Proceeding IAC (2019).

The technology of shrimp larvae transportation: ecophysiology and bioeconomy effects on high stocking density shrimp *Litopenaeus vannamei*

Henry Kasmanhadi Saputra[1*], *Muhammad Subhan* Hamka[2], *Ardana* Kurniaji[3], *Lily* Susanti[4], *Sri Wahyuni* Firman[5], *Agus* Dwiarto[6], and *Hilman Syaeful* Alam[7]

[1]College of Vocational Studies, IPB-University, Bogor, West Java, Indonesia
[2]Freshwater Fish Farming, Rejang Lebong State Community College, Rejang Lebong, Bengkulu, Indonesia
[3]Aquaculture, Polytechnic of Marine and Fisheries Bone, Bone, South Sulawesi, Indonesia
[4]School of Environment, University of Indonesia, Jakarta, Indonesia
[5]Aquaculture, Faculty of Science and Technology, Muhammadiyah University of Education Sorong, Sorong, West Papua, Indonesia
[6]PT Nanobubble Karya Indonesia, Serpong, South Tangerang, Indonesia
[7]Instrumentation Development Center, National Research and Innovation Agency, Bandung, West Java, Indonesia

Abstract. Intensification is part of modern shrimp farming technology, feed, and high stocking density. High density could save time and production costs. This research aimed to determine the optimal tolerance of shrimp larvae with high stocking density per bag unit without oxygen liquid in terms of environmental physiology (ecophysiology) and economic feasibility aspects (bioeconomics). The shrimp larvae used were post-larvae 10 with a weight of 0.0026 ± 0.0021 g, which were previously fasted. The method used was a simulation of a three-hour trip with treatments: 200 larvae/packing (control), 400 larvae/packing, and 600 larvae/packing. The treatment for each packing was 450 ml of water and oxygen gas and added ice cubes in each styrofoam during delivery at night. The results showed that the stocking density of 600 larvae/packing had the best value. In terms of ecophysiology, the survival rate was highest when arriving at the location, with 98.20% of total larvae. The lowest oxygen consumption in the metabolic process was 0.76 ± 0.10 mg/L. Reviewing a bioeconomic, the potential profit per packing reaches IDR 995,528. It can be concluded that the use of oxygen gas and the addition of ice cubes in transportation increased the survival of shrimp larvae.

1. Introduction

Modern shrimp culture is characterized by intensification, feed technology, pellets, and high stocking density in one pond area. One of the leading aquaculture products is the white shrimp commodity. One of the leading aquaculture products is the white shrimp

* Corresponding author: henry_khs@apps.ipb.ac.id

commodity, *Litopenaeus vannamei*. Many sectors are related to vaname shrimp farming, both in grower and hatchery culture. It is essential to pay attention to the live transportation from the seller to the buyer's location in the hatchery sector. If there are many tourist trips, the tourist is the seller or entrepreneur of shrimp larvae at a loss. Generally, industrial-scale shrimp entrepreneurs are not worried about this because they use aeration technology such as adding liquid oxygen to the larval packing bag or aeration technology in micro/nanobubble.

Industrial-scale entrepreneurs are currently able to apply high-density cultivation. It can be achieved if the need for dissolved oxygen through aeration in shrimp culture ponds is met because it is the absolute main factor needed in creating an atmosphere that supports the growth of this type of biota [1,2]. This has minimal impact on mortality during the journey from the harvest location to the stocking location. Packaging of shrimp larvae to consumers is one of the critical supply chains for shrimp larvae hatchery businesses, especially for conventional farmers. Reducing costs in this process is mandatory to support the increase in profit for producers, especially for small-scale hatcheries such as traditional non-corporate system cultivators.

Especially for conventional shrimp larvae cultivators, one effort to reduce costs or capital costs when sending larvae is to reduce the number of larvae densities per bag even though oxygen gas has been added. Stunning the traditional way with the addition of ice cubes and oxygen gas is expected to increase the number of larvae in the bag. Therefore, this study aimed to develop an optimal tolerance system for shrimp larvae with high stocking densities per bag unit and environmental, physiological aspects (ecophysiology) and economic feasibility aspects of high stocking densities (bioeconomics).

2. Methods

The research was conducted at a company that provides quality larvae around the Province of Banten, and this activity began on August 22-23, 2020. The Covid 19 pandemic caused research to be carried out outside the production site to avoid contamination by cultivation activities. In this study, oxygen gas (liquid oxygen was generally used) was added through an oxygen tube at packing. The addition was to the habits of shrimp larvae farmers. The methods in this study were divided into several series, starting from the preparation of shrimp *Litopenaeus vannamei* larvae, preparation of raw water for packaging, determination of research treatment, and transportation simulation. The first step is by weighing the weight of the larvae, PL10, with a weight of 0.0026 ± 0.0021 g. The second step is preparing raw water by adding an aeration technology such as bubbles or bubbling [3].

The preparation of raw water for packing media in shrimp larvae bags is transported as much as 5 tons of water or approximately 5,000 L. The third step was the provision of larvae with research treatment in plastic bags. The treatments applied were 200 larvae/bag, 400 larvae/bag, and 600 larvae/bag, with the volume of water/bag being 450 mL. The density was based on the minimum standard in white shrimp larvae supplying companies, 2,000 shrimp larvae/45,000 mL of media water in the bag. This study was compressed (empirically) to 200 shrimp larvae/450 mL of media water in the bag. The stocking density of each bag that has been suitable for treatment is then given oxygen gas and then tied with rubber, then put in styrofoam and ice cubes inserted in plastic. The fourth step was a short-distance transportation simulation process with a three hours-trip from 10.00 pm to 01.00 am. The mileage division was divided into three based on the length of time the oxygen gas ran out once oxygen gas was given to the bag. Short distances are from one hour to six hours, medium distances are from seven hours to twelve hours, and finally, long distances are at least twelve hours to twenty-four hours (a day).

In this research, shrimp larvae that have been packaged according to treatment in bags are then shaken in a box container randomly every 60 minutes which are divided into several times. This activity was considered the same as in the field when transporting using a pickup car. Water quality was measured every hour from the beginning to the end of the study for three hours. The parameters measured were water quality parameters in physical and chemical parameters: temperature, TAN (Total Ammonia Nitrogen), and dissolved oxygen (DO). Production performance parameters in the form of OCR (Oxygen Consumption Rate) and survival rate of larvae.

2.1 Water Quality

The following are the water quality parameters and tools used in this study, as well as the measurement times listed in the table below:

Table 1. Water quality parameters and specifications for measuring water quality

Parameter	Unit	Method/Tools	Specification	Measurement Time
Temperature	°C	Thermometer	Lutron 5510	Every hour
TAN	mg/L	Spectrophotometer	Lutron 5510	Every hour
pH	-	pH-meter	ATC pH tester	Every hour
DO	mg/L	DO-meter	Lutron 5510	Every hour

2.2 Oxygen Consumption

The level of oxygen consumption was measured by obtaining the difference between the DO of the medium after and before transportation. The formula used was as follows [4]:

$$OCR = \frac{V \times DO_{t0} - DO_{tn}}{W \times T}$$

Description :
OCR : Oxygen consumption rate (mgO_2/g/hour)
V : Volume of water in tank (L)
DO_{t0} : Dissolved oxygen concentration at the beginning of observation (mg/L)
DO_{tn} : Dissolved oxygen concentration at time in n (mg/L)
W : Biota weight (g)
T : Observation period (hour)

2.3 Survival Rate

The survival rate is calculated using the formula [5]:

$$SR = \frac{N_t}{N_0} \times 100$$

Description:
SR : Survival rate (%)
N_t : Calculated using the formula t (larvae)
N_0 : Number of live shrimp on day 0 (larvae)

2.4 Data Analysis

The design used in this study was a Completely Randomized Design (CRD) with three replications. Water quality data were analyzed descriptively, while production performance

https://doi.org/10.1051/e3sconf/202234800038

data were analyzed using analysis of variance with a 90% confidence level. If there was a significant difference between treatments, then further tests with *Duncan's Multiple Range* using SPSS 16.

3. Results and Discussion

3.1 Water Quality

Temperature, dissolved oxygen, TAN, and pH are the main water qualities that play an essential role in conditioning a good aquatic environment. These parameters will affect other chemical and physical parameters. Water quality in the form of temperature, TAN content, pH levels, and oxygen consumption rate in shrimp larvae media was presented in Figure 1. The results of observations of water quality show different fluctuations in each treatment and observation time. The results of the water temperature measurement show an increase in the value from the beginning to the end of the observation. The temperature increase was slower in the treatment of the density of 200 larvae, especially during the observation at the first hour. The time of transportation carried out at night causes the temperature to be maintained in the range of 20.37-22.97°C.

Temperature is closely related to dissolved oxygen. The higher the temperature, the smaller the dissolved oxygen (DO) in the water and biological and chemical processes will increase, so oxygen consumption will also increase [6]. This increase in oxygen consumption results in a decrease in dissolved oxygen in the water. This can be observed at DO, which decreases from the beginning to the end of the observation. Dissolved oxygen was an essential factor in aquaculture activities. Not only do shrimp consume oxygen for respiration, but oxygen is also used to stabilize water quality. Metabolic waste can be high along with shrimp density. The observations showed that the density treatment of 600 larvae had the lowest DO value compared to other treatments. The highest DO levels were observed at the treatment density of 200 larvae at each observation time. Reduced oxygen in aquaculture is high density and larger shrimp sizes [7]. The leading cause of reduced dissolved oxygen in water is waste materials that consume oxygen [8].

Shrimp was a poikilothermic biota, so any increase in water temperature around the environment will cause the temperature of cold-blooded animals to increase due to increased metabolism in their bodies, and stress plus abnormal stocking density has an impact on increasing oxygen demand. This causes a decrease in the concentration of oxygen in the water medium. Converting oxygen into energy (metabolism) and producing ATP or energy in cells also produces urea or feces rich in ammonia. This was the cause of the increase in ammonia, which was measured using the TAN parameter. The results of the TAN measurement showed an increase from the beginning to the end of the observation. The proportion of unionized ammonia increases as the temperature and pH increase. Un-ionized ammonia dominates at high pH and is toxic, whereas ammonium is relatively non-toxic at low pH [9]. The presence of residual respiration impacts the balance of hydrogen ions in the water medium. This was evident from measuring the pH value periodically, experiencing fluctuations in each treatment every hour of observation.

The treatment with a density of 600 larvae was observed to decrease lower than the other treatments in the second hour and increase in the 3rd hour of observation. Likewise, the 200 and 400 larvae density decreased in the 2nd hour and increased in the 3rd hour. The increase in TAN occurred more slowly at the treatment density of 200 larvae observed in the 1st hour. This indicates that the low density has a lower TAN value than the higher density. TAN increases due to the secretion of shrimp which also uses oxygen, the higher the density, the higher the ammonia secretion produced [10]. TAN is the most toxic form of un-ionized ammonia nitrogen and has a negative impact on shrimp health, so its presence needs to be controlled [11].

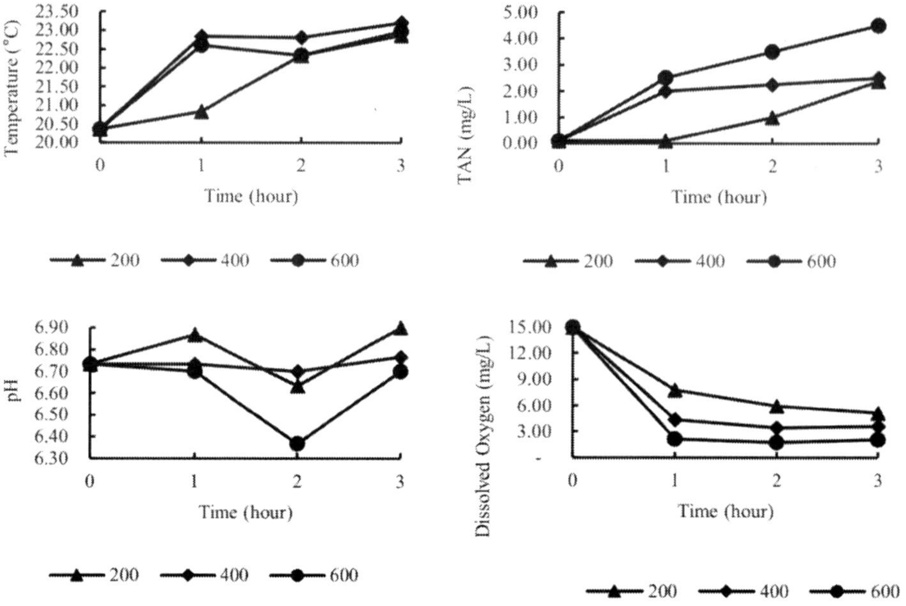

Fig. 1. Fluctuations in Temperature, TAN, pH, and oxygen consumption rate in Vaname Shrimp Larvae Media with Different Densities

3.2 Ecophysiology

Ecophysiology was a study of the physiology of organisms related to the surrounding environment. In this case, two parameters describe the interaction between biota (larvae) and the environment (water media in bags) in this research, the level of oxygen consumption and the survival rate during treatment.

3.2.1 Oxygen Consumption Rate

The level of oxygen consumption was a parameter to determine the metabolic rate closely related to growth [12]. The level of oxygen consumption was an illustration of the amount of oxygen consumed by biota from the oxygen consumption rate available in aquaculture media. Based on Figure 2, it was known that the highest value in the treatment of 200 larvae/bag was 3.66 ± 0.44 mgO_2/g/hour and the lowest value in the treatment of 600 larvae/bag was 0.76 ± 0.14 mgO_2/g/hour. The low level of oxygen consumption in the treatment of 600 larvae/bag indicated that the larvae on high stocking density media could not consume much oxygen 0.0022 mgO_2/g/hour for each PL 10 shrimp larvae. The lowest temperature in the treatment supported this fact, which tends to be colder in the range of 22.97±0.31°C compared to other treatments, 23.20±0.20°C.

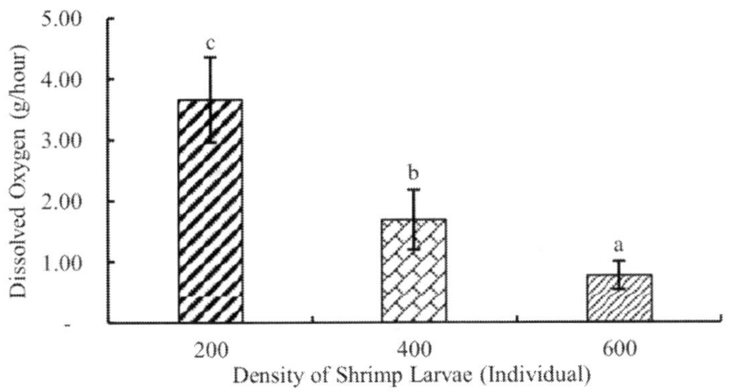

Fig. 2. The level of oxygen consumption of vaname shrimp larvae on short-distance transportation with different densities

Low-temperature levels can affect the metabolic rate in the body of vaname shrimp larvae, including being able to inhibit the movement of larvae. As stated, temperature affects the level of oxygen consumed by shrimp and the salinity and activity level of biota [13]. Shrimp that are slow to move affect the lack of oxygen consumed, converted into energy [14]. If the shrimp larvae are slow to move, the potential for ammonia or waste was also low, so the chance of environmental stress/pressure and death due to poisoning was low. The energy produced will be converted to growth, not to metabolism.

3.2.2 Survival Rate

All treatments were not statistically significantly different. Based on the results of the research, it was found that the highest survival rate or survival rate was found in the treatment with a stocking density of 600 larvae/bag (98.20±0.67%), followed by a stocking density of 400 larvae/bag, 97.12±1.76%, and the lowest was at 200 larvae/bag (97.08 ± 0.72%). From a numerical point of view, the best choice in treating 600 shrimp larvae/bag, but according to [15], statistical significance was a direct result of mathematical equations and not necessarily relevant to describe all influencing factors in the field. A possible approach for making decisions regarding statistically non-significant outcomes was to draw relationships with key parameters closer to the parameters being analyzed (Figure 3) oxygen consumption level. Judging from the level of oxygen consumption parameters, the best value was the treatment of 600 larvae/bag. The value of the frequency of media shaking, the volume of media water, the health status of the test biota, and other parameters are some of the factors that affect significantly or are not related to the statistical test.

The survival rate of shrimp larvae was calculated based on the number of shrimp larvae. Live at the end of the treatment so that the statistically significant difference was more towards the label and the potential impact in the study itself was not an external impact, for example, economic impact or other impacts that are not the purpose of the research. Based on the above, the recommended survival rate in this study was the treatment of 600 shrimp larvae/bag with a survival rate of 98.20±0.67%.

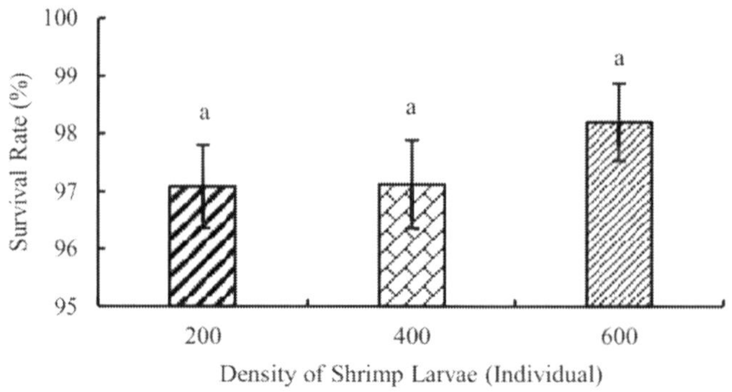

Fig. 3. The survival rate of vaname shrimp larvae at different density transport

3.3 Bioeconomics

Bioeconomics is an activity with economic value based on data on the production and growth of living organisms that can adapt to a fluctuating environment in the context of applied science to achieve prosperity. One of the activities that included bioeconomics was aquaculture activities. Each cultivation activity related to stocking density or density can be compared between the amount of stocking density and the constant output [16]. This activity was inseparable from production, including production costs and investment costs to make profits from the business. It was impossible for good cultivation results, but it was still a loss from the economy of scale. It includes businesses related to vaname shrimp larvae PL 10. In larvae cultivation business activities on a traditional business scale by prioritizing the use of oxygen gas.

Based on table 2, the production cost was the same in all treatments (200 larvae/bag, 400 larvae/bag, and 600 larvae/bag) during the process of transporting larvae from the farmer to the buyer, which was IDR 19,497,000. This was because the supporting material for the transportation of shrimp larvae and the employee costs incurred were the same because the weight of the tail was not the primary consideration but the number of larvae. The unit observed was per unit of transport capacity, one pickup car with a travel time of three hours or a short distance. The capacity of the pickup car was 1,305 litres which could be filled with 58 boxes, and each box contained 50 bags with a volume of 450 mL of water. At the beginning of the departure conditions, it was assumed that all larvae were 100% alive, and during the trip, there were no obstacles such as accidents or traffic tickets on the way or other events.

Table 2. Production costs of vaname shrimp larvae with different densities

Stocking Density (Larvae/Bag)	Total Gas Cost / Bag (IDR)	Packaging Fee / Bag (IDR)	Cost of Ice Cubes and Styrofoam / Bag (IDR)	Pick Up Rental L 300 (IDR)	Total Number of Fry (Larvae)	Total Cost (IDR)
200	400	5,500	600	300,000	580,000	19,150,000
400	400	5,500	600	300,000	1,160,000	19,150,000
600	400	5,500	600	300,000	1,740,000	19,150,000

There was the most considerable total profit value among them in the treatment of 600 larvae/bag (Table 3). The profit value obtained from this treatment was IDR 995,528 or ten times the value with a stocking density of 200 larvae/bag, which was IDR 106,688.

Although not statistically significant, the number of survivors living in the treatment was 98.20% or the lowest compared to the other treatments. The cause of the considerable profit value was the larvae's selling price and the bag's capacity to accommodate the larvae stocking density. On the other hand, this was an indication that the density of 600 larvae/bag was still categorized into the optimal tolerance for the number of biota/bag in secure packing with oxygen gas and ice cubes in styrofoam. The survival rate of larvae was still above 85%, which means it was still good in terms of pressure resistance during the trip.

Table 3. Acceptance of results on different density vaname shrimp larvae

Stocking Density (Larvae/Bag)	Survival Rate of Larvae (%)	Selling Price Per Larva (IDR)	Total Number of Live (Larva)	Revenue (IDR)	Total Profit (IDR)	Profit / Bag (IDR)
200	97.08	45	563,064	25,337,880	6,187,880	106,688
400	97.12	45	1,126,592	50,696,640	31,546,640	543,908
600	98.20	45	1,708,680	76,890,600	57,740,600	995,528

4. Conclusion

The best treatment that can be used at short distances was a density of 600 larvae/bag with a survival rate of 98.20%. The results support this that the level of oxygen consumed by each larva was more efficient at 0.0022 mgO_2/g/hour, which is lower than other treatments and in terms of economic potential, the most significant net profit is IDR 57,740,600 compared to other treatments.

References

1. R. Arantes, R. Schveitzer, C. Magnotti, K.R.Lapa, L. Vinatea. Aquac. Res. **48(4)**, 1359–1980 (2020)
2. M. Jayanthi, A.A.K. Balasubramaniam, S. Suryaprakash, N. Veerapandian, T. Ravisankar, K.K.Vijayan. J. Aquaeng, **92**, 81–90 (2021)
3. C.E. Boyd. J. Aquaeng, **18(1)**, 9–40 (1998)
4. I.C. Liao, H.J. Huang. J. Fish. Soc. Taiwan, **4(1)**, 33–50 (1975)
5. H. Zokaeifar, J.L.Balcázar, C.R. Saad, M.S. Kamarudin, K. Sijam, A. Arshad, N. Nejat. Fish Shellfish Immun., **33(4)**, 683–689 (2012).
6. R. Mallekh, J.P. Lagardere. J. Fish. Biol., **60**, 1105-1115 (2002)
7. S.Z.H. Shah. Departement of Zoology. University of Gurajat (2019)
8. R.V. Anandasari, E. Supriyono, O. Carman, K. Adiyana. J. Akuakultur Indonesia, **14(1)**, 42–49 (2015)
9. J.A. Hargreave, C.S.Tucker. Louisiana State University Agricultural Center Mississippi State University (2004)
10. G. Palomino, F. Contreras, A. Sanchez, C. Rosas. J. of the World Aquacult. Soc., **32(2)**, 167-176 (2002)
11. S.M. Levit. A literature review of effects of ammonia on fish. The Nature Conservancy, Center for Science in Public Participation, Bozeman, Montana. (2010)

12. W. Nurussalam, K. Nirmala, E. Supriyono, Y.P. Hastuti. 2017. J. Akuakultur Indonesia, **16(2)**, 144–153 (2017)
13. R.G. Chittleborough. Aust J. Mar and Freshwater Res, **26(2)**, 177–196 (1975)
14. Li E, Chen L, Zeng C, Chen X, Yu N, Lai Q, Qin JG. Aquaculture, **265(1–4)**, 385–390 (2007)
15. J.D. Harris, J.C.Brand, M.P.Cote, S.C.Faucett, A.Dhawan. Arthroscopy, **33(6)**, 1102–1112 (2017)
16. B. Muangkeow, K. Ikejima, S. Powtongsook S, Y. Yi. Aquaculture, **269(1–4)**, 363–376 (2007)

The Importance of Halal Certification for the Processed Food by SMEs to Increase Export Opportunities

Made Gayatri Anggarkasih[1*] and *Prima* Sukmana Resma[2]

[1]College of Vocational Studies IPB University, Jl. Kumbang No.14 Bogor, Indonesia.
[2]The Assessment Institute for Foods, Drugs, and Cosmetics Majelis Ulama Indonesia, Global Halal Center Building, Jl. Pemuda No.5 Bogor, Indonesia.

Abstract. As the largest Muslim population country, Indonesia potentially becomes the largest exporter of halal food products globally, especially to OIC countries with demands for the fulfillment of halal product guarantees and food safety. Currently, BPJPH is assigned as the organizer of halal product guarantees in Indonesia, and MUI remains the party authorized to issue a fatwa on the determination of halalness. This fatwa will be submitted to BPJPH as the basis for issuing halal certificates. The halal certification and assurance systems designed and implemented by MUI have also been recognized and even adopted by halal certifier bodies abroad. Until 2019, only 0.11% of SMEs had halal certificates. Promoting halal certification for SMEs is an important strategy in increasing food exports. In addition, to the obligation to comply with Law number 33 of 2014 on Halal Products Guarantee (Law of JPH). Halal certification significantly enhances the marketability of SMEs products. This paper describes the importance of halal certification of processed foods to increase export opportunities and reviews the government's concrete steps in implementing the Halal Certification for SMEs.

1 Background

Indonesia is a country with the largest Muslim population in the world. Currently, halal products are necessary for the community, even for non-Muslim communities. Indonesia has a large Muslim population but has not become the world's main producer of halal products. Many exporting countries for halal food products come from countries with a non-Muslim majority, such as Brazil and Thailand. At the same time, Indonesia is in the 20th position [1].

Halal-certified products can assure consumers that the products are processed with halal materials and with halal processes. Besides, they also meet good hygiene sanitation standards so that the products consumed are guaranteed halal and have their food safety. Producers who have halal certificates can market their products to Muslims and to the world because halal food is not only part of Islamic law but also because of its emphasis on

* Corresponding author: madegayatri@apps.ipb.ac.id

hygiene and health. The main halal products in this paper focus on processed foods. Indonesia enforces the Halal Product Assurance Law, whereby products that enter, circulate and are traded in Indonesian territory must be certified halal (Law Number 33 of 2014, article 4). This law is also applied to SMEs. Halal certification can increase competitiveness and marketing for SMEs.

The benefits of halal certification on processed food products are:
- Ensuring that the products produced by manufacturers are guaranteed to be halal and safe.
- Providing peace for Muslim consumers for the products they consume.
- Increasing sales value and expanding marketing
- Increasing product competitiveness
- Being consumed not only consumed by Muslims but also by non-Muslims
- Meeting export requirements, especially to Muslim-majority countries

Halal certificate is a written fatwa of the Indonesian Ulama Council (MUI) which states the halalness of a product following Islamic law. This certificate is exceptionally important for SMEs and can become an added value to products and facilitate wider market access. Halal certificates for SME products can be an added value and reach greater market access since it provides certainty of the product's halal status so that it soothes consumers in consuming them [2].

The government has made several efforts to assist SME halal certification. Through halal certification, Indonesia has the potential to become the largest exporter of halal products in the world, especially to OIC countries that have strict halal and food safety requirements. This paper aims to explain the importance of halal certification, especially for processed food products, to increase export opportunities and review the government's concrete steps in implementing halal certification for SMEs.

2 Literature Review

The literature review in this paper was carried out to obtain conclusions about the importance of halal certification for SMEs, especially in increasing the export opportunities of processed food products. This literature review contains information from the literature such as books, journals, encyclopedias, documents, news, and so on to find various theories and ideas that can be formulated according to the paper's aims.

2.1 Halal and haram

The concepts of halal and haram are written in the holy Qur'an QS 2:168 and QS 2:172, which explain that a Muslim is required to consume halal food. The word halal, taken from Arabic, means "permissible," while the definition of halal food and drink is food and drink permitted to be consumed according to Islam. In addition to the term halal, the term toyyiban also means "good". A Muslim should consume food that is halal and *toyyiban*, which is permissible and has good quality, and does not damage health. As opposed to halal and *toyyiban*, there is haram food. Haram food is food that is strictly prohibited from being consumed [3]. Someone will sin if they still consume it. In consuming foods, several conditions must be met and considered so that humans avoid various diseases that come from food [4].

In Islamic law, pork, blood, and carrion are forbidden to be consumed. In addition, animals with fangs and tusks, poisonous animals, animals that live in two realms, animals that eat carrion and dung are also forbidden to eat. The plant-based ingredients are halal if they are not toxic (the poisons can be eliminated during the processing) and are not processed into haram foods such as alcoholic beverages. *Khamr* or alcoholic beverages are unlawful because it is intoxicating. The use of ethanol in food processing (e.g., being used as a solvent in extraction) should be eliminated as much as possible in the final product [5].

Food derived from animals that are permitted are halal if the animals are slaughtered according to Islamic law: (1) The animals slaughtered by an adult Muslim, healthy and knowledgeable about halal slaughter; (2) The tools used must be sharp and not in the form of nails, fangs or bones; (3) Slaughter is based on the intention of slaughtering and by mentioning the name of Allah; (4) Slaughter is carried out by draining the blood through the cutting of the food channel (oesophagus), respiratory/throat (trachea), and two blood vessels (jugular vein and carotid artery); (5) After the slaughter, the blood is allowed to flow out until it stops flowing; (6) Slaughter is carried out hygienically and maintains a clean environment; (7) Storage is carried out separately between halal and non-halal [6].

2.2 Processed Food

Food processing is converting raw materials into intermediate products and ready-to-eat products that can be directly consumed and have nutritional values [7]. Food should be prepared, processed, packaged, transported, and stored according to the general principles of the Codex on Food Hygiene and other related Codex Standards [8]. In processed foods, there is a change in the form of raw materials into products through several processes with the addition of food additives. The criteria for halal food are as follows (1) it must not contain anything that is considered haram according to Islamic law; (2) At the stage of preparation, processing, transportation, and storage, the equipment used must be free from anything that is considered haram according to Islamic law; (3) At the stage of preparation, processing, transportation, and storage there is no direct contact with food that is haram according to Islamic law [9].

In food processing, the ingredients that can be eaten cannot be confirmed as halal. Chicken and beef can be eaten, but they must also follow the procedures for slaughtering animals according to Islamic law regulated in the MUI Fatwa number 12 of 2009 concerning Halal Slaughter Certification Standards [10]. In addition, the use of food additives also creates ambiguity about the halalness of the product because it is obtained from various processes. In bread and pastries, L-cysteine, commonly used in production, is made from feathers, pig bristles, and sometimes even human hair [11]. Most of the gelatine used in sweets and marshmallows is obtained from pigskin collagen (46%) and bovine bone for halal gelatine alternatives [12]. Rennet and pepsin are the two enzymes commonly used in cheese production. Pepsin, an enzyme derived from pigs' stomachs, is also present in some cheeses available at supermarkets [13]. Analysis of the halal status of a product is very important for detecting raw materials or the detection of mixed materials. Counterfeiting of halal materials with haram materials is very likely to be carried out by irresponsible parties [14]. Therefore, a system is needed to guarantee the production of processed food according to halal criteria, namely the Halal Assurance System.

2.3 Halal Assurance System

Halal Assurance System (HAS) is a work tool composed of management commitments, resources, and interconnected procedures to ensure the halalness of products following the requirements so that the halal status is consistent and sustainable. The Halal Assurance System can adopt management system principles developed previously, such as TQM, ISO 9000, and HACCP [9]. There are 11 criteria for the Halal Assurance System that are covered in HAS 23000, which include (1) Halal Policy; (2) Halal Management Team; (3) Training; (4) Materials; (5) Production Facilities; (6) Products; (7) Written Procedure for Critical Activities; (8) Traceability; (9) Handling of Products that do not meet the criteria;

(10) Internal Audit; (11) Management Review [15]. HAS 23000 is a halal certification requirement set by LPPOM MUI for halal certification of a product

The halal assurance system contains requirements applicable to all business categories, including food processing, pharmaceutical and cosmetic industries based on agricultural products, slaughterhouses, restaurants or caterings, and service industries (distributors, warehouses, transportation, retailers). HAS 23101 is a guideline for fulfilling the criteria for a halal assurance system in the processing industry. The food processing industry in Indonesia must implement a Halal Assurance System. The Halal Assurance System is a management system compiled, implemented, and maintained by companies holding halal certificates to maintain the continuity of the halal production process according to the provisions [16]. The Halal certificate owned by each producer is a way for Muslim consumers to get guarantees that the products they consume are halal. Meanwhile, the halal assurance system is one way to convince the public that the product is consistently halal during the validity period of the halal certificate.

2.4 Halal food regulation

The Indonesian government responded positively to the importance of halal certification through several regulations. Halal certification and labeling aim to protect domestic consumers in the AEC and increase the competitiveness of Indonesian halal product producers and exporters in the international market. The entrepreneurs' products, including Indonesian exporters who already have a halal certificate, will be more easily accepted by consumers of halal products in other countries by looking at the halal label listed on the product [17]. On October 17, 2014, Law number 33 of 2014 concerning Halal Product Guarantee (Law of JPH) stated that halal certification is required for all products traded in Indonesia. In addition, starting October 17, 2019, the government has started to enforce Government Regulation number 31 of 2019 concerning the implementation regulations of Law Number 13 of 2014, which regulates the strengthening of previous regulations. It starts the actual implementation from what was originally voluntary to become mandatory, whereas Government Regulation number 31 of 2019 provides five years from the date of promulgation [18].

The enactment of the Law of JPH brings changes regarding halal certification providers. The Halal Product Guarantee Agency (BPJPH) cooperates with the Ministry, the Halal Inspection Agency (LPH), and the Indonesian Ulama Council (MUI) in realizing the Law of JPH. LPH carries out halal product inspection, and MUI carries out the halal fatwa determination by issuing a decision to determine the halal product through the Fatwa session [19]. Halal Certificate is valid for 4 (four) years from the date of issuance by BPJPH, unless there is a change in the composition of the ingredients. Producers are required to extend the Halal Certificate by applying for an extension of Halal Certificate no later than three months before the validity period of the Halal Certificate expires [20].

Implementing the mandatory halal certification under the Law of JPH began to be implemented on October 17, 2019. Food and beverage certification will be carried out in gradually until October 17, 2024. BPJPH will be more persuasive in conducting socialization and guidance to producers during these five years and has not yet started law enforcement. Next, it will start for non-food products on October 17, 2021 [21].

2.5 Halal certification procedures for SMEs

The flow of the halal certification mechanism through BPJPH consists of five main stages, namely (1) Registering yourself and attaching the required documents; (2) Waiting for BPJPH will check the attached requirements; (3) Conducting product audits and inspections; (4) Submit the results of the examination to MUI to issue a fatwa; (5) Issue a halal certificate if it has passed the audit. Halal certification was originally a work area of MUI through LPPOM and transferred to BPJPH, an official government agency under the Ministry of Religious Affairs. This transition was also followed by several changes to the flow of the certification process and the parties involved [22]. Currently, LPPOM is one of the Halal Inspection Agencies (LPH) in charge of inspecting and testing products' halalness.

2.6 The Indonesia government programs to build a halal ecosystem

2.6.1 Establishment of halal regulations

A halal certificate is government protection for Muslim consumers and non-Muslims. With halal certification, consumers will be safer in consuming or using products and avoiding products containing haram elements. The government's protection is manifested in the form of Law of JPH signed by President Susilo Bambang Yudhoyono on October 17, 2014, in the form of Law No. 33 of 2014 concerning Halal Product Assurance mandates the establishment of the Halal Product Guarantee Agency (BPJPH). BPJPH is directly responsible to the Ministry of Religious Affairs [19]. The halal certificate issued by BPJPH based on the results of the examination by the LPH and the determination of the MUI Fatwa is valid for four years. It has a strong guarantee of legal certainty. There are both criminal sanctions and fines for business actors who do not maintain the halalness of products that have been certified halal [22].

2.6.2 The free halal certification program

The free halal certification program has been carried out as a form of implementation of the memorandum of understanding (MoU) of 2020 between Ministry of Religious Affairs and Ministry of Industry, Ministry of Trade, State Minister for Cooperatives Small and Medium Enterprises, Ministry of Tourism and Creative Economy/Tourism and Creative Economy Agency, Ministry of State-Owned Enterprises, Ministry of Home Affairs, Amil Zakat National Agency, the Indonesian Waqf Agency, National Sharia Economic and Finance Committee (KNEKS) on the implementation of halal certification for MSEs.

This MoU functions as a guideline for the parties to facilitate the implementation of halal certification for micro and small enterprises (MSEs), including halal certification costs. In addition, this collaboration is also facilitating halal supervisors for MSE, starting with providing prospective halal supervisors, halal supervisor data, financing for halal supervisor certification training, financing for participation in halal supervisor certification competency tests to providing socialization, information, and education of halal supervisors in the context of mentoring MSEs [23]. Apart from the ministry, free halal certification is also carried out by LPPOM MUI through the Syawal Festival, in which 644 of qualified SMEs take apart in this program [24]. Likewise, PT Indofood Sukses Makmur Tbk Bogasari Division provided free halal certification for 50 chicken noodle SME partners in 2020 [25].

2.6.3 Establishing a halal industrial area

The development of halal industrial areas (KIH) in the National Medium-Term Development Plan (RPJMN) in 2020-2024 was achieved in 2021 including Modern Cikande Industrial Estate in Serang (Banten), Safe n Lock Halal Industrial Park in Sidoarjo

(East Java), and the Halal Industrial Park in Sidoarjo (East Java), and Bintan Industry Inti Halal Hub (Bintan Regency). KIH Cikande is an area designed with systems and facilities to develop industries that produce halal products according to sharia principles [26]. KIH will be developed into a halal industry cluster as a halal ecosystem from upstream to downstream, including a logistics system with the hope of becoming an international halal hub in Indonesia. In addition, the government also encourages large and small and medium industries (SMEs) to synergize well [27].

2.6.4 Establishing a national sharia economy and finance committee

The National Committee for Sharia Economy and Finance (KNEKS) aims to improve the development of the sharia economic and financial ecosystem to support national economic development and make Indonesia a World Halal Centre [28]. In line with KNEKS, the Islamic Economic Community (MES) is a non-profit organization that aims to develop and ground the sharia economic system as an economic system that is just and based on sharia principles [29]. The vision and mission and the direction of the MES policy are in line with the KNEKS work plan in developing the sharia economy according to the sharia economic master plan launched in 2019. In strengthening the halal value chain on the food and beverage cluster in the Sharia economic master plan, there is a strategy for developing the halal food and beverage industry with a diversification of export destinations and product specialization [30].

3 Discussions

3.1 The importance of halal certification for SMEs in Indonesia

The halal food market is one of the markets with the largest consumers, according to the Global Islamic Economy Report 2014. This is because Muslims worldwide spend 16.6% of the total expenditure on global food, making the global halal food one of the largest markets. The world's Muslim population is predicted to reach 2.2 billion in 2030 and 2.6 billion in 2050, and Muslims should consume halal food, causing the halal food market to dominate the global food market [31]. According to the Global Islamic Economy Report 2019/2020, Muslim spending on food and beverages was $1.4 trillion in 2018 and is expected to reach $2.0 trillion by 2024. Indonesia is currently in 20th place as an exporter of halal food with a 1.86% market share, much lower than that in the non-Muslim majority countries such as Brazil and Thailand. Based on the data from the Ministry of Trade, in 2020, Indonesia's export performance for the food sector reached US$ 454.16 million, with superior products being sauces and their processed products (20.03%); pasta (17.71%); processed fish (13.5%); extracts, essences, and concentrates of coffee (7.78%) and processed foods (7.32%). Indonesia's export destinations in the food sector include Malaysia, Saudi Arabia, Nigeria, United Arab Emirates (UAE), and Jordan. OIC countries' imports for the processed food category showed a rapid growth rate in 10 years, from US$ 540 million (2005) to US$ 2.19 billion (2014). It may reflect the impact of increased purchasing power and changes in eating habits. Younger people are favored by pre-served or prepared products, more likely by international trends and advertising [32]. Indonesia potentially becomes the largest exporter of halal food globally, including OIC countries.

The condition of global halal product trade opens opportunities for producers and SME exporters in Indonesia and poses challenges not only for SMEs but also for the government as a regulator. Two challenges for the halal food industry include bureaucracy and regulation. The Indonesian Food and Beverage Entrepreneurs Association (GAPMMI) complains that food regulations constantly change. On the other hand, GAPMMI considers halal certification labeling in Indonesia is very good so that products produced in Indonesia are easy to market, especially in the Middle East [31]. Halal certification is required to export products to OIC countries, especially category A countries such as the UAE, Saudi Arabia, Bahrain, Yemen, and 15 other countries. In exporting their food products to Islamic countries, the manufacturers must comply with halal manufacturing standards. There is an incentive to choose an internationally recognized halal certification authority to certify products because several countries have implemented a cross-certification system [33]. In the GCC guide for control imported goods, the required documents for products that will enter the GCC countries member include a health certificate (must be issued by the appropriate and competent authority), list of ingredients, halal certificate, supporting documents for label claims and temperature records where required. Food products are handled in places that have been inspected by the competent authority and or officially recognized bodies and implement a food safety management system based on HACCP principles or an equivalent system. This shows that in addition to being halal, the food safety of food products exported to GCC countries, including the UAE, must also be guaranteed.

Since June 15, 2020, in the halal certification process, LPPOM MUI has conducted additional audits and audits according to the criteria for the halal assurance system according to HAS 23000. The audit is related to food safety. This regulation refers to SNI ISO/IEC 17065: 2012 and UAE 2055:2 for halal certification bodies, namely the addition of food safety requirements in the halal certification process for the food and beverage industry imposed by the National Accreditation Committee (KAN). Previously, LPPOM MUI obtained an accreditation certificate from the National Accreditation Committee (KAN) with the SNI ISO/IEC 17065: 2012 and UAE 2055:2. With that, LPPOM MUI can run a certification body according to the international standards and the acceptance of products certified by MUI to countries with the same standard references. LPPOM MUI has also received recognition from the Emirates Authority for Standardization and Metrology (ESMA) to accept MUI-certified products in the UAE [34].

3.2 Review the government's concrete steps

The government, through BPJPH carries out the mandate in the JPH Law in holding halal certification together with LPH and MUI. The current flow of halal bag certification is carried out through the registration process to BPJPH to check for completeness of documents. After the documents are complete, BPJPH will issue a registration receipt (STTD). Next, the applicant registers with the LPH that has been selected by attaching documents, including STTD from BPJPH. LPH will check the adequacy of documents and implement the criteria for the existing halal assurance system in the company. LPPOM MUI is one of the LPH that BPJPH has recognized. Registration for LPPOM MUI halal certification is done online through CEROL SS-23000. The next process is a fatwa trial by the MUI to determine the product's halal status. If it is proven that it meets the halal requirements, the MUI will issue a halal decree that is the basis for issuing a halal certificate by BPJPH. The flow of the halal certification process can be seen in Figure 1. The flow of halal certification in Indonesia is carried out to ensure that products that pass the inspection are halal and have no doubts. Products that are declared halal will then obtain a halal certificate and can include a halal logo on the packaging. The Halal logo attached to the packaging can reassure Muslim consumers that the products produced are prepared according to Islamic requirements. As for non-Muslim consumers, the Halal logo recognizes that food is prepared hygienically and cleanly for consumption [35]. The presence of the Halal certification is expected to be able to give some confidence to Muslim consumers and save them from time-consuming, tedious, and somehow problematic ingredients checking [36].

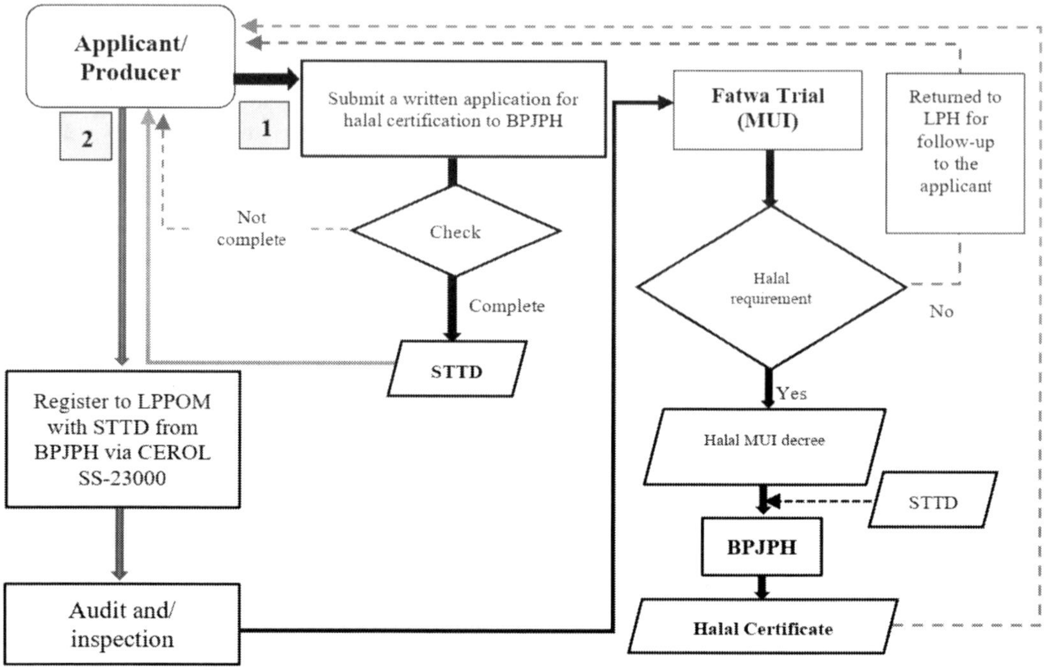

Fig. 1. Halal certification process by BPJPH, LPPOM (LPH), dan MUI

In the halal certification process, SMEs face several obstacles, including the limited cost of certification by SMEs. The Indonesian government has issued a regulation regarding the tariff for halal certification fees through Regulation of Minister of Finance of Republic Indonesia Number 57 of 2021 as a follow-up to Act Number 11 of 2020 on Job Creation. In addition, subsidized programs and even free halal certification can be obtained through several ministries, such as State Minister for Cooperatives Small and Medium Enterprises and the provincial government. This program is a manifestation of the MoU carried out by ten ministries and institutions in Indonesia to accelerate the implementation of halal certification for SMEs. Until mid-2021, the halal certification assistance program was still running. Therefore, the government still requires more massive socialization through BPJPH and LPH for SMEs that have not registered themselves in the halal certification program.

Another challenge faced by SMEs in halal certification is fulfilling the criteria for the halal product guarantee system, especially related to materials and traceability. The use of materials is not only permitted in Islam but must be able to trace its halalness, especially for critical materials. If SMEs use critical materials that are not halal, then SMEs cannot use these materials. This is often an obstacle for SMEs because substitute materials are difficult to obtain or more expensive price. In addition to materials, traceability criteria are often obstacles for SMEs. The traceability of certified products comes from materials that meet material criteria (already approved by LPPOM/listed in the ingredients list) and are produced in production facilities that meet criteria such as being free of impurities [37]. Each producer must guarantee traceability by having records or documentation, including documentation of the origin of the raw material for chicken used by producers from clear and halal slaughterhouses. Most SME does not yet have a good documentation system

The obstacles SMEs face are a challenge for the government in realizing the implementation of halal products in Indonesia. Based on the conditions, several important points that the Indonesian government must immediately realize are as follow:

1. SMEs mapping
2. Halal supply chain

SME mapping aims to classify SMEs based on the type of product. SMEs with the same type of product will find it easier to coordinate to find solutions to their problems. In addition, with this mapping, a supply chain map can be drawn up that will benefit many parties. The halal supply chain will develop all business links in the system to support the halal food industry so that the economic benefits generated will be greater. The halal food industry does not develop independently but also drives various other business units that support the process. The traceability aspect of raw materials is important in determining the critical point of halal food products. Therefore, a potential solution is to map all raw material supply units with a halal supply chain approach and ensure all products are managed with the principles to maintain halal integrity. The obligation of halal certification for food products in Indonesia and applying the principle of traceability make Indonesian halal food products acceptable to importing countries, especially the OIC countries [38].

The implementation of halal certification, which will be carried out gradually until October 17, 2024, has not been carried out optimally. Based on data from the ministry of cooperatives and SMEs, in 2019, as many as 69,577 companies (total of SMEs and large industries) were certified halal, while the number of SMEs in Indonesia reached 65 million SMEs so only 0.11% of businesses were registered. Halal certification has not been maximized in Indonesia due to unintegrated politics and inadequate facilities and infrastructure [39]. The Indonesian government is currently preparing to support Indonesia as the world's center for halal products. The supporting infrastructure for the halal industry has been prepared in the form of developing a halal industrial area and halal logistics. The development of the halal industrial area (KIH) is a flagship program to strengthen the Halal Value Chain. The concept of developing KIH must provide the location required for the halal industry with a halal product guarantee system including the selection of raw materials, production processes, packaging, storage, and distribution. Industrial products from factories located in halal industrial areas are halal products that have been certified. A halal industrial area can be part or all of an industrial area designed with systems and facilities to develop industries that produce halal products by the halal assurance system [31]. KIH will be developed into a halal industry cluster as an integrated halal ecosystem, including logistics to ports.

In addition to the halal industrial area, other infrastructure to increase demand for halal products includes halal logistics. The halal logistics system must ensure that products are guaranteed to be halal during logistics activities, both in warehouses, depots, terminals, transportation equipment, and packaging. Halal logistics manages the procurement, movement, storage, and handling of materials and supplies of semi-finished goods, both food and non-food, along with related information and documentation flow-through company organizations and supply chains that comply with general Sharia principles [30]. The principles in halal logistics are almost the same as those of handling halal products, namely separating halal products from non-halal products to avoid cross-contamination and ensuring the consistency of halal products until they reach consumers.

The Indonesian government continues to increase SME awareness of the importance of halal certification in Indonesia through socialization by BPJPH together with various ministries, institutions, and agencies through virtual meetings. Before the BPJPH era, LPPOM MUI had carried out socialization and education through the INDHEX (Indonesia International Halal Expo) program, Halal Olympics, Halal Food Goes to School, Halal Tourism (Halal Tour), Halal Talk show Seminar, Halal Competition, Halal Socialization to

SMEs/Large Companies, Halal Competition, Facilitation of Free Halal Certification, Halal Socialization through Media and social media and merchandise. In addition to internal socialization, the Indonesian government also conducts external socialization as a strategy to strengthen the market in the international arena by displaying Indonesian halal products in exhibitions abroad by the ministry of trade of the Republic of Indonesia, such as the Malaysia International Halal Showcase (MIHAS) in 2015, 2018 and 2019; Taiwan International Halal Expo 2016 and 2017; and Russia Halal Expo 2020; World Expo 2020 Dubai in 2021.

4 Conclusion and Suggestion

The condition of the global halal food market opens opportunities for producers and SME exporters in Indonesia, especially for processed food. Food processing creates ambiguity about the halalness of products due to the process and use of food additives, so a system that can guarantee product halalness is required, i.e., a halal assurance system. Based on the JPH Law, all products in Indonesia must be certified halal, including MSE products. This opens more opportunities for MSEs to export their products. Food products certified by the MUI have been internationally recognized and accepted by the UAE because they have guaranteed halal and food safety. Halal certification is important for MSEs to increase product competitiveness.

In the halal certification process, SMEs face several obstacles, including the limited cost of certification and the fulfillment of the criteria for the halal product guarantee system, especially related to materials and traceability. Based on the conditions, the Indonesian government must immediately realize several important points: SMEs mapping and implementing the halal supply chain. Therefore, mapping MSEs with a halal supply chain approach will ensure that all processes are managed to maintain the integrity of halal starting from raw materials to transportation and distribution to consumers, which will facilitate the traceability system and is a potential solution. The obligation of halal certification for food products in Indonesia and applying the principle of traceability make Indonesian halal food products acceptable to importing countries, especially OIC countries. The Indonesian government continues to increase SME awareness of the importance of halal certification in Indonesia through socialization. They provide a free halal certification facility program for SMEs and promote Indonesian halal products abroad; as a result, Indonesia may become the world's halal center.

References

1. O. Andrianita, *Peluang dan Tantangan Produk Halal Indonesia dalam Memasuki Pasar Global* (Ditjen Pengembangan Ekspor Nasional, Kementerian Perdagangan RI, 2020).
2. Small and Medium Enterprises and Cooperatives, *Label Halal Bikin Produk UMKM Makin Dikenal* (Portal SMESCO, Kementerian Koperasi dan UKM RI, 2021).
3. Indonesia Halal Training and Education Center, *Pengertian Makanan dan Minuman Halal* (Portal IHATEC, 2021).
4. A. Andriyani, J. Kedok. Kes. **15**, 178 (2019).
5. S. Zulaekah, Y. Kusumawati, SUHUF **17**, 25 (2005).
6. Indonesia Ulema Council, *Standar Spesifikasi Penyembelihan Halal* (Fatwa Majelis Ulama Indonesia No 12, 2009).
7. E. Sobari, Tim Agrotekuin13, *Dasar-dasar Proses Pengolahan Bahan Pangan* (Polsub Press, Subang, Indonesia, 2019).

https://doi.org/10.1051/e3sconf/202234800039

8. D.M. Rangkuty, *Indonesia's halal commodity supply for international trade market,* in Proceedings 1st International Halal Conference and Exhibitions, IHCE-2019, 29-30 April 2019, Medan, Indonesia (2019).
9. A. Apriyantono, J. Hermanianto, N. Wahid, *Pedoman Produksi Pangan Halal* (Khairul Bayan Press, Jakarta, 2007).
10. M. Solek, DIMAS **17**, 297 (2017).
11. Y. Teinaz. *Does your daily bread contain human hair, or duck and chicken feathers which are from animals not Halal slaughtered in accordance with the Islamic Shariah Law?* (The London Central Mosque Trust & The Islamic Culture Centre, London, 2006).
12. A. Bahar, Rusijono, N. Kusumawati, *Extraction and characterization of the base halal gelatin based on bovine bone*, in Proceedings of National Seminar on Chemistry, SNK, 3-4 September 2018, Samarinda, Indonesia (2018).
13. N.Y. Farkye, Int. J. Dairy Technol. **57**(2-3), 91 (2004).
14. A. Rohman, Y.B. Che Man, Food Rev. Int. **28**, 97 (2012).
15. The Assessment Institute for Foods, Drugs, and Cosmetics Majelis Ulama Indonesia, *Pedoman Pemenuhan Kriteria Sistem Jaminan Halal di Industri Pengolahan* (LPPOM MUI, 2020).
16. The Assessment Institute for Foods, Drugs, and Cosmetics Majelis Ulama Indonesia, *Panduan Umum Sistem Jaminan Halal* LPPOM-MUI (LPPOM MUI, 2008).
17. The Ministry of Trade Republic of Indonesia, Warta Ekspor Juli: *Hidup Sehat dengan Produk Halal* (Directorate General for National Export Development, Jakarta, 2015).
18. M. Muhammad, JIEBI, **2**, 1 (2020).
19. S. Suparto, Djanurdi, D. Yuanitasari, A. Suwandono, Mim. Huk. **28**(3), 427 (2016).
20. Law of The Republic of Indonesia Number 33 Year 2014 On Halal Product Assurance (2014).
21. CNN Indonesia, *Kemenag Belum Hukum Produsen yang Tak Urus Sertifikat Halal*, (Portal CNN Indonesia, 2019).
22. H.D Faridah, J. Hal. Prod. Res. **2(2)**, 68 (2019).
23. Ministry of Religious Affairs, *Pemerintah Percepat Fasilitasi Sertifikasi Halal Pelaku* UMK (Portal Badan Penyelenggara Jaminan Produk Halal/BPJPH, Kementerian Agama RI, 2020).
24. The Assessment Institute for Foods, Drugs, and Cosmetics Majelis Ulama Indonesia, *Festival Syawal LPPOM MUI 1442 H Dorong Kemajuan UMK Indonesia* (LPPOM MUI, 2021).
25. Bogasari, *Saat Pandemi Bogasari Urus Halal 50 UKM* (Bogasari, PT Indofood Sukses Makmur Tbk, 2020).
26. E. Sutrisno, *Menyiapkan 3 Kawasan Industri Halal di Indonesia* (Portal Informasi Indonesia, 2020).
27. The Ministry of Industry, *Siaran Pers: Pembangunan Kawasan Industri Halal Harus Dipercepat* (Portal Kementerian Perindustrian RI, 2021).
28. Komite Nasional Ekonomi dan Keuangan Syariah, *Profil KNEKS* (Portal KNEKS, 2020).
29. Masyarakat Ekonomi Syariah, *Profil Organisasi* (Porta MES, 2020).
30. The Ministry of National Development Planning of Indonesia, *Masterplan Ekonomi Syariah Indonesia 2019-2024* (Kementerian Perencanaan Pembangunan Nasional/ Badan Perencanaan Pembangunan Nasional, 2018).
31. Sukoso, A. Wiryawan, J. Kusnadi, Sucipto, *Ekosistem Industri Halal* (Bank Indonesia, Jakarta, 2020).
32. International Trade Centre, *Halal goes global* (ITC, Geneva, 2015).
33. E.K. Yun, H.Y. Lee, D.H. Kim, Int. J. Philos, Cult. Axio. **17(1)**, 173 (2020).

34. The Assessment Institute for Foods, Drugs, and Cosmetics Majelis Ulama Indonesia, *Pemenuhan Kriteria Keamanan Pangan pada Proses Sertifikasi Halal* (Portal LPPOM MUI, 2020).
35. R. Nurrachmi, TIFBR **11**, 39 (2017).
36. F.F.P. Perdana, B. Kartika, *the importance of the halal certification based on Imam Nursi's concept of frugality* in Proceedings International Conference on Halal Innovation in Product and Services, i-CHIPS, 13-14 July 2018, Songkhla, Thailand (2018).
37. S. Prabowo, A.A. Rahman, For. Penelit. Agro Eko. **34**, 57 (2016).
38. N. Masruroh, J. Ekon. Isl. **11**, 25 (2020).
39. M.C. Nafis, J. Hal. Prod. Res. **2(1)**, 1 (2019).

Improving the overall equipment effectiveness (OEE) on the chicken bowl printing machine by using the theory of change perspective

Suhendi Irawan [1*], *Chandra* Ayu Kurniawati [1], and *Sherly* Dea Febiola [1]

[1]Industrial Management Program Study, College of Vocational Studies, IPB University, Jl. Kumbang No 14 Bogor 16151, Indonesia

Abstract. This article discusses how to improve the overall equipment effectiveness (OEE) of the chicken bowl printing machine. OEE is important to measure machine performance in the production process. OEE will affect the costs incurred during production activities, such as product defects, how effective the performance of machines is, and how much time is available for production. The failure-tags analysis was used to identify the category factors causing engine failure. The why-why analysis was used to identify the root cause of the problem, and the theory of change was used to explain what should be implemented by the company to improve the OEE value. In 2020, only December had a value above standard by 95%, based on the Seiichi Nakajima OEE has a good value if the value >85%. The OEE values From January to May were fell from 73% to 34% and then kept increasing to 73% in November. The increase of the OEE value can be achieved by implementing a communication schedule, monitoring and evaluating the performance of maintenance staff to achieve a fast maintenance machine process, and applying maintenance rules and guidelines.

1. Introduction

Industry in Indonesia has progressed over time. Intense competition forces companies to present their advantages compared to other companies. One way that must be done is by producing output that accordance the quality and quantity with the market needs. One of the success rates of the manufacturing industry is determined by the smoothness of the production process. [1] Said that increasing productivity in production is a challenge for companies. The smooth production process will be run well if the machine in the production process runs according to the planned schedule, but unfortunately, in the production process, downtime machine is often happening and make the production stop.

To overcome the occurrence of downtime machines, various efforts have been made by the company, one of which is implementing Total Productive Maintenance (TPM) to prevent and reduce breakdown machines. [2] TPM is a philosophy that originated and was

* Corresponding author: Suhendiirawan1@apps.ipb.ac.id

developed in Japan and then continues to spread and be applied throughout the world. The purpose of this research is as additional information for companies related to production, to find out the possible causes of the occurrence of machines, so the company knows what the actions needed to obtain OEE values above the standard are. In addition, this study is a continuation of knowing the OEE value of a machine using the theory of change perspective.

2. Literature Review

Total Productive Maintenance (TPM), an important approach to lean maintenance, has been widely used in the manufacturing industry since the 1950s [3]. Manufacturing as a company that makes production keeps maintaining the quality of its TPM handling to keep it above the OEE value standard. [4] explain that all manufacturers are very concerned about the maintenance strategies used in the workshop to avoid downtime because failures directly lead to reduced capacity utilization and therefore reduced productivity. The organization's primary goal is to improve efficiency, lifecycle, and productivity by maintaining better delivery times so that customer satisfaction can be guaranteed. One way to achieve this goal is to implement the concept of Total Productive Maintenance [5].

The implementation of TPM can be measured by using the OEE. OEE is the most common method for measuring production performance, especially for assessing the condition of equipment [6]. It also helps measure losses and corrective actions [7]. Today industrial companies are constantly improving their processes and increasing their efficiency to achieve higher profits [8] by maintaining the machine's performance. Machinery is a key factor in successful production, especially the ability to use production time to produce products that match quality and quantity, [9] and all of those are measured by using the OEE. Therefore, the OEE value can be an effective monitoring and evaluation tool [9].

Interventions are needed to make the machine has the OEE value above the standard. Therefore, the theory of change can be used to package several interventions to improve the OEE value of the machine. The Theory of Change describes how activities are understood and produce different outcomes that help achieve the final intended effect [10].

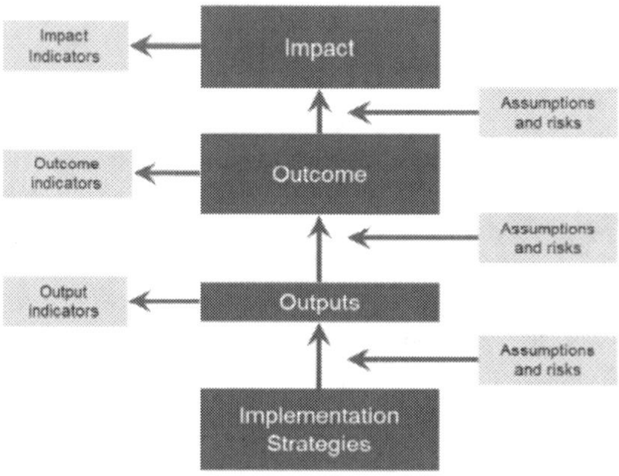

Fig. 1. The principle of the theory of change

3. Methodology

The collection of data and information was carried out in three ways, direct observation in the field, interviews, and literature study through recorded data stored by the company. Data identification and analysis were carried out to solve the problem under study. The steps taken in the data identification and analysis include:

1. F-tags analysis
2. Why-why analysis
3. The theory of change

Failure tags are systems on machines that are downtime. The downtime is caused by a malfunction of a component or a break in a circuit of the machine. Damaged components caused system failure resulting in machine downtime [11]. The Five Whys Technique helps to start at the result, reflect on what caused that, and ask the answer five times. It is a basic and often effective approach to problem-solving that promotes deep thinking by questioning and can be adapted quickly and applied to most problems [12]. The theory of change can be used for strategic planning or program to identify the current condition, the intended situation, and what needs to be done to move from one condition/situation to another condition/situation. It can help design more realistic goals and establish a common understanding of the strategies to achieve the goals. It can also be used to identify which indicators must be monitored and explained to staff [10].

4. Results

4.1 OEE value identification

The ideal condition of the engine can be determined by performing OEE calculations to see the engine performance. The OEE calculation is calculated for one year, from January to December 2020. The Chicken Bowl Printing Machine is classified as a critical unit machine, so OEE calculations must be carried out.

Table 1. OEE value of the chicken bowl 6.5 inch molding machine

Month	Availability (%)	Performance (%)	Quality Yield (%)	OEE (%)
January	98	77	97	73
February	98	85	97	81
March	98	74	95	69
April	98	69	95	64
May	98	36	95	34
June	98	51	96	48
July	96	60	99	57
August	98	60	99	59
September	85	57	100	49
Oktober	98	69	100	67
November	96	76	100	73
Desember	98	98	99	95

The results of the OEE calculation show that for almost one year, the OEE value has been below the established standard based on the Seiichi Nakajima. OEE has a good value if the value > 85%. The company must continue to make improvements to make the production activities run well. Based on table one, the main factor that causes the OEE value to be below the standard is because the performance value is below the average, so this causes the OEE value to be below the standard.

4.2 Root cause analysis of the low value of OEE

Table one shows that the cause of many below standard OEE values is because many performance values are below standard. The flow shop production process caused it. When a machine has downtime, it will cause the other machine not to work. Several elements that cause engine damage are identified in table two.

Table 2. F-tags category

No	Damage Type	F-tags category					
		BCN	OSF	UD	ISL	DW	U
1.	*V Belt on ball mill machine stiff due to exposure to water*			√			
2.	The stripper spring on the press machine lacks oil			√			
3.	wind up on the extruder machine is clogged	√		√			
4.	the oil in the print engine pulley starts to dry			√			
5.	oil on the dryer machine rail that starts to dry			√			
6.	The rubber on the press is torn	√					

Keterangan:
BCN : Basic Condition Neglect
OSF : Operating Standart not Followed
UD : Uncheck Deteriration
ISL : Inadequate Skill Level
DW : Design Weakness
UN : Unknown

The F-tag category indicates that most of the causes of damage are caused by unchecking deterioration which indicates that the check was not carried out properly. The parts that must be replaced continue to work until the damage occurs. In addition, the damage caused by basic condition neglect indicates that the maintenance of the machine is not running properly, so the damage occurs.

4.3 The root causes analysis of the long and frequent of breakdowns machine

After knowing the cause of the small value of performance that impacts the OEE value, the root cause is searched using root cause analysis tools. Based on the root cause analysis, it is known that the root cause of the pretty high downtime on critical machines is because the company pays less attention to employees in the machine repair process.

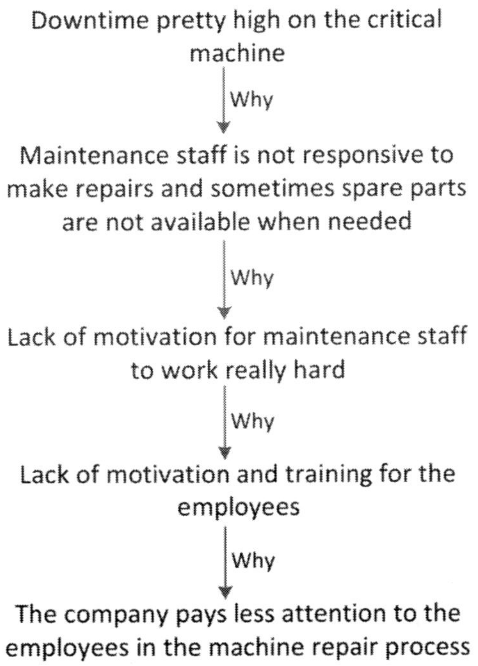

Fig. 2. Root cause analysis of downtime machine

In addition to the problems above, uncheck deterioration and basic condition neglect, which dominate the category of causing breakdown is also analyzed using root cause analysis. From the analysis results, it is known that unchecked deterioration and basic condition neglect that dominate the damage category are caused by a lack of attention and company regulations in the engine repair process.

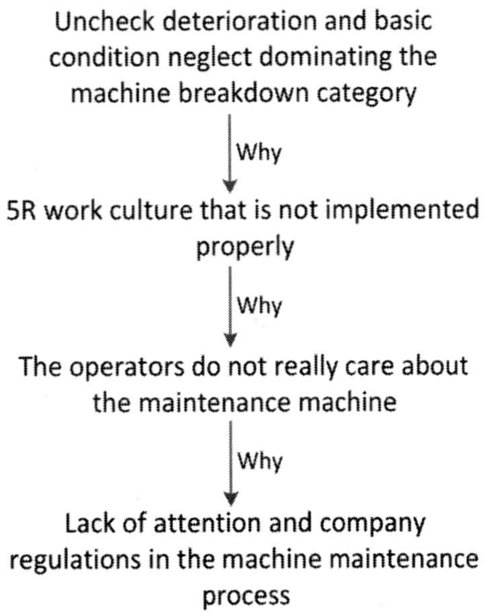

Fig. 3. Root cause analysis of uncheck deterioration and condition neglect dominating

4.4 The theory of change as an effort to increase the OEE value

To increase the OEE value to be above the standard, actions must be taken by the company in an effort for improvement. It starts with scheduling communication and monitoring and evaluating the performance of the maintenance staff. With this implementation, it is hoped that various information to improve the performance of repairing machines will be obtained so that the company will implement the improvement based on the information earned. In addition, the application of rules and engine maintenance guidelines also needs to be implemented. With the implementation of these rules, over time, the employees will get used to running the rules and guidelines for machine maintenance, so a sense of ownership will arise, making the machine more durable (long-lived) and not easily downtime.

Table 3. The theory of change

Intervention	Activities	Output	Outcome	Impact
Scheduling communication as well as monitoring and evaluation of the performance maintenance staff	Communicate with employees on the obstacles faced and monitor and evaluate in carrying out the machine repair process	Information to improve the performance of the maintenance process	Implementation of the improvement of the maintenance machine	Fast maintenance machine process
The implementation of rules and guidelines for maintenance of the machine	Posting the machine maintenance rules and guidelines in the workplace area	Employees follow the rules and guidelines of maintenance machine	Employees are used to following the rules and guidelines for maintenance machines and a sense of ownership arises over the machines being cared for	The machine becomes more durable (long life) and is not easily breakdown

5. Conclusion

To increase the OEE value of the chicken bowl machine, the company must increase the machine's performance value. It can be achieved by implementing a communication schedule and monitoring and evaluating the performance of maintenance staff to achieve a fast maintenance machine process, and applying maintenance rules and guidelines to achieve a machine that has a long life and does not easily downtime.

References

1. A. Amrussalam, P. Budi Santoso, and I. Pambudi Tama, “Pengukuran Dan Perbaikan Total Productive Maintenance (Tpm) Menggunakan Overall Equipment Effectiveness (Oee) Dan Root Cause Failure Analysis (Rcfa),” J. Eng. Manag. Industial Syst., vol. 4, no. 2, pp. 102–108 (2016)

2. R. Rahmad, P. Pratikto, and S. Wahyudi, "Penerapan Overall Equipment Effectiveness (Oee) Dalam Implementasi Total Productive Maintenance (TPM) (Studi Kasus Di Pabrik Gula PT. 'Y'.)," J. Rekayasa Mesin, vol. 3, no. 3, pp. 431–437 (2012)
3. G. L. Tortorella, F. S. Fogliatto, P. A. Cauchick-Miguel, S. Kurnia, and D. Jurburg, "Integration of Industry 4.0 technologies into Total Productive Maintenance practices," Int. J. Prod. Econ., vol. 240, no. February, p. 108224 (2021)
4. R. Rathi, M. Singh, M. Sabique, M. Al Amin, S. Saha, and M. Hari Krishnaa, "Identification of total productive maintenance barriers in Indian manufacturing industries," Mater. Today Proc., no. xxxx (2021)
5. R. P. Mishra, G. Gupta, and A. Sharma, "Development of a Model for Total Productive Maintenance Barriers to Enhance the Life Cycle of Productive Equipment," Procedia CIRP, vol. 98, pp. 241–246 (2021)
6. T. Djatna and M. R. Ihsan, "A Fuzzy Associative Memory Modeling for Production Equipment Status Assessment," Procedia Manuf., vol. 4, no. Iess, pp. 163–167 (2015)
7. A. S. Relkar and K. N. Nandurkar, "Optimizing & analysing overall equipment effectiveness (OEE) through design of experiments (DOE)," Procedia Eng., vol. 38, pp. 2973–2980 (2012)
8. P. Dobra and J. Jósvai, "Enhance of OEE by hybrid analysis at the automotive semi-automatic assembly lines," Procedia Manuf., vol. 54, no. Det 2020, pp. 184–190 (2020)
9. S. H. Santosa, S. Irawan, and I. Ardani, "Determination of Overall Equipment Effectiveness Superflex Machine Using Fuzzy Approach," Int. J. Artif. Intell. Res., vol. 4, no. 2 (2021)
10. D. Gangloff, "Theory of change," Am. For., vol. 113, no. 3, p. 7 (2007)
11. Borris. Total Productive Maintenance. New York (US): Mc Graw-Hill Companies (2006)
12. O. Serrat, "The Five Whys Technique," Knowl. Solut., no. February 2009, pp. 307–310 (2017)

AUTHOR INDEX